Adolf Grünbaum

Particle Physics and Inflationary Cosmology

Contemporary Concepts in Physics

A series edited by
Herman Feshbach
Massachusetts Institute of Technology

Associate Editors
N. Bloembergen
Harvard University
Mildred S. Dresselhaus
Massachusetts Institute of Technology
Mal Ruderman
Columbia University
S.B. Treiman
Princeton University

Founding Editor
Henry Primakoff
(1914–1983)

This book is part of a series. The publisher will accept continuation orders which may be cancelled at any time and which provide for automatic billing and shipping of each title in the series upon publication. Please write for details.

Particle Physics and Inflationary Cosmology

by
Andrei Linde

Lebedev Physical Institute, Moscow, USSR
and CERN, Geneva, Switzerland

Translated from the Russian by
Marc Damashek

h•
ap harwood academic publishers
chur • london • paris • new york • melbourne

Harwood Academic Publishers

Post Office Box 197
London WC2E 9PX
United Kingdom

58, rue Lhomond
75005 Paris
France

Post Office Box 786
Cooper Station
New York, New York 10276
United States of America

Private Bag 8
Camberwell, Victoria 3124
Australia

Library of Congress Cataloging-in-Publication Data

Linde, A. D.
Particle physics and inflationary cosmology / by Andrei D. Linde
translated from the Russian by Marc Damashek.
p. cm.—(Contemporary concepts in physics; v. 5)
Bibliography:
Includes index.
ISBN 3-7186-0489-2. ISBN 3-7186-0490-6 (pbk.).
1. Cosmology 2. Astrophysics. 3. Particles (Nuclear physics)
I. Title. II. Series.
QB981.L56 1990 89-11037
523.1—dc20 CIP

Reproduced from camera-ready copy supplied by the translator.

Contents

Preface to the Series

The series of volumes, *Contemporary Concepts in Physics*, is addressed to the professional physicist and to the serious graduate student of physics. The subjects to be covered will include those at the forefront of current research. It is anticipated that the various volumes in the series will be rigorous and complete in their treatment, supplying the intellectual tools necessary for the appreciation of the present status of the areas under consideration and providing the framework upon which future developments may be based.

Introduction

With the invention and development of unified gauge theories of weak and electromagnetic interactions, a genuine revolution has taken place in elementary particle physics in the last 15 years. One of the basic underlying ideas of these theories is that of spontaneous symmetry breaking between different types of interactions due to the appearance of constant classical scalar fields φ over all space (the so-called Higgs fields). Prior to the appearance of these fields, there is no fundamental difference between strong, weak, and electromagnetic interactions. Their spontaneous appearance over all space essentially signifies a restructuring of the vacuum, with certain vector (gauge) fields acquiring high mass as a result. The interactions mediated by these vector fields then become short-range, and this leads to symmetry breaking between the various interactions described by the unified theories.

The first consistent description of strong and weak interactions was obtained within the scope of gauge theories with spontaneous symmetry breaking. For the first time, it became possible to investigate strong and weak interaction processes using high-order perturbation theory. A remarkable property of these theories — asymptotic freedom — also made it possible in principle to describe interactions of elementary particles up to center-of-mass energies $E \sim M_P \sim 10^{19}$ GeV, that is, up to the Planck energy, where quantum gravity effects become important.

Here we will recount only the main stages in the development of gauge theories, rather than discussing their properties in detail. In the 1960s, Glashow, Weinberg, and Salam proposed a unified theory of the weak and electromagnetic interactions [1], and real progress was made in this area in 1971–1973 after the theories were shown to be renormalizable [2]. It was proved in 1973 that many such theories, with quantum chromodynamics in particular serving as a description of strong interactions, possess the property of asymptotic freedom (a decrease in the coupling constant with increasing energy [3]). The first unified gauge theories of strong, weak, and electromagnetic interactions with a simple symmetry group, the so-called grand unified theories [4], were proposed in 1974. The first theories to unify all of the fundamental interactions, including gravitation, were proposed in 1976 within the context of supergravity theory. This was followed by the development of Kaluza–Klein theories, which maintain that our four-dimensional space-time results from the spontaneous compactification of a higher-dimensional space [6]. Finally, our most recent hopes for a unified theory of all interactions have been invested in superstring theory [7]. Modern theories of elementary particles are covered in a number of excellent reviews and monographs (see [8–17], for example).

The rapid development of elementary particle theory has not only led to great advances in our understanding of particle interactions at superhigh energies, but also (as a consequence) to significant progress in the theory of superdense matter. Only fifteen years ago, in fact, the term *superdense matter* meant matter with a density somewhat higher than nuclear values, $\rho \sim 10^{14}$–10^{15} $g \cdot cm^{-3}$ and it was virtually impossible to conceive of how one might describe matter with $\rho \gg 10^{15}$ $g \cdot cm^{-3}$. The main problems involved strong-interaction theory, whose typical coupling constants at $\rho \gtrsim 10^{15}$ $g \cdot cm^{-3}$ were large, making standard perturbation-theory predictions of the properties of such matter unreliable. Because of asymptotic freedom in quantum chromodynamics, however, the corresponding coupling constants decrease with increasing temperature (and density). This enables one to describe the behavior of matter at temperatures approaching $T \sim M_P \sim 10^{19}$ GeV, which corresponds to a density $\rho_P \sim M_P^4 \sim 10^{94}$ $g \cdot cm^{-3}$. Present-day elementary particle theories thus make it possible, in principle, to describe the properties of matter more than 80 orders of magnitude denser than nuclear matter!

The study of the properties of superdense matter described by unified gauge theories began in 1972 with the work of Kirzhnits [18], who showed that the classical scalar field φ responsible for symmetry breaking should disappear at a high enough temperature T. This means that a phase transition (or a series of phase transitions) occurs at a sufficiently high temperature $T > T_c$, after which symmetry is restored between various types of interactions. When this happens, elementary particle properties and the laws governing their interaction change significantly.

This conclusion was confirmed in many subsequent publications [19–24]. It was found that similar phase transitions could also occur when the density of cold matter was raised [25–29], and in the presence of external fields and currents [22, 23, 30, 33]. For brevity, and to conform with current terminology, we will hereafter refer to such processes as phase transitions in gauge theories.

Such phase transitions typically take place at exceedingly high temperatures and densities. The critical temperature for a phase transition in the Glashow–Weinberg–Salam theory of weak and electromagnetic interactions [1], for example, is of the order of 10^2 GeV ~ 10^{15} K. The temperature at which symmetry is restored between the strong and electroweak interactions in grand unified theories is even higher, $T_c \sim 10^{15}$ GeV ~ 10^{28} K. For comparison, the highest temperature attained in a supernova explosion is about 10^{11} K. It is therefore impossible to study such phase transitions in a laboratory. However, the appropriate extreme conditions could exist at the earliest stages of the evolution of the universe.

According to the standard version of the hot universe theory, the universe could have expanded from a state in which its temperature was at least $T \sim 10^{19}$ GeV [34, 35], cooling all the while. This means that in its earliest stages, the symmetry between the strong, weak, and electromagnetic interactions should have been intact. In cooling, the universe would have gone through a number of phase transitions, breaking the symmetry between the different interactions [18–24].

This result comprised the first evidence for the importance of unified theories of elementary particles and the theory of superdense matter for the development of the theory of the evolution of the universe. Cosmologists became particularly interested in recent theories of elementary particles after it was found that grand unified theories provide a natural framework within which the observed baryon asymmetry of the universe (that is, the lack of antimatter in the observable part of the universe) might arise [36–38]. Cosmology has likewise turned out to be an important source of information for elementary particle theory. The recent rapid development of the latter has resulted in a somewhat unusual situation in that branch of theoretical physics. The reason is that typical elementary particle energies required for a direct test of grand unified theories are of the order of 10^{15} GeV, and direct tests of supergravity, Kaluza–Klein theories, and superstring theory require energies of the order of 10^{19} GeV. On the other hand, currently planned accelerators will only produce particle beams with energies of about 10^4 GeV. Experts estimate that the largest accelerator that could be built on earth (which has a radius of about 6000 km) would enable us to study particle interactions at energies of the order of 10^7 GeV, which is typically the highest (center-of-mass) energy encountered in cosmic ray experiments. Yet this is twelve orders of magnitude lower than the Planck energy $E_P \sim M_P \sim 10^{19}$ GeV.

The difficulties involved in studying interactions at superhigh energies can be highlighted by noting that 10^{15} GeV is the kinetic energy of a small car, and 10^{19} GeV is the kinetic energy of a medium-sized airplane. Estimates indicate that accelerating particles to energies of the order of 10^{15} GeV using present-day technology would require an accelerator approximately one light-year long.

It would be wrong to think, though, that the elementary particle theories currently being developed are totally without experimental foundation — witness the experiments on a huge scale which are under way to detect the decay of the proton, as predicted by grand unified theories. It is also possible that accelerators will enable us to detect some of the lighter particles (with mass $m \sim 10^2$–10^3 GeV) predicted by certain versions of supergravity and superstring theories. Obtaining information solely in this way, however, would be similar to trying to discover a unified theory of weak and electromagnetic interactions using only radio telescopes, detecting radio waves with an energy E_γ no greater than 10^{-5} eV (note that $\frac{E_P}{E_W} \sim \frac{E_W}{E_\gamma}$, where $E_W \sim 10^2$ GeV is the characteristic energy in the unified theory of weak and electromagnetic interactions).

The only laboratory in which particles with energies of 10^{15}–10^{19} GeV could ever exist and interact with one another is our own universe in the earliest stages of its evolution.

At the beginning of the 1970s, Zeldovich wrote that the universe is the poor man's accelerator: experiments don't need to be funded, and all we have to do is collect the experimental data and interpret them properly [39]. More recently, it has become quite clear that the universe is the only accelerator that could ever produce particles at energies high enough to test unified theories of all fundamental interactions directly, and in that sense it is not just the poor man's accelerator but the richest man's as well. These days, most new elementary particle theories must first take a "cosmological validity" test — and only a very few pass.

It might seem at first glance that it would be difficult to glean any reasonably definitive or reliable information from an experiment performed more than ten billion years ago, but recent studies indicate just the opposite. It has been found, for instance, that phase transitions, which should occur in a hot universe in accordance with the grand unified theories, should produce an abundance of magnetic monopoles, the density of which ought to exceed the observed density of matter at the present time, $\rho \sim 10^{-29}$ g • cm^{-3}, by approximately fifteen orders of magnitude [40]. At first, it seemed that uncertainties inherent in both the hot universe theory and the grand unified theories, being very large, would provide an easy way out of the primordial monopole problem. But many attempts to resolve this problem within the context of the standard hot universe theory have not led to final success. A similar situation has arisen in dealing with theories involving spontaneous breaking of a discrete symmetry (spontaneous CP-invariance breaking, for example). In such models, phase transitions ought to give rise to supermassive domain walls, whose existence would sharply conflict with the astrophysical data [41–43]. Going to more complicated theories such as N = 1 supergravity has engendered new problems rather than resolving the old ones. Thus it has turned out in most theories based on N = 1 supergravity that the decay of gravitinos (spin=3/2 superpartners of the graviton) which existed in the early stages of the universe leads to results differing from the observational data by about ten orders of magnitude [44, 45]. These theories also predict the existence of so-called scalar Polonyi fields [15, 46]. The energy density that would have been accumulated in these fields by now differs from the cosmological data by fifteen orders of magnitude [47, 48]. A number of axion theories [49] share this difficulty, particularly in the simplest models based on superstring theory [50]. Most Kaluza–Klein theories based on supergravity in an 11-dimensional space lead to vacuum energies of order $-M_P^4 \sim -10^{94}$ g • cm^{-3} [16], which differs from the cosmological data by approximately 125 orders of magnitude....

This list could be continued, but as it stands it suffices to illustrate why elementary particle theorists now find cosmology so interesting and important. An even more general reason is that no real unification of all interactions including gravitation is possible without an analysis of the most important manifestation of that unification, namely the existence of the universe itself. This is illustrated especially clearly by Kaluza–Klein and superstring theories, where one must simultaneously investigate the properties of the space-time formed by compactification of "extra" dimensions, and the phenomenology of the elementary particles.

It has not yet been possible to overcome some of the problems listed above. This places important constraints on elementary particle theories currently under development. It is all the more surprising, then, that many of these problems, together with a number of others that predate the hot universe theory, have been resolved in the context of one fairly simple scenario for the development of the universe — the so-called inflationary universe scenario [51–57]. According to this scenario, the universe, at some very early stage of its evolution, was in an unstable vacuum-like state and expanded exponentially (the stage of inflation). The vacuum-like state then decayed, the universe heated up, and its subsequent evolution can be described by the usual hot universe theory.

Since its conception, the inflationary universe scenario has progressed from something akin to science fiction to a well-established theory of the evolution of the universe accepted by most cosmologists. Of course this doesn't mean that we have now finally achieved total enlightenment as to the physical processes operative in the early universe. The incompleteness of the current picture is reflected by the very word *scenario*, which is not normally found in the working vocabulary of a theoretical physicist. In its present form, this scenario only vaguely resembles the simple models from which it sprang. Many details of the inflationary universe scenario are changing, tracking rapidly changing (as noted above) elementary particle theories. Nevertheless, the basic aspects of this scenario are now well- developed, and it should be possible to provide a preliminary account of its progress.

Most of the present book is given over to discussion of inflationary cosmology. This is preceded by an outline of the general theory of spontaneous symmetry breaking and a discussion of phase transitions in superdense matter, as described by present-day theories of elementary particles. The choice of material has been dictated by both the author's interests and his desire to make the contents useful both to quantum field theorists and astrophysicists. We have therefore tried to concentrate on those problems that yield an understanding of the basic aspects of the theory, referring the reader to the original papers for further details.

In order to make this book as widely accessible as possible, the main exposition has been preceded by a long introductory chapter, written at a relatively elementary level. Our hope is that by using this chapter as a guide to the book, and the book itself as a guide to the original literature, the reader will gradually be able to attain a fairly complete and accurate understanding of the present status of this branch of science. In this regard, he might also be assisted by an acquaintance with the books ***Cosmology of the Early Universe***, by A. D. Dolgov, Ya. B. Zeldovich, and M. V. Sazhin; ***How the Universe Exploded***, by I. D. Novikov; ***A Brief History of Time: From the Big Bang to Black Holes***, by S. W. Hawking; and ***An Introduction to Cosmology and Particle Physics***, by R. Dominguez–Tenreiro and M. Quiros. A good collection of early papers on inflationary cosmology and galaxy formation can also be found in the book ***Inflationary Cosmology***, edited by L. Abbott and S.-Y. Pi. We apologize in advance to those authors whose work in the field of inflationary cosmology we have not been able to treat adequately. Much of the material in this book is based on the ideas and work of S. Coleman, J. Ellis, A. Guth, S. W. Hawking, D. A. Kirzhnits, L. A. Kofman, M. A. Markov, V. F. Mukhanov, D. Nanopoulos, I. D. Novikov, I. L. Rozental', A. D. Sakharov, A. A. Starobinsky, P. Steinhardt, M. Turner, and many other scientists whose contribution to modern cosmology could not possibly be fully reflected in a single monograph, no matter how detailed.

I would like to dedicate this book to the memory of Yakov Borisovich Zeldovich, who should by rights be considered the founder of the Soviet school of cosmology.

1

Overview of Unified Theories of Elementary Particles and the Inflationary Universe Scenario

1.1 The scalar field and spontaneous symmetry breaking

Scalar fields φ play a fundamental role in unified theories of the weak, strong, and electromagnetic interactions. Mathematically, the theory of these fields is simpler than that of the spinor fields ψ describing electrons or quarks, for instance, and it is simpler than the theory of the vector fields A_μ which describes photons, gluons, and so on. The most interesting and important properties of these fields for both elementary particle theory and cosmology, however, were grasped only fairly recently.

Let us recall the basic properties of such fields. Consider first the simplest theory of a one-component real scalar field φ with the Lagrangian[1]

$$L = \frac{1}{2}(\partial_\mu \varphi)^2 - \frac{m^2}{2}\varphi^2 - \frac{\lambda}{4}\varphi^4. \tag{1.1.1}$$

---

[1] In this book we employ units such that $\hbar = c = 1$, the system commonly used in elementary particle theory. In order to transform expressions to conventional units, corresponding terms must be multiplied by appropriate powers of $\hbar$ or c to give the correct dimensionality (note that $\hbar = 6.6 \cdot 10^{-22}$ MeV $\cdot$ sec $\sim 10^{-27}$ erg $\cdot$ sec, $c \sim 3 \cdot 10^{10}$ cm $\cdot$ sec^{-1}) . Thus, for example, Eq. (1.1.1) would acquire the form

$$L = \frac{1}{2}(\partial_\mu \varphi)^2 - \frac{m^2 c^2}{2\hbar^2}\varphi^2 - \frac{\lambda}{4}\varphi^4.$$

In this equation, m is the mass of the scalar field, and λ is its coupling constant. For simplicity, we assume throughout that $\lambda \ll 1$. When φ is small and we can neglect the last term in (1.1.1), the field satisfies the Klein–Gordon equation

$$(\Box + m^2)\varphi \equiv \ddot{\varphi} - \Delta\varphi + m^2\varphi = 0, \tag{1.1.2}$$

where a dot denotes differentiation with respect to time. The general solution of this equation is expressible as a superposition of plane waves, corresponding to the propagation of particles of mass m and momentum k [58]:

$$\begin{aligned}\varphi(x) &= (2\pi)^{-3/2}\int d^4k\,\delta(k^2-m^2)\,[e^{ikx}\varphi^+(k) + e^{-ikx}\varphi^-(k)] \\ &= (2\pi)^{-3/2}\int \frac{d^3k}{\sqrt{2k_0}}[e^{ikx}a^+(\boldsymbol{k}) + e^{-ikx}a^-(\boldsymbol{k})],\end{aligned} \tag{1.1.3}$$

where $a^\pm(\boldsymbol{k}) = \frac{1}{\sqrt{2k_0}}\varphi^\pm(k)$, $k_0 = \sqrt{\boldsymbol{k}^2 + m^2}$, $kx = k_0t - \boldsymbol{k}\cdot\boldsymbol{x}$. According to (1.1.3), the field $\varphi(x)$ will oscillate about the point $\varphi = 0$, the reason being that the minimum of the potential-energy density for the field φ (the so-called effective potential)

$$V(\varphi) = \frac{1}{2}(\nabla\varphi)^2 + \frac{m^2}{2}\varphi^2 + \frac{\lambda}{4}\varphi^4 \tag{1.1.4}$$

occurs at $\varphi = 0$ (see Fig. 1a).

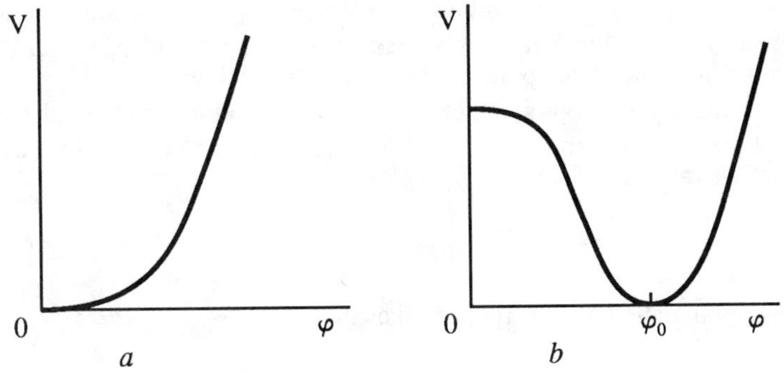

Figure 1. Effective potential V(φ) in the simplest theories of the scalar field φ. *a*) V(φ) in the theory (1.1.1), and *b*) in the theory (1.1.5).

Fundamental advances in the unification of the weak, strong, and electromagnetic interactions were finally achieved when simple theories based on Lagrangians like (1.1.1) with $m^2 > 0$ gave way to what were at first glance somewhat strange-looking theories with negative mass squared:

$$L = \frac{1}{2}(\partial_\mu\varphi)^2 + \frac{\mu^2}{2}\varphi^2 - \frac{\lambda}{4}\varphi^4. \tag{1.1.5}$$

Instead of oscillations about $\varphi = 0$, the solution corresponding to (1.1.3) gives modes that grow exponentially near $\varphi = 0$ when $\boldsymbol{k}^2 < \mu^2$:

$$\delta\varphi(\boldsymbol{k}) \sim \exp\left(\pm\sqrt{\mu^2 - \boldsymbol{k}^2}\, t\right)\cdot \exp\,(\pm\, i\boldsymbol{k}\boldsymbol{x})\,. \tag{1.1.6}$$

What this means is that the minimum of the effective potential

$$V(\varphi) = \frac{1}{2}(\nabla\varphi)^2 - \frac{\mu^2}{2}\varphi^2 + \frac{\lambda}{4}\varphi^4 \tag{1.1.7}$$

will now occur not at $\varphi = 0$, but at $\varphi_c = \pm\,\mu/\sqrt{\lambda}$ (see Fig. 1b).[2] Thus, even if the field φ is zero initially, it soon undergoes a transition (after a time of order μ^{-1}) to a stable state with the classical field $\varphi_c = \pm\,\mu/\sqrt{\lambda}$, a phenomenon known as spontaneous symmetry breaking.

After spontaneous symmetry breaking, excitations of the field φ near $\varphi_c = \pm\,\mu/\sqrt{\lambda}$ can also be described by a solution like (1.1.3). In order to do so, we make the change of variables

$$\varphi \to \varphi + \varphi_0. \tag{1.1.8}$$

The Lagrangian (1.1.5) thereupon takes the form

$$\begin{aligned} L(\varphi+\varphi_0) &= \frac{1}{2}\left(\partial_\mu(\varphi+\varphi_0)\right)^2 + \frac{\mu^2}{2}(\varphi+\varphi_0)^2 - \frac{\lambda}{4}(\varphi+\varphi_0)^4 \\ &= \frac{1}{2}(\partial_\mu\varphi)^2 - \frac{3\lambda\varphi_0^2 - \mu^2}{2}\varphi^2 - \lambda\varphi_0\varphi^3 - \frac{\lambda}{4}\varphi^4 \\ &\quad + \frac{\mu^2}{2}\varphi_0^2 - \frac{\lambda}{4}\varphi_0^4 - \varphi(\lambda\varphi_0^2 - \mu)\varphi_0. \end{aligned} \tag{1.1.9}$$

We see from (1.1.9) that when $\varphi_0 \neq 0$, the effective mass squared of the field φ is not equal to $-\mu^2$, but rather

$$m^2 = 3\lambda\varphi_0^2 - \mu^2, \tag{1.1.10}$$

and when $\varphi_0 = \pm\,\mu/\sqrt{\lambda}$, at the minimum of the potential $V(\varphi)$ given by (1.1.7), we have

$$m^2 = 2\lambda\varphi_0^2 = 2\mu^2 > 0; \tag{1.1.11}$$

in other words, the mass squared of the field φ has the correct sign. Reverting to the original variables, we can write the solution for φ in the form

$$\varphi(x) = \varphi_0 + (2\pi)^{-3/2}\int \frac{d^3k}{\sqrt{2k_0}}\left[e^{ikx}a^+(\boldsymbol{k}) + e^{-ikx}a^-(\boldsymbol{k})\right]. \tag{1.1.12}$$

[2] $V(\varphi)$ usually attains a minimum for homogeneous fields φ, so gradient terms in the expression for $V(\varphi)$ are often omitted.

The integral in (1.1.12) corresponds to particles (quanta) of the field φ with mass given by (1.1.11), propagating against the background of the constant classical field φ_0.

The presence of the constant classical field φ_0 over all space will not give rise to any preferred reference frame associated with that field: the Lagrangian (1.1.9) is covariant, irrespective of the magnitude of φ_0. Essentially, the appearance of a uniform field φ_0 over all space simply represents a restructuring of the vacuum state. In that sense, the space filled by the field φ_0 remains "empty." Why then is it necessary to spoil the good theory (1.1.1)?

The main point here is that the advent of the field φ_0 changes the masses of those particles with which it interacts. We have already seen this in considering the example of the sign "correction" for the mass squared of the field φ in the theory (1.1.5). Similarly, scalar fields can change the mass of both fermions and vector particles.

Let us examine the two simplest models. The first is the simplified σ-model, which is sometimes used for a phenomenological description of strong interactions at high energy [26]. The Lagrangian for this model is a sum of the Lagrangian (1.1.5) and the Lagrangian for the massless fermions ψ, which interact with φ with a coupling constant h:

$$L = \frac{1}{2}(\partial_\mu\varphi)^2 + \frac{\mu^2}{2}\varphi^2 - \frac{\lambda}{4}\varphi^4 + \bar{\psi}\left(i\,\partial_\mu\gamma_\mu - h\varphi\right)\psi. \tag{1.1.13}$$

After symmetry breaking, the fermions will clearly acquire a mass

$$m_\psi = h|\varphi_0| = h\,\frac{\mu}{\sqrt{\lambda}}. \tag{1.1.14}$$

The second is the so-called Higgs model [59], which describes an Abelian vector field A_μ (the analog of the electromagnetic field) that interacts with the complex scalar field $\chi = (\chi_1 + \chi_2)/\sqrt{2}$. The Lagrangian for this theory is given by

$$\begin{aligned} L = -\frac{1}{4}(\partial_\mu A_\nu - \partial_\nu A_\mu)^2 + (\partial_\mu + ieA_\mu)\,\chi^*\,(\partial_\mu - ieA_\mu)\,\chi \\ + \mu^2\chi^*\chi - \lambda(\chi^*\chi)^2. \end{aligned} \tag{1.1.15}$$

As in (1.1.7), when $\mu^2 < 0$ the scalar field χ acquires a classical component. This effect is described most easily by making the change of variables

$$\chi(x) \to \frac{1}{\sqrt{2}}\left(\varphi(x) + \varphi_0 \exp\frac{i\zeta(x)}{\varphi_0}\right),$$
$$A_\mu(x) \to A_\mu(x) - \frac{1}{e\varphi_0}\partial_\mu \zeta(x), \tag{1.1.16}$$

whereupon the Lagrangian (1.1.15) becomes

$$L = -\frac{1}{4}(\partial_\mu A_\nu - \partial_\nu A_\mu)^2 + \frac{e^2}{2}(\varphi + \varphi_0)^2 A_\mu^2 + \frac{1}{2}(\partial_\mu \varphi)^2$$
$$-\frac{3\lambda\varphi_0^2 - \mu^2}{2}\varphi^2 - \lambda\varphi_0\varphi^3 - \frac{\lambda}{4}\varphi^4 + \frac{\mu^2}{2}\varphi_0^2 - \frac{\lambda}{4}\varphi_0^4 \tag{1.1.17}$$
$$-\varphi(\lambda\varphi_0^2 - \mu^2)\varphi_0.$$

Notice that the auxiliary field $\zeta(x)$ has been entirely cancelled out of (1.1.17), which describes a theory of vector particles of mass $m_A = e\varphi_0$ that interact with a scalar field having the effective potential (1.1.7). As before, when $\mu^2 > 0$, symmetry breaking occurs, the field $\varphi_0 = \mu/\sqrt{\lambda}$ appears, and the vector particles of A_μ acquire a mass $m_A = e\mu/\sqrt{\lambda}$. This scheme for making vector mesons massive is called the Higgs mechanism, and the fields χ, φ are known as Higgs fields. The appearance of the classical field φ_0 breaks the symmetry of (1.1.15) under U(1) gauge transformations:

$$A_\mu \to A_\mu + \frac{1}{e}\partial_\mu \zeta(x),$$
$$\chi \to \chi \exp[i\zeta(x)]. \tag{1.1.18}$$

The basic idea underlying unified theories of the weak, strong, and electromagnetic interactions is that prior to symmetry breaking, all vector mesons (which mediate these interactions) are massless, and there are no fundamental differences among the interactions. As a result of the symmetry breaking, however, some of the vector bosons do acquire mass, and their corresponding interactions become short-range, thereby destroying the symmetry between the various interactions. For example, prior to the appearance of the constant scalar Higgs field H, the Glashow–Weinberg–Salam model [1] has SU(2) × U(1) symmetry, and electroweak interactions are mediated by massless vector bosons. After the appearance of the constant scalar field H, some of the vector bosons ($W_\mu^\pm$ and Z_μ^0) acquire masses of order eH ~ 100 GeV, and the corresponding interactions become short-range (weak interactions), whereas the electromagnetic field A_μ remains massless.

The Glashow–Weinberg–Salam model was proposed in the 1960's [1], but the real explosion of interest in such theories did not come until 1971–1973, when it was shown that gauge theories with spontaneous symmetry breaking are renormalizable, which means that there is a regular method for dealing with the ultraviolet divergences, as in ordinary quantum electrodynamics [2]. The proof of renormalizability for unified field theories is rather complicated, but the basic physical idea behind it is quite simple. Before the appearance of the scalar field φ_0, the unified theories are renormalizable, just like ordinary quantum electrodynamics. Naturally, the appearance of a classical scalar field φ_0 (like the presence of the ordinary classical electric and magnetic fields) should not affect the high-energy properties of the theory; specifically, it should not destroy the original renormalizability of the theory. The creation of unified gauge theories with spontaneous symmetry breaking and the proof that they are renormalizable carried elementary particle theory in the early 1970's to a qualitatively new level of development.

The number of scalar field types occurring in unified theories can be quite large. For example, there are two Higgs fields in the simplest theory with SU(5) symmetry [4]. One of these, the field Φ, is represented by a traceless 5 × 5 matrix. Symmetry breaking in this theory results from the appearance of the classical field

$$\Phi_0 = \sqrt{\frac{2}{15}}\,\varphi_0 \begin{pmatrix} 1 & & & & 0 \\ & 1 & & & \\ & & 1 & & \\ & & & -3/2 & \\ 0 & & & & -3/2 \end{pmatrix}, \qquad (1.1.19)$$

where the value of the field φ_0 is extremely large — $\varphi_0 \sim 10^{15}$ GeV. All vector particles in this theory are massless prior to symmetry breaking, and there is no fundamental difference between the weak, strong, and electromagnetic interactions. Leptons can then easily be transformed into quarks, and vice versa. After the appearance of the field (1.1.19), some of the vector mesons (the X and Y mesons responsible for transforming quarks into leptons) acquire enormous mass: $m_{X,Y} = (5/3)^{1/2} g\varphi_0/2 \sim 10^{15}$ GeV, where $g^2 \sim 0.3$ is the SU(5) gauge coupling constant. The transformation of quarks into leptons thereupon becomes strongly inhibited, and the proton becomes almost stable. The original SU(5) symmetry breaks down into SU(3) × SU(2) × U(1); that is, the strong interactions (SU(3)) are separated from the electroweak (SU(2) × U(1)). Yet another classical scalar field $H \sim 10^2$ GeV then makes its appearance, breaking the symmetry between the weak and electromagnetic interactions, as in the Glashow–Weinberg–Salam theory [4, 12].

The Higgs effect and the general properties of theories with spontaneous symmetry breaking are discussed in more detail in Chapter 2. The elementary theory of spontaneous symmetry breaking is discussed in Section 2.1. In Section 2.2, we further study this phenomenon, with quantum corrections to the effective potential $V(\varphi)$ taken into consideration. As will be shown in Section 2.2, quantum corrections can in some cases significantly modify the general form of the potential (1.1.7). Especially interesting and unexpected properties of that potential will become apparent when we study it in the $1/N$ approximation.

1.2 Phase transitions in gauge theories

The idea of spontaneous symmetry breaking, which has proven so useful in building unified gauge theories, has an extensive history in solid-state theory and quantum statistics, where it has been used to describe such phenomena as ferromagnetism, superfluidity, superconductivity, and so forth.

Consider, for example, the expression for the energy of a superconductor in the phenomenological Ginzburg–Landau theory [60] of superconductivity:

$$E = E_0 + \frac{H^2}{2} + \frac{1}{2m} |(\nabla - 2ieA)\Psi|^2 - \alpha |\Psi|^2 + \beta |\Psi|^4 . \qquad (1.2.1)$$

Here E_0 is the energy of the normal metal without a magnetic field H, Ψ is the field describing the Cooper-pair Bose condensate, and α and β are positive parameters.

Bearing in mind, then, that the potential energy of a field enters into the Lagrangian with a negative sign, it is not hard to show that the Higgs model (1.1.15) is simply a relativistic generalization of the Ginzburg–Landau theory of superconductivity (1.2.1), and the classical field φ in the Higgs model is the analog of the Cooper-pair Bose condensate.[3]

The analogy between unified theories with spontaneous symmetry breaking and theories of superconductivity has been found to be extremely useful in studying the properties of superdense matter described by unified theories. Specifically, it is well known that when the temperature is raised, the Cooper-pair condensate shrinks to zero and superconductivity disappears. It turns out that the uniform scalar field φ should also disappear when the temperature of matter is raised; in other words, at superhigh temperatures, the symmetry between the weak, strong, and electromagnetic interactions ought to be restored [18–24].

A theory of phase transitions involving the disappearance of the classical field φ is discussed in detail in Ref. 24. In gross outline, the basic idea is that the equilibrium value of the field φ at fixed temperature $T \neq 0$ is governed not by the location of the minimum of the potential energy density $V(\varphi)$, but by the location of the minimum of the free energy density $F(\varphi, T) \equiv V(\varphi, T)$, which equals $V(\varphi)$ at $T = 0$. It is well-known that the temperature-dependent contribution to the free energy F from ultrarelativistic scalar particles of mass m at temperature $T \gg m$ is given [61] by

$$\Delta F = \Delta V(\varphi, T) = -\frac{\pi}{90}T^4 + \frac{m^2}{24}T^2\left(1 + O\left(\frac{m}{T}\right)\right). \tag{1.2.2}$$

If we then recall that

$$m^2(\varphi) = \frac{d^2V}{d\varphi^2} = 3\lambda\varphi^2 - \mu^2$$

---

[3] Where this does not lead to confusion, we will simply denote the classical scalar field by φ, rather than φ_0. In certain other cases, we will also denote the initial value of the classical scalar field φ by φ_0. We hope that the meaning of φ and φ_0 in each particular case will be clear from the context.

in the model (1.1.5) (see Eq. (1.1.10)), the complete expression for $V(\varphi, T)$ can be written in the form

$$V(\varphi, T) = -\frac{\mu^2}{2}\varphi^2 + \frac{\lambda\varphi^4}{4} + \frac{\lambda T^2}{8}\varphi^2 + \ldots, \tag{1.2.3}$$

where we have omitted terms that do not depend on φ. The behavior of $V(\varphi, T)$ is shown in Fig. 2 for a number of different temperatures.

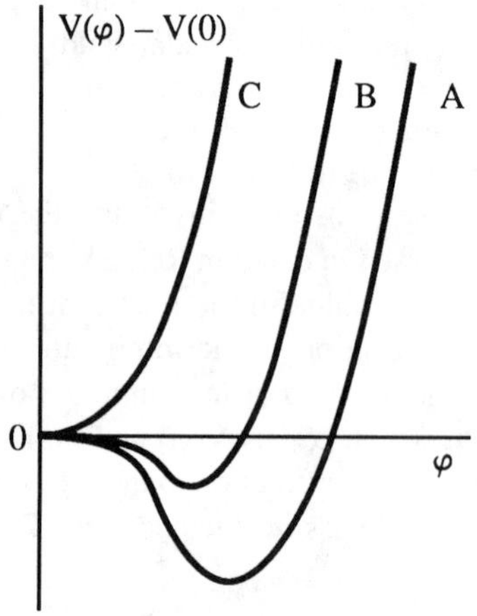

Figure 2. Effective potential $V(\varphi, T)$ in the theory (1.1.5) at finite temperature. A) $T = 0$; B) $0 < T < T_c$; C) $T > T_c$. As the temperature rises, the field φ varies smoothly, corresponding to a second-order phase transition.

It is clear from (1.2.3) that as T rises, the equilibrium value of φ at the minimum of $V(\varphi, T)$ decreases, and above some critical temperature

$$T_c = \frac{2\mu}{\sqrt{\lambda}}, \tag{1.2.4}$$

the only remaining minimum is the one at $\varphi = 0$, i.e., symmetry is restored (see Fig. 2). Equation (1.2.3) then implies that the field φ decreases continuously to zero with rising temperature; the restoration of symmetry in the theory (1.1.5) is a second-order phase transition.

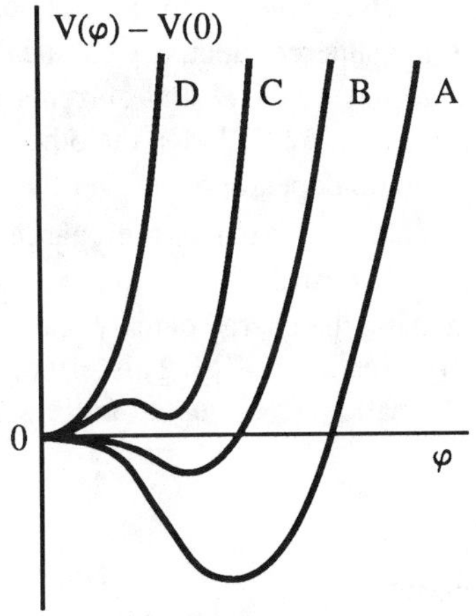

Figure 3. Behavior of the effective potential $V(\varphi, T)$ in theories in which phase transitions are first-order. Between T_{c_1} and T_{c_2}, the effective potential has two minima; at $T = T_c$, these minima have the same depth. A) $T = 0$; B) $T_{c_1} < T < T_c$; C) $T_c < T < T_{c_2}$; D) $T > T_{c_2}$.

Note that in the case at hand, when $\lambda \ll 1$, $T_c \gg m$ over the entire range of values of φ that is of interest ($\varphi \lesssim \varphi_c$), so that a high-temperature expansion of $V(\varphi, T)$ in powers of m/T in (1.2.2) is perfectly justified. However, it is by no means true that phase transitions take place only at $T \gg m$ in all theories. It often happens that at the instant of a phase transition, the potential $V(\varphi, T)$ has two local minima, one giving a stable state and the other an unstable state of the system (Fig. 3). We then have a first-order phase transition, due to the formation and subsequent expansion of bubbles of a stable phase within an unstable one, as in boiling water. Investigation of the first-order phase transitions in gauge theories [62] indicates that such transitions are sometimes considerably delayed, so that the transition takes place (with rising temperature) from a strongly superheated state, or (with falling temperature) from a strongly supercooled one. Such processes are explosive, which can lead to many important and interesting effects in an expanding universe. The formation of bubbles of a new phase is typically a barrier tunnelling process; the theory of this process at a finite temperature was given in [62].

It is well known that superconductivity can be destroyed not only by heating, but also by external fields **H** and currents j; analogous effects exist in unified gauge theories [22, 23]. On the other hand, the value of the field φ, being a scalar, should depend not just on the currents j, but on the square of current $j^2 = \rho^2 - \mathbf{j}^2$, where ρ is the charge density. Therefore, while increasing the current j usually leads to the restoration of symmetry in gauge theories, increasing the charge density ρ usually results in the enhancement of symmetry breaking [27]. This effect and others that may exist in superdense cold matter are discussed in Refs. 27–29.

1.3 Hot universe theory

There have been two important stages in the development of twentieth-century cosmology. The first began in the 1920's, when Friedmann used the general theory of relativity to create a theory of a homogeneous and isotropic expanding universe with metric [63–65]

$$ds^2 = dt^2 - a^2(t)\left[\frac{dr^2}{1 - kr^2} + r^2(d\theta^2 + \sin^2\theta\, d\varphi^2)\right], \qquad (1.3.1)$$

where $k = +1, -1$, or 0 for a closed, open, or flat Friedmann universe, and $a(t)$ is the "radius" of the universe, or more precisely, its scale factor (the total size of the universe may be infinite). The term *flat universe* refers to the fact that when $k = 0$, the metric (1.3.1) can be put in the form

$$ds^2 = dt^2 - a^2(t)\,(dx^2 + dy^2 + dz^2). \qquad (1.3.2)$$

At any given moment, the spatial part of the metric describes an ordinary three-dimensional Euclidean (flat) space, and when $a(t)$ is constant (or slowly varying, as in our universe at present), the flat-universe metric describes Minkowski space.

For $k = \pm 1$, the geometrical interpretation of the three-dimensional space part of (1.3.1) is somewhat more complicated [65]. The analog of a closed world at any given time t is a sphere S^3 embedded in some auxiliary four-dimensional space (x, y, z, τ). Coordinates on this sphere are related by

$$x^2 + y^2 + z^2 + \tau^2 = a^2(t). \tag{1.3.3}$$

The metric on the surface can be written in the form

$$dl^2 = a^2(t)\left[\frac{dr^2}{1 - r^2} + r^2(d\theta^2 + \sin^2\theta\, d\varphi^2)\right], \tag{1.3.4}$$

where r, ϑ, and φ are spherical coordinates on the surface of the sphere S^3.

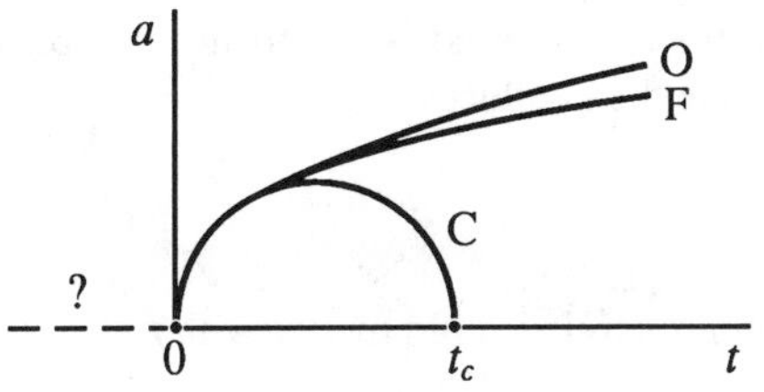

Figure 4. Evolution of the scale factor $a(t)$ for three different versions of the Friedmann hot universe theory: open (O), flat (F), and closed (C).

The analog of an open universe at fixed t is the surface of the hyperboloid

$$x^2 + y^2 + z^2 + \tau^2 = -a^2(t). \tag{1.3.5}$$

The evolution of the scale factor $a(t)$ is given by the Einstein equations

$$\ddot{a} = -\frac{4\pi}{3} G(\rho + 3p)a, \tag{1.3.6}$$

$$H^2 + \frac{k}{a^2} \equiv \left(\frac{\dot{a}}{a}\right)^2 + \frac{k}{a^2} = \frac{8\pi}{3} G\rho. \tag{1.3.7}$$

Here ρ is the energy density of matter in the universe, and p is its pressure. The gravitational constant $G = M_P^{-2}$, where $M_P = 1.2 \cdot 10^{19}$ GeV is the Planck mass, and $H = \frac{\dot{a}}{a}$ is the Hubble "constant," which in general is a function of time. Equations (1.3.6) and (1.3.7) imply an energy conservation law, which can be written in the form

$$\dot{\rho} a^3 + 3(\rho + p)a^2\dot{a} = 0. \tag{1.3.8}$$

To find out how this universe will evolve in time, one also needs to know the so-called equation of state, which relates the energy density of matter to its pressure. One may assume, for instance, that the equation of state for matter in the universe takes the form $p = \alpha\rho$. From the energy conservation law, one then deduces that

$$\rho \sim a^{-3(1+\alpha)}. \tag{1.3.9}$$

In particular, for nonrelativistic cold matter with $p = 0$,

$$\rho \sim a^{-3}, \tag{1.3.10}$$

and for a hot ultrarelativistic gas of noninteracting particles with $p = \frac{\rho}{3}$,

$$\rho \sim a^{-4}. \tag{1.3.11}$$

In either case (and in general for any medium with $p > -\frac{\rho}{3}$), when a is small, the quantity $\frac{8\pi}{3} G\rho$ is much greater than $\frac{k}{a^2}$. We then find from (1.3.7) that for small a, the expansion of the universe goes as

$$a \sim t^{\frac{2}{3(1+\alpha)}}. \tag{1.3.12}$$

In particular, for nonrelativistic cold matter

$$a \sim t^{2/3}, \tag{1.3.13}$$

and for the ultrarelativistic gas

$$a \sim t^{1/2}. \tag{1.3.14}$$

Thus, regardless of the model used ($k = \pm 1, 0$), the scale factor vanishes at some time $t = 0$, and the matter density at that time becomes infinite. It can also be shown that at that time, the curvature tensor $R_{\mu\nu\alpha\beta}$ goes to infinity as well. That is why the point $t = 0$ is known as the point of the initial cosmological singularity (Big Bang).

An open or flat universe will continue to expand forever. In a closed universe with $p > -\frac{\rho}{3}$, on the other hand, there will be some point in the expansion when the term $\frac{1}{a^2}$ in (1.3.7) becomes equal to $\frac{8\pi}{3}G\rho$. Thereafter, the scale constant a decreases, and it vanishes at some time t_c (Big Crunch). It is straightforward to show [65] that the lifetime of a closed universe filled with a total mass M of cold nonrelativistic matter is

$$t_c = \frac{4M}{3}G = \frac{4M}{3M_P^2} \sim \frac{M}{M_P} \cdot 10^{-43}\ \text{sec}. \tag{1.3.15}$$

The lifetime of a closed universe filled with a hot ultrarelativistic gas of particles of a single species may be conveniently expressed in terms of the total entropy of the universe, $S = 2\pi^2 a^3 s$, where s is the entropy density. If the total entropy of the universe does not change (adiabatic expansion), as is often assumed, then

$$t_c = \left(\frac{32}{45\pi^2}\right)^{\frac{1}{6}} \frac{S^{2/3}}{M_P} \sim S^{2/3} \cdot 10^{-43}\ \text{sec}. \tag{1.3.16}$$

These estimates will turn out to be useful in discussing the difficulties encountered by the standard theory of expansion of the hot universe.

Up to the mid-1960's, it was still not clear whether the early universe had been hot or cold. The critical juncture marking the beginning of the second stage in the development of modern cosmology was Penzias and Wilson's 1964–65 discovery of the 2.7 K microwave background radiation arriving from the farthest reaches of the universe. The existence of the microwave background had been predicted by the hot universe theory [66, 67], which gained immediate and widespread acceptance after the discovery.

According to that theory, the universe, in the very early stages of its evolution, was filled with an ultrarelativistic gas of photons, electrons, positrons, quarks, antiquarks, etc. At that epoch, the excess of baryons over antibaryons was but a small fraction (at most 10^{-9}) of the total number of particles. As a result of the decrease of the effective coupling constants for weak, strong, and electromagnetic interactions with increasing density, effects related to interactions among those particles affected the equation of state of the superdense matter only slightly, and the quantities s, ρ, and p were given [61] by

$$\rho = 3\,p = \frac{\pi^2}{30}\,N(T)\,T^4, \tag{1.3.17}$$

$$s = \frac{2\pi^2}{45}\,N(T)\,T^3, \tag{1.3.18}$$

where the effective number of particle species N(T) is $N_B(T) + \frac{7}{8}N_F(T)$, and N_B and N_F are the number of boson and fermion species[4] with masses $m \ll T$.

In realistic elementary particle theories, N(T) increases with increasing T, but it typically does so relatively slowly, varying over the range 10^2 to 10^4. If the universe expanded adiabatically, with $sa^3 \approx$ const, then (1.3.18) implies that during the expansion, the quantity aT also remained approximately constant. In other words, the temperature of the universe dropped off as

$$T(t) \sim a^{-1}(t). \tag{1.3.19}$$

The background radiation detected by Penzias and Wilson is a result

[4] To be more precise, N_B and N_F are the number of boson and fermion degrees of freedom. For example, $N_B = 2$ for photons, $N_F = 1$ for neutrinos, $N_F = 2$ for electrons, etc.

of the cooling of the hot photon gas during the expansion of the universe. The exact equation for the time-dependence of the temperature in the early universe can be derived from (1.3.7) and (1.3.17):

$$t = \frac{1}{4\pi}\sqrt{\frac{45}{\pi N(T)}}\,\frac{M_P^2}{T^2}\,. \tag{1.3.20}$$

In the later stages of the evolution of the universe, particles and antiparticles annihilate each other, the photon-gas energy density falls off relatively rapidly (compare (1.3.10) and (1.3.11)), and the main contribution to the matter density starts to come from the small excess of baryons over antibaryons, as well as from other fields and particles which now comprise the so-called hidden mass in the universe.

The most detailed and accurate description of the hot universe theory can be found in the fundamental monograph by Zeldovich and Novikov [34] (see also [35]).

Several different avenues were pursued in the 1970's in developing this theory. Two of these will be most important in the subsequent discussion: the development of the hot universe theory with regard to the theory of phase transitions in superdense matter [18–24], and the theory of formation of the baryon asymmetry of the universe [36–38].

Specifically, as just stated in the preceding paragraph, symmetry should be restored in grand unified theories at superhigh temperatures. As applied to the simplest SU(5) model, for instance, this means that at a temperature $T \gtrsim 10^{15}$ GeV, there was essentially no difference between the weak, strong, and electromagnetic interactions, and quarks could easily transform into leptons; that is, there was no such thing as baryon number conservation. At $t_1 \sim 10^{-35}$ sec after the Big Bang, when the temperature had dropped to $T \sim T_{c_1} \sim 10^{14}$–$10^{15}$ GeV, the universe underwent the first symmetry-breaking phase transition, with SU(5) perhaps being broken into SU(3) × SU(2) × U(1). After this transition, strong interactions were separated from electroweak and leptons from quarks, and superheavy-meson decay processes ultimately leading to the baryon asymmetry of the universe were initiated. Then, at $t_2 \sim 10^{-10}$ sec, when the temperature had dropped to $T_{c_2} \sim 10^2$ GeV, there was a second phase transition, which broke the symmetry between the weak and electromagnetic interactions, SU(3) × SU(2) × U(1) → SU(3) × U(1). As the temperature dropped still further to $T_{c_3} \sim 10^2$ MeV, there was yet another phase transition (or perhaps two distinct ones), with the formation of baryons and mesons from quarks and the breaking of chiral invariance in strong interaction theory. Physical processes taking place at later stages in the evolution of the universe were much less dependent on the specific features of unified gauge theories (a description of these processes can be found in the books cited above [34, 35]).

Most of what we have to say in this book will deal with events that transpired approximately 10^{10} years ago, in the time up to about 10^{-10} seconds after the Big Bang. This will make it possible to examine the global structure of the universe, to derive a more adequate understanding of the present state of the universe and its future, and finally, even to modify considerably the very notion of the Big Bang.

1.4 Some properties of the Friedmann models

In order to provide some orientation for the problems of modern cosmology, it is necessary to present at least a rough idea of typical values of the quantities appearing in the equations, the relationships among these quantities, and their physical meaning.

We start with the Einstein equation (1.3.7), which we will find to be particularly important in what follows. What can one say about the Hubble parameter $H = \frac{\dot{a}}{a}$, the density ρ, and the quantity k?

At the earliest stages of the evolution of the universe (not long after the singularity), H and ρ might have been arbitrarily large. It is usually assumed, though, that at densities $\rho \gtrsim M_P^4 \sim 10^{94}$ g/cm³, quantum gravity effects are so significant that quantum fluctuations of the metric exceed the classical value of $g_{\mu\nu}$, and classical space-time does not provide an adequate description of the universe [34]. We therefore restrict further discussion to phenomena for which $\rho \lesssim M_P^4$, $T \lesssim M_P \sim 10^{19}$ GeV, $H < M_P$, and so on. This restriction can easily be made more precise by noting that quantum corrections to the Einstein equations in a hot universe are already significant for $T \sim \frac{M_P}{\sqrt{N}} \sim 10^{17}$–$10^{18}$ GeV and $\rho \sim \frac{M_P^4}{N} \sim 10^{90}$–$10^{92}$ g/cm³. It is also worth noting that in an expanding universe, thermodynamic equilibrium cannot be established immediately, but only when the temperature T is sufficiently low. Thus in SU(5) models, for example, the typical time for equilibrium to be established is only comparable to the age t of the universe from (1.3.20) when $T \lesssim T^* \sim 10^{16}$ GeV (ignoring hypothetical graviton processes that might lead to equilibrium even before the Planck time has elapsed, with $\rho \gg M_P^4$).

The behavior of the nonequilibrium universe at densities of the order of the Planck density is an important problem to which we shall return again and again. Notice, however, that $T^* \sim 10^{16}$ GeV exceeds the typical critical temperature for a phase transition in grand unified theories, $T_c \lesssim 10^{15}$ GeV.

At the present time, the values of H and ρ are not well-determined. For example,

$$H = 100\, h \frac{\text{km}}{\text{sec} \cdot \text{Mpc}} \sim h \cdot (3 \cdot 10^{17})^{-1}\ \text{sec}^{-1} \sim h \cdot 10^{-10}\ \text{yr}^{-1}, \qquad (1.4.1)$$

where the factor h lies somewhere in the range 1/2 to 1 (1 megaparsec (Mpc) equals $3.09 \cdot 10^{24}$ cm or $3.26 \cdot 10^{6}$ light years). For a flat universe, H and ρ are uniquely related by Eq. (1.3.7); the corresponding value $\rho = \rho_c(H)$ is known as the critical density, since the universe must be closed (for given H) at higher density, and open at lower:

$$\rho_c = \frac{3H^2}{8\pi G} = \frac{3H^2 M_P^2}{8\pi}, \qquad (1.4.2)$$

and at present, the critical density of the universe is

$$\rho_c \approx 2 \cdot 10^{-29}\, h^2\ \text{g/cm}^3. \qquad (1.4.3)$$

The ratio of the actual density of the universe to the critical density is given by the quantity Ω,

$$\Omega = \frac{\rho}{\rho_c}. \qquad (1.4.4)$$

Contributions to the density ρ come both from luminous baryon matter, with $\rho_{LB} \sim 10^{-2}\, \rho_c$, and from dark (hidden, missing) matter, which should have a density at least an order of magnitude higher. The observational data imply a current ratio

$$0.1 \lesssim \Omega \lesssim 2. \qquad (1.4.5)$$

The present-day universe is thus not too far from being flat (while according to the inflationary universe scenario, $\Omega = 1$ to high accuracy; see below). Furthermore, as we remarked previously, the early universe not far from being spatially flat because of the relatively small value of $\frac{k}{a^2}$ compared to $\frac{8\pi G}{3}\rho$ in (1.3.7). From here on, therefore, we confine our estimates to those for a flat universe ($k = 0$).

Equations (1.3.13) and (1.3.14) imply that the age of a universe filled with ultrarelativistic gas is related to the quantity $H = \frac{\dot{a}}{a}$ by

$$t = \frac{1}{2H}, \qquad (1.4.6)$$

and for a universe with the equation of state $p = 0$,

$$t = \frac{2}{3H}. \qquad (1.4.7)$$

If, as is often supposed, the major contribution to the missing mass comes from nonrelativistic matter, the age of the universe will presently be given by Eq. (1.4.7):

$$t \sim \frac{2}{3h} \cdot 10^{10}\ \text{yr}; \qquad \frac{1}{2} \lesssim h \lesssim 1. \qquad (1.4.8)$$

H(t) not only determines the age, but the distance to the horizon as well, that is, the radius of the observable part of the universe.

To be more precise, one must distinguish between two horizons — the particle horizon and the event horizon [35].

The particle horizon delimits the causally connected part of the universe that an observer can see in principle *at a given time t*. Since light propagates on the light cone $ds^2 = 0$, we find from (1.3.1) that the rate at which the radius r of a wavefront changes is

$$\frac{dr}{dt} = \frac{\sqrt{1 - kr^2}}{a(t)}, \qquad (1.4.9)$$

and the physical distance travelled by light in time t is

$$R_p(t) = a(t) \int_0^{r(t)} \frac{dr}{\sqrt{1-kr^2}} = a(t) \int_0^t \frac{dt'}{a(t')}. \qquad (1.4.10)$$

In particular, for $a(t) \sim t^{2/3}$ (1.3.13),

$$R_p(t) = 3t = 2[H(t)]^{-1}. \qquad (1.4.11)$$

The quantity R_p gives the size of the observable part of the universe at time t. From (1.4.1) and (1.4.11), we obtain the present-day value of R_p (i.e., the distance to the particle horizon)

$$R_p = 0.9\ h^{-1} \cdot 10^{28}\ \text{cm}. \qquad (1.4.12)$$

In a certain conceptual sense, the event horizon is the complement of the particle horizon: it delimits that part of the universe from which we can ever (up to some time t_{max}) receive information about events taking place *now* (at time t):

$$R_e(t) = a(t) \int_t^{t_{max}} \frac{dt'}{a(t')}. \qquad (1.4.13)$$

For a flat universe with $a(t) \sim t^{2/3}$, there is no event horizon: $R_e(t) \to \infty$ as $t_{max} \to \infty$. In what follows, we will be particularly interested in the case $a(t) \sim e^{Ht}$, where H = const. This corresponds to the de Sitter metric, and gives

$$R_e(t) = H^{-1}. \qquad (1.4.14)$$

The thrust of this result is that an observer in an exponentially expanding universe sees only those events that take place at a distance no farther away than H^{-1}. This is completely analogous to the situation for a black hole, from whose surface no information can escape. The difference is that an observer in de Sitter space (in an exponentially expanding universe) will find himself effectively *surrounded* by a "black hole" located at a distance H^{-1}.

In closing, let us note one more rather perplexing circumstance. Consider two points separated by a distance R at time t in a flat Friedmann universe. If the spatial coordinates of these points remain unchanged (and in that sense, they remain stationary), the distance between them will nevertheless increase, due to the general expansion of the universe, at a rate

$$\frac{dR}{dt} = \frac{\dot{a}}{a} R = H R. \tag{1.4.15}$$

What this means, then, is that two points more than a distance H^{-1} apart will move away from one another faster than the speed of light $c = 1$. But there is no paradox here, since what we are concerned with now is the rate at which two objects subject to the general cosmological expansion separate from each other, and not with a signal propagation velocity at all, which is related to the local variation of particle spatial coordinates. On the other hand, it is just this effect that provides the foundation for the existence of an event horizon in de Sitter space.

1.5 Problems of the standard scenario

Following the discovery of the microwave background radiation, the hot universe theory immediately gained widespread acceptance. Workers in the field have indeed pointed out certain difficulties which, over the course of many years, have nevertheless come to be looked upon as only temporary. In order to make the changes now taking place in cosmology more comprehensible, we list here some of the problems of the standard hot universe theory.

1.5.1. *The singularity problem*

Equations (1.3.9) and (1.3.12) imply that for all "reasonable" equations of state, the density of matter in the universe goes to infinity as $t \to \infty$, and the corresponding solutions cannot be formally continued to the domain $t < 0$.

One of the most distressing questions facing cosmologists is whether anything existed *before* $t = 0$; if not, then where did the universe come from? The birth and death of the universe, like the birth and death of a human being, is one of the most worrisome problems facing not just cosmologists, but all of contemporary science.

At first, there seemed to be some hope that even if the problem could not be solved, it might at least be possible to circumvent it by considering a more general model of the universe than the Friedmann model — perhaps an inhomogeneous, anisotropic universe filled with matter having some exotic equation of state. Studies of the general structure of space-time near a singularity [68] and several important theorems on singularities in the general theory of relativity [69, 70] proven by topological methods, however, demonstrated that it was highly unlikely that this problem could be solved within the framework of classical gravitation theory.

1.5.2. *The flatness of space*

This problem admits of several equivalent or almost equivalent formulations, differing somewhat in the approach taken.

a. THE EUCLIDICITY PROBLEM. We all learned in grade school that our world is described by Euclidean geometry, in which the angles of a triangle sum to 180° and parallel lines never meet (or they "meet at infinity"). In college, we were told that it was Riemannian geometry that described the world, and that parallel lines *could* meet or diverge at infinity. But nobody ever explained why what we learned in school was also true (or almost true) — that is, why the world is Euclidean to such an incredible degree of accuracy. This is even more surprising when one realizes that there is but one natural scale length in general relativity, the Planck length $l_P \sim M_P^{-1} \sim 10^{-33}$ cm.

One might expect that the world would be close to Euclidean except perhaps at distances of the order of l_P or less (that is, less than the characteristic radius of curvature of space). In fact, the opposite is true: on small scales $l \lesssim l_P$, quantum fluctuations of the metric make it impossible in general to describe space in classical terms (this leads to the concept of space-time foam [71]). At the same time, for reasons unknown, space is almost perfectly Euclidean on large scales, up to $l \sim 10^{28}$ cm — 60 orders of magnitude greater than the Planck length.

b. THE FLATNESS PROBLEM. The seriousness of the preceding problem is most easily appreciated in the context of the Friedmann model (1.3.1). We have from Eq. (1.3.7) that

$$|\Omega - 1| = \frac{|\rho(t) - \rho_c|}{\rho_c} = [\dot{a}(t)]^{-2}, \tag{1.5.1}$$

where ρ is the energy density in the universe, and ρ_c is the critical density for a flat universe with the same value of the Hubble parameter $H(t)$.

As already mentioned in Section 1.4, the present-day value of Ω is known only roughly, $0.1 \lesssim \Omega \lesssim 2$, or in other words our universe could presently show a fairly sizable departure from flatness. On the other hand, $(\dot{a})^{-2} \sim t$ in the early stages of evolution of a hot universe (see (1.3.14)), so the quantity $|\Omega - 1| = \left|\frac{\rho}{\rho_c} - 1\right|$ was extremely small. One can show that in order for Ω to lie in the range $0.1 \lesssim \Omega \lesssim 2$ now, the early universe must have had $|\Omega - 1| \lesssim 10^{-59}\frac{M_P}{T^2}$, so that at $T \sim M_P$,

$$|\Omega - 1| = \left|\frac{\rho}{\rho_c} - 1\right| \lesssim 10^{-59}. \tag{1.5.2}$$

This means that if the density of the universe were initially (at the Planck time $t_P \sim M_P^{-1}$) greater than ρ_c, say by $10^{-55}\rho_c$, it would be closed, and the limiting value t_c would be so small that the universe would have collapsed long ago. If on the other hand the density at the Planck time were $10^{-55}\rho_c$ less than ρ_c, the present energy density in the universe would be vanishingly low, and the life could not exist. The question of why the energy density ρ in the early universe was so fantastically close to the critical density (Eq. (1.5.2)) is usually known as the flatness problem.

c. THE TOTAL ENTROPY AND TOTAL MASS PROBLEM. The question here is why the total entropy S and total mass M of matter in the observable part of the universe, with $R_P \sim 10^{28}$ cm, is so large. The total entropy S is of order $(R_P T_\gamma)^3 \sim 10^{87}$, where $T_\gamma \sim 2.7$ K is the temperature of the primordial background radiation. The total mass is given by

$M \sim R_P^3 \rho_c \sim 10^{55}$ g $\sim 10^{49}$ tons.

If the universe were open and its density at the Planck time had been subcritical, say, by $10^{-55}\rho_c$, it would then be easy to show that the total mass and entropy of the observable part of the universe would presently be many orders of magnitude lower.

The corresponding problem becomes particularly difficult for a closed universe. We see from (1.3.15) and (1.3.16) that the total lifetime t_c of a closed universe is of order $M_P^{-1} \sim 10^{-43}$ sec, and this will be a long timespan ($\sim 10^{10}$ yr) only when the total mass and energy of the entire universe are extremely large. But why is the total entropy of the universe so large, and why should the mass of the universe be tens of orders of magnitude greater than the Planck mass M_P, the only parameter with the dimension of mass in the general theory of relativity? This question can be formulated in a paradoxically simple and apparently naïve way: Why are there so many different things in the universe?

d. THE PROBLEM OF THE SIZE OF THE UNIVERSE. Another problem associated with the flatness of the universe is that according to the hot universe theory, the total size l of the part of the universe currently accessible to observation is proportional to $a(t)$; that is, it is inversely proportional to the temperature T (since the quantity aT is practically constant in an adiabatically expanding hot universe — see Section 1.3). This means that at $T \sim M_P \sim 10^{19}$ GeV $\sim 10^{32}$ K, the region from which the observable part of the universe (with a size of 10^{28} cm) formed was of the order of 10^{-4} cm in size, or 29 orders of magnitude greater than the Planck length $l_P \sim M_P^{-1} \sim 10^{-33}$ cm. Why, when the universe was at the Planck density, was it 29 orders of magnitude bigger than the Planck length? Where do such large numbers come from?

We discuss the flatness problem here in such detail not only because an understanding of the various aspects of this problem turns out to be important for an understanding of the difficulties inherent in the standard hot universe theory, but also in order to be able to understand later which versions of the inflationary universe scenario to be discussed in this book can resolve this problem.

1.5.3. *The problem of the large-scale homogeneity and isotropy of the universe*

In Section 1.3, we assumed that the universe was initially absolutely homogeneous and isotropic. In actuality, or course, it is not completely homogeneous and isotropic even now, at least on a relatively small scale, and this means that there is no reason to believe that it was homogeneous *ab initio*. The most natural assumption would be that the initial conditions at points sufficiently far from one another were chaotic and uncorrelated [72]. As was shown by Collins and Hawking [73] under certain assumptions, however, the class of initial conditions for which the universe tends asymptotically (at large t) to a Friedmann universe (1.3.1) is one of measure zero among all possible initial conditions. This is the crux of the problem of the homogeneity and isotropy of the universe. The subtleties of this problem are discussed in more detail in the book by Zeldovich and Novikov [34].

1.5.4. *The horizon problem*

The severity of the isotropy problem is somewhat ameliorated by the fact that effects connected with the presence of matter and elementary particle production in an expanding universe can make the universe locally isotropic [34, 74]. Clearly, though, such effects cannot lead to global isotropy, if only because causally disjoint regions separated by a distance greater than the particle horizon (which in the simplest cases is given by $R_p \sim t$, where t is the age of the universe) cannot influence each other. In the meantime, studies of the microwave background have shown that at $t \sim 10^5$ yr, the universe was quite accurately homogeneous and isotropic on scales orders of magnitude greater than t, with temperatures T in different regions differing by less than $O(10^{-4})$T. Inasmuch as the observable part of the universe presently consists of about 10^6 regions that were causally unconnected at $t \sim 10^5$ yr, the probability of the temperature T in these regions being fortuitously correlated to the indicated accuracy is at most 10^{-24}–10^{-30}. It is exceedingly difficult to come up with a convincing explanation of this fact within the scope of the standard scenario. The corresponding problem is known as the horizon problem or the causality problem [48, 56].

There is one more aspect of the horizon problem which will be important for our purposes. As we mentioned in the earlier discussion of the flatness problem, at the Planck time $t_P \sim M_P^{-1} \sim 10^{-43}$ sec, when the size (the radius of the particle horizon) of each causally connected region of the universe was $l_P \sim 10^{-33}$ cm, the size of the overall region from which the observable part of the universe formed was of order 10^{-4} cm. The latter thus consisted of $(10^{29})^3 \sim 10^{87}$ causally unconnected regions. Why then should the expansion of the universe (or its emergence from the space-time foam with the Planck density $\rho \sim M_P^4$) have begun simultaneously (or nearly so) in such a huge number of causally unconnected regions? The probability of this occurring at random is close to $\exp(-10^{90})$.

1.5.5. *The galaxy formation problem*

The universe is of course not perfectly homogeneous. It contains such important inhomogeneities as stars, galaxies, clusters of galaxies, etc. In explaining the origin of galaxies, it has been necessary to assume the existence of initial inhomogeneities [75] whose spectrum is usually taken to be almost scale-invariant [76]. For a long time, the origin of such density inhomogeneities remained completely obscure.

1.5.6. *The baryon asymmetry problem*

The essence of this problem is to understand why the universe is made almost entirely of matter, with almost no antimatter, and why on the other hand baryons are many orders of magnitude scarcer than photons, with $\frac{n_B}{n_\gamma} \sim 10^{-9}$.

Over the course of time, these problems have taken on an almost metaphysical flavor. The first is self-referential, since it can be restated by asking “What was there before there was anything at all?” or “What was at the time at which there was no space-time at all?” The others could always be avoided by saying that by sheer good luck, the initial conditions in the universe were such as to give it precisely the form it finally has now, and that it is meaningless to discuss initial conditions. Another possible answer is based on the so-called Anthropic Principle, and seems almost purely metaphysical: we live in a homogeneous, isotropic universe containing an excess of matter over antimatter simply because in an inhomogeneous, anisotropic universe with equal amounts of matter and antimatter, life would be impossible and these questions could not even be asked [77].

Despite its cleverness, this answer is not entirely satisfying, since it explains neither the small ratio $\frac{n_B}{n_\gamma} \sim 10^{-9}$, nor the high degree of homogeneity and isotropy in the universe, nor the observed spectrum of galaxies. The Anthropic Principle is also incapable of explaining why all properties of the universe are approximately uniform over its entire observable part ($l \sim 10^{28}$ cm) — it would be perfectly possible for life to arise if favorable conditions existed, for example, in a region the size of the solar system, $l \sim 10^{14}$ cm. Furthermore, the Anthropic Principle rests on an implicit assumption that either universes are constantly created, one after another, or there exist many different universes, and that life arises in those universes which are most hospitable. It is not clear, however, in what sense one can speak of different universes if ours is in fact unique. We shall return to this question later and provide a basis for a version of the Anthropic Principle in the context of inflationary cosmology [57, 78, 79].

The first breach in the cold-blooded attitude of most physicists toward the foregoing "metaphysical" problems appeared after Sakharov discovered [36] that the baryon asymmetry problem could be solved in theories in which baryon number is not conserved by taking account of nonequilibrium processes with C and CP-violation in the very early universe. Such processes can occur in all grand unified theories [36–38]. The discovery of a way to generate the observed baryon asymmetry of the universe was considered to be one of the greatest successes of the hot universe cosmology. Unfortunately, this success was followed by a whole series of disappointments.

1.5.7. *The domain wall problem*

As we have seen, symmetry is restored in the theory (1.1.5) when $T > 2\mu/\sqrt{\lambda}$. As the temperature drops in an expanding universe, the symmetry is broken. But this symmetry breaking occurs independently in all causally unconnected regions of the universe, and therefore in each of the enormous number of such regions comprising the universe at the time of the symmetry-breaking phase transition, both the field $\varphi = +\mu/\sqrt{\lambda}$ and the field $\varphi = -\mu/\sqrt{\lambda}$ can arise. Domains filled by the field $\varphi = +\mu/\sqrt{\lambda}$ are separated from those with the field $\varphi = -\mu/\sqrt{\lambda}$ by domain walls. The energy density of these walls turns out to be so high that the existence of just one in the observable part of the universe would lead to unacceptable cosmological consequences [41]. This implies that a theory with spontaneous breaking of a discrete symmetry is inconsistent with the cosmological data. Initially, the principal theories fitting this description were those with spontaneously broken CP invariance [80]. It was subsequently found that domain walls also occur in the simplest version of the SU(5) theory, which has the discrete invariance $\Phi \to -\Phi$ [42], and in most axion theories [43]. Many of these theories are very appealing, and it would be nice if we could find a way to save at least some of them.

1.5.8. *The primordial monopole problem*

Other structures besides domain walls can be produced following symmetry-breaking phase transitions. For example, in the Higgs model with broken U(1) symmetry and certain others, strings of the Abrikosov superconducting vortex tube type can occur [81]. But the most important effect is the creation of superheavy t'Hooft—Polyakov magnetic monopoles [82, 83], which should be copiously produced in practically all of the grand unified theories [84] when phase transitions take place at $T_{c_1} \sim 10^{14}$–10^{15} GeV. It was shown by Zeldovich and Khlopov [40] that monopole annihilation proceeds very slowly, and that the monopole density at present should be comparable to the baryon density. This would of course have catastrophic consequences, as the mass of each monopole is perhaps 10^{16} times that of the proton, giving an energy density in the universe about 15 orders of magnitude higher than the critical density $\rho_c \sim 10^{29}$ g/cm^3. At that density, the universe would have collapsed long ago. The primordial monopole problem is one of the sharpest encountered thus far by elementary particle theory and cosmology, since it relates to practically all unified theories of weak, strong, and electromagnetic interactions.

1.5.9. *The primordial gravitino problem*

One of the most interesting directions taken by modern elementary particle physics is the study of supersymmetry, the symmetry between fermions and bosons [85]. Here we will not list all the advantages of supersymmetric theories, referring the reader instead to the literature [13, 14]. We merely point out that phenomenological supersymmetric theories, and N = 1 supergravity in particular, may provide a way to solve the mass hierarchy problem of unified field theories [15]; that is, they may explain why there exist such drastically differing mass scales $M_P \gg M_X \sim 10^{15}$ GeV and $M_X \gg m_W \sim 10^2$ GeV.

One of the most interesting attempts to resolve the mass hierarchy problem for N = 1 supergravity is based on the suggestion that the gravitino (the spin-3/2 superpartner of the graviton) has mass $m_{3/2} \sim m_W \sim 10^2$ GeV [15]. It has been shown [86], however, that gravitinos with this mass should be copiously produced as a result of high-energy particle collisions in the early universe, and that gravitinos decay rather slowly.

Most of these gravitinos would only have decayed by the later stages of evolution of the universe, after helium and other light elements had been synthesized, which would have led to many consequences that are inconsistent with the observations [44, 45]. The question is then whether we can somehow rescue the universe from the consequences of gravitino decay; if not, must we abandon the attempt to solve the hierarchy problem?

Some particular models [87] with superlight or superheavy gravitinos manage to avoid these difficulties. Nevertheless, it would be quite valuable if we could somehow avoid the stringent constraints imposed on the parameters of N = 1 supergravity by the hot universe theory.

1.5.10. *The problem of Polonyi fields*

The gravitino problem is not the only one that arises in phenomenological theories based on N = 1 supergravity (and superstring theory). The so-called scalar Polonyi fields χ are one of the major ingredients of these theories [46, 15]. They are relatively low-mass fields that interact weakly with other fields. At the earliest stages of the evolution of the universe they would have been far from the minimum of their corresponding effective potential $V(\chi)$. Later on, they would start to oscillate about the minimum of $V(\chi)$, and as the universe expanded, the Polonyi field energy density ρ_χ would decrease in the same manner as the energy density of nonrelativistic matter ($\rho_\chi \sim a^{-3}$), or in other words much more slowly than the energy density of hot plasma. Estimates indicate that for the most likely situations, the energy density presently stored in these fields should exceed the critical density by about 15 orders of magnitude [47, 48]. Somewhat more refined models give theoretical predictions of the density ρ_X that no longer conflict with the observational data by a factor of 10^{15}, but only by a factor of 10^6 [48], which of course is also highly undesirable.

1.5.11. *The vacuum energy problem*

As we have already mentioned, the advent of a constant homogeneous scalar field φ over all space simply represents a restructuring of the vacuum, and in some sense, space filled with a constant scalar field φ remains "empty" — the constant scalar field does not carry a preferred reference frame with it, it does not disturb the motion of objects passing through the space that it fills, and so forth. But when the scalar field appears, there is a change in the vacuum energy density, which is described by the quantity $V(\varphi)$. If there were no gravitational effects, this change in the energy density of the vacuum would go completely unnoticed. In general relativity, however, it affects the properties of space-time. $V(\varphi)$ enters into the Einstein equation in the following way:

$$R_{\mu\nu} - \frac{1}{2} g_{\mu\nu} R = 8\pi G T_{\mu\nu} = 8\pi G (\tilde{T}_{\mu\nu} + g_{\mu\nu} V(\varphi)) , \qquad (1.5.3)$$

where $T_{\mu\nu}$ is the total energy-momentum tensor, $\tilde{T}_{\mu\nu}$ is the energy-momentum tensor of substantive matter (elementary particles), and $g_{\mu\nu} V(\varphi)$ is the energy-momentum tensor of the vacuum (the constant scalar field φ). By comparing the usual energy-momentum tensor of matter

$$\tilde{T}_{\mu}^{\ \nu} = \begin{pmatrix} \rho & & & \\ & -p & & \\ & & -p & \\ & & & -p \end{pmatrix} \qquad (1.5.4)$$

with $g_{\mu}^{\ \nu} V(\varphi)$, one can see that the "pressure" exerted by the vacuum and its energy density have opposite signs, $p = -\rho = -V(\varphi)$.

The cosmological data imply that the present-day vacuum energy density ρ_v is not much greater in absolute value than the critical density $\rho_c \sim 10^{29}$ g/cm^3:

$$|\rho_{vac}| = |V(\varphi_0)| \lesssim 10^{-29}\ \mathrm{g/cm^3}. \tag{1.5.5}$$

This value of $V(\varphi)$ was attained as a result of a series of symmetry-breaking phase transitions. In the SU(5) theory, after the first phase transition $SU(5) \to SU(3) \times SU(2) \times U(1)$, the vacuum energy (the value of $V(\varphi)$) decreased by approximately 10^{80} g/cm^3. After the $SU(3) \times SU(2) \times U(1) \to SU(3) \times U(1)$ transition, it was reduced by about another 10^{25} g/cm^3. finally, after the phase transition that formed the baryons from quarks, the vacuum energy again decreased, this time by approximately 10^{14} g/cm^3, and surprisingly enough after all of these enormous drops, it turned out to equal zero to an accuracy of $\pm 10^{-29}$ g/cm^3! It seems unlikely that the complete (or almost complete) cancellation of the vacuum energy should occur merely by chance, without some deep physical reason. The vacuum energy problem in theories with spontaneous symmetry breaking [88] is presently deemed to be one of the most important problems facing elementary particle theories.

The vacuum energy density multiplied by $8\pi G$ is usually called the cosmological constant Λ [89]; in the present case, $\Lambda = 8\pi G V(\varphi)$ [88]. The vacuum energy problem is therefore also often called the cosmological constant problem.

Note that by no means do all theories ensure, even in principle, that the vacuum energy at the present epoch will be small. This is one of the most difficult problems encountered in Kaluza–Klein theories based on $N = 1$ supergravity in 11-dimensional space [16]. According to these theories, the vacuum energy would now be of order $-M_P^4 \sim -10^{94}$ g·cm^{-3}. On the other hand, indications that the vacuum energy problem may be solvable in superstring theories [17] have stimulated a great deal of interest in the latter.

1.5.12. *The problem of the uniqueness of the universe*

The essence of this problem was most clearly enunciated by Einstein, who said that "we wish to know not just the structure of Nature (and how natural phenomena are played out), but insofar as we can, we wish to attain a daring and perhaps utopian goal — to learn why Nature is just the way it is, and not otherwise" [90]. As recently as a few years ago, it would have seemed rather meaningless to ask why our space-time is four-dimensional, why there are weak, strong, and electromagnetic interactions and no others, why the fine-structure constant $\alpha = \frac{e^2}{4\pi}$ equals 1/137, and so on. Of late, however, our attitude toward such questions has changed, since unified theories of elementary particles frequently provide us with many different solutions of the relevant equations that in principle could describe our universe.

In theories with spontaneous symmetry breaking, for example, the effective potential will often have several local minima — in the theory (1.1.5), for instance, there are two, at $\varphi = \pm\mu/\sqrt{\lambda}$. In the minimal supersymmetric SU(5) grand unification theory, there are three local minima of the effective potential for the field Φ that have nearly the same depth [91]. The degree of degeneracy of the effective potential in supersymmetric theories (the number of different types of vacuum states having the same energy) becomes even greater when one takes into account other Higgs fields H which enter into the theory [92].

The question then arises as to how and why we come to be in a minimum in which the broken symmetry is SU(3) × U(1) (this question becomes particularly complicated if we recall that the early high-temperature universe was at an SU(5)-symmetric minimum $\Phi = H = 0$ [93], and there is no apparent reason for the entire universe to jump to the SU(3) × U(1) minimum upon cooling).

It is assumed in the Kaluza–Klein and superstring theories that we live in a space with $d > 4$ dimensions, but that $d - 4$ of these dimensions have been compactified — the radius of curvature of space in the corresponding directions is of order M_P^{-1}. That is why we cannot move in those directions, and space is apparently four-dimensional.

Presently, the most popular theories of that kind have $d = 10$ [17], but others with $d = 26$ [94] and $d = 506$ [95, 96] have also been considered. One of the most fundamental questions that comes up in this regard is why precisely $d - 4$ dimensions were compactified, and not $d - 5$ or $d - 3$. Furthermore, there are usually a great many ways to compactify $d - 4$ dimensions, and each results in its own peculiar laws of elementary particle physics in four-dimensional space. A frequently asked question is then why Nature chose just that particular vacuum state which leads to the strong, weak, and electromagnetic interactions with the coupling constants that we measure experimentally. As the dimension d of the parent space rises, this problem becomes more and more acute. Thus, it has variously been estimated that in $d = 10$ superstring theory, there are perhaps 10^{1500} ways of compactifying the ten-dimensional space into four dimensions (some of which may lead to unstable compactification), and there are many more ways to do this in space with $d > 10$. The question of why the world that surrounds us is structured just so, and not otherwise, has therefore lately turned into one of the most fundamental problems of modern physics.

We could continue this list of problems facing cosmologists and elementary particle theorists, of course, but here we are only interested in those that bear some relation to our basic theme.

The vacuum energy problem has yet to be solved definitively. There are many interesting attempts to do so, some of which are based on quantum cosmology and on the inflationary universe scenario. A solution to the baryon asymmetry problem was proposed by Sakharov long before the advent of the inflationary universe scenario [36], but the latter also introduces much that is new [97–99]. As for the other ten problems, they can all be solved either partially or completely within the framework of inflationary cosmology, and we now turn to a description of that theory.

1.6 A sketch of the development of the inflationary universe scenario

The main idea underlying all existing versions of the inflationary universe scenario is that in the very earliest stages of its evolution, the universe could be in an unstable vacuum-like state having high energy density. As we have already noted in the preceding section, the vacuum pressure and energy density are related by Eq. (1.5.4), $p = -\rho$. This means, according to (1.3.8), that the vacuum energy density does not change as the universe expands (a "void" remains a "void", even if it has weight). But (1.3.7) then implies that at large times t, the universe in an unstable vacuum state $\rho > 0$ should expand exponentially, with

$$a(t) = \mathrm{H}^{-1} \cosh \mathrm{H} t \qquad (1.6.1)$$

for $k = +1$ (a closed Friedmann universe),

$$a(t) = \mathrm{H}^{-1} e^{\mathrm{H}t} \qquad (1.6.2)$$

for $k = 0$ (a flat universe), and

$$a(t) = \mathrm{H}^{-1} \sinh \mathrm{H}t \tag{1.6.3}$$

for $k = -1$ (an open universe). Here $\mathrm{H} = \sqrt{\frac{8\pi}{3} G\rho} = \sqrt{\frac{8\pi\rho}{3M_P^2}}$. More generally, during expansion the magnitude of H in the inflationary universe scenario changes, but very slowly,

$$\dot{\mathrm{H}} \ll \mathrm{H}^2. \tag{1.6.4}$$

Over a characteristic time $\Delta t = \mathrm{H}^{-1}$ there is little change in the magnitude of H, so that one may speak of a quasiexponential expansion of the universe,

$$a(t) = a_0 \exp\left[\int_0^t \mathrm{H}(t)\, dt\right] \sim a_0 e^{\mathrm{H}t}, \tag{1.6.5}$$

or of a quasi-de Sitter stage in its expansion; just this regime of quasiexponential expansion is known as inflation.

Inflation comes to an end when H begins to decrease rapidly. The energy stored in the vacuum-like state is then transformed into thermal energy, and the universe becomes extremely hot. From that point onward, its evolution is described by the standard hot universe theory, with the important refinement that the initial conditions for the expansion stage of the hot universe are determined by processes which occurred at the inflationary stage, and are practically unaffected by the structure of the universe prior to inflation. As we shall demonstrate below, just this refinement enables us to solve many of the problems of the hot universe theory discussed in the preceding section.

The space (1.6.1)–(1.6.3) was first described in the 1917 papers of de Sitter [100], well before the appearance of Friedmann's theory of the expanding universe. However, de Sitter's solution was obtained in a form differing from (1.6.1)–(1.6.3), and for a long time its physical meaning was somewhat obscure. Before the advent of the inflationary universe scenario, de Sitter space was employed principally as a convenient staging area for developing the methods of general relativity and quantum field theory in curved space.

The possibility that the universe might expand exponentially during the early stages of its evolution, and be filled with superdense matter with the equation of state $p = -\rho$, was first suggested by Gliner [51]; see also [101–103]. When they appeared, however, these papers did not arouse much interest, as they dealt mainly with superdense baryonic matter, which, as we now believe, has an equation of state close to $p = \frac{\rho}{3}$, according to asymptotically free theories of weak, strong, and electromagnetic interactions.

It was subsequently realized that the constant (or almost constant) scalar field φ appearing in unified theories of elementary particles could play the role of a vacuum state with energy density $V(\varphi)$ [88]. The magnitude of the field φ in an expanding universe depends on the temperature, and at times of phase transitions that change φ, the energy stored in the field is transformed into thermal energy [21–24]. If, as sometimes happens, the phase transition takes place from a highly supercooled metastable vacuum state, the total entropy of the universe can increase considerably afterwards [23, 24, 104], and in particular, a cold Friedmann universe can become hot. The corresponding model of the universe was developed by Chibisov and the present author (in this regard, see [24, 105]).

In 1979–80, a very interesting model of the evolution of the universe was proposed by Starobinsky [52]. His model was based on the observation of Dowker and Critchley [106] that the de Sitter metric is a solution of the Einstein equations with quantum corrections. Starobinsky noted that this solution is unstable, and after the initial vacuum-like state decays (its energy density is related to the curvature of space R), de Sitter space transforms into a hot Friedmann universe [52].

Starobinsky's model proved to be an important step on the road towards the inflationary universe scenario. However, the principal advantages of the inflationary stage had not yet been recognized at that time. The main objective pursued in [52] was to solve the problem of the initial cosmological singularity. The goal was not reached at that time, and the question of initial conditions for the model remained unclear. In that model, furthermore, the density inhomogeneities that appeared after decay of de Sitter space turned out to be too large [107]. All of these considerations required that the foundations of the model be significantly altered [108–110]. In its modified form, the Starobinsky model has become one of the most actively developed versions of the inflationary universe scenario (or, to be more precise, the chaotic inflation scenario; see below).

The necessity of considering models of the universe with a stage of quasiexponential expansion was fully recognized only after the work of

Guth [53], who suggested using the exponential expansion (inflation) of the universe in a supercooled vacuum state $\varphi = 0$ to solve three of the problems discussed in Section 1.5, namely the flatness problem, the horizon problem, and the primordial monopole problem (a similar possibility for solving the flatness problem was independently suggested by Lapchinsky, Rubakov, and Veryaskin [111]). The scenario suggested by Guth was based on three fundamental propositions:

1. The universe initially expands in a state with superhigh temperature and restored symmetry, $\varphi_{(T)} = 0$.

2. One considers theories in which the potential $V(\varphi)$ retains a local minimum at $\varphi = 0$ even at a low temperature T. As a result, the evolving universe remains in the supercooled metastable state $\varphi = 0$ for a long time. Its temperature in this state falls off, the energy-momentum tensor gradually becomes equal to $T_{\mu\nu} = g_{\mu\nu} V(0)$, and the universe expands exponentially (inflates) for a long time.

3. Inflation continues until the end of a phase transition to a stable state $\varphi_0 \neq 0$. This phase transition proceeds by forming bubbles containing the field $\varphi = \varphi_0$. The universe heats up due to bubble-wall collisions, and its subsequent evolution is described by the hot universe theory.

The exponential expansion of the universe in stage (2) is introduced to make the term $\frac{k}{a^2}$ in the Einstein equation (1.3.7) vanishingly small as compared with $\frac{8\pi G}{3}\rho$, i.e., in order to make the universe flatter and flatter. This same process is invoked to ensure that the observable part of the universe, some 10^{28} cm in size, came about as the result of inflation of a very small region of space that was initially causally connected. In this scenario, monopoles are created at places where the walls of several exponentially large bubbles collide, and they therefore have exponentially low density.

The main idea behind the Guth scenario is very simple and extremely attractive. As noted by Guth himself [53], however, collisions of the walls of very large bubbles should lead to an unacceptable destruction of homogeneity and isotropy in the universe after inflation. Attempts to improve this situation were unsuccessful [112, 113] until cosmologists managed to surmount a certain psychological barrier and renounce all three of the aforementioned assumptions of the Guth scenario, while retaining the idea that the universe might have undergone inflation during the early stages of its evolution.

The invention of the so-called new inflationary universe scenario [54, 55] marked the departure from assumptions (2) and (3). This scenario is based on the fact that inflation can occur not only in a supercooled state $\varphi = 0$, but also during the process of growth of the field φ if this field increases to its equilibrium value φ_0 slowly enough, so that the time t for φ to reach the minimum of $V(\varphi)$ is much longer than H^{-1}. This condition can be realized if the effective potential of the field φ has a sufficiently flat part near $\varphi = 0$. If inflation during the stage when φ is rolling downhill is large enough, the walls of bubbles of the field φ (if they are formed) will, after inflation, be separated from one another by much more than 10^{28} cm, and will not engender any inhomogeneities in the observable part of the universe. In this scenario, the universe is heated after inflation not because of collisions between bubble walls, but because of the creation of elementary particles by the classical field φ, which executes damped oscillations about the minimum of $V(\varphi)$.

The new inflationary universe scenario is free of the major shortcomings of the old scenario. In the context of this scenario, it is possible to propose solutions not just to the flatness, horizon, and primordial monopole problems, but also to the homogeneity and isotropy problems, as well as many of the others referred to in Section 1.5. It has been found, in particular, that at the time of inflation in this scenario, density inhomogeneities are produced with a spectrum that is virtually independent of the logarithm of the wavelength (a so-called flat, scale-free, or Harrison–Zeldovich spectrum [214, 76]). This marked an important step on the road to solving the problem of the origin of the large-scale structure of the universe.

The successes of the new inflationary universe scenario were so impressive that even now, many scientists who speak of the inflationary universe scenario mean this new scenario [54, 55]. In our opinion, however, this scenario is still far from perfect; there are at least three problems that stand in the way of its successful implementation:

1. The new scenario requires a realistic theory of elementary particles in which the effective potential satisfies many constraints that are rather unnatural. For example, the potential $V(\varphi)$ must be very close to flat ($V(\varphi) \sim$ const) for values of the field close to $\varphi = 0$. If, for instance, the behavior of $V(\varphi)$ at small φ is close to $V(0) - \frac{\lambda}{4}\varphi^4$, then in order for density inhomogeneities generated at the time of inflation to have the required amplitude

$$\frac{\delta\rho}{\rho} \sim 10^{-4}\text{–}10^{-5}, \tag{1.6.6}$$

the constant λ must be extremely small [114],

$$\lambda \sim 10^{-12}\text{–}10^{-14}. \tag{1.6.7}$$

On the other hand, the curvature of the effective potential $V(\varphi)$ near its minimum at $\varphi=\varphi_0$ must be great enough to make the field φ oscillate at high frequency after inflation, thereby heating the universe to a rather high temperature T. It has turned out to be rather difficult to suggest a natural yet realistic theory of elementary particles that satisfies all the necessary requirements.

2. The second problem is related to the fact that the weakly interacting field φ (see (1.6.7)) is most likely not to be in a state of thermodynamic equilibrium with the other fields present in the early universe. But even if it were in equilibrium, if λ is small, high-temperature corrections to $V(\varphi)$ of order $\lambda T^2\varphi^2$ cannot alter the initial value of the field φ and make it zero in the time between the birth of the universe and the assumed start of inflation [115, 116].

3. Yet another problem relates to the fact that in both the old and new scenarios, inflation will only begin when the temperature of the universe has dropped sufficiently far, $T^4 \lesssim V(0)$. However, the condition (1.6.6) implies not only the constraint (1.6.7) on λ, but also (in most models) a constraint on the value of $V(\varphi)$ in the last stages of inflation, which in the new inflationary universe scenario is practically equal to V(0) [116, 117]:

$$V(0) \lesssim 10^{-13}\, M_P^4. \tag{1.6.8}$$

This means that inflation will start when $T^2 \lesssim 10^{-7} M_P^2$, i.e., at a time t following the beginning of expansion of the universe that exceeds the Planck time $t_P \sim M_P^{-1}$ (1.3.21) by 6 orders of magnitude. But for a hot, closed universe to live that long, its total entropy must at the very outset be greater than $S \sim 10^9$ (1.3.16). Thus, the flatness problem for a closed universe has not been solved [116], either in the context of the Guth scenario or the new inflationary universe scenario. One could look upon this result as an argument in favor of the universe being either open or flat. We think, however, that this is not a problem of the theory of a closed universe; rather, it is just one more shortcoming of the new inflationary universe scenario.

Fortunately, there is another version of the inflationary universe scenario, the so-called chaotic inflation scenario [56, 57], which does not share these problems. Rather than being based on the theory of high-temperature phase transitions, it is simply concerned with the evolution of a universe filled with a chaotically (or almost chaotically — see below) distributed scalar field φ. In what follows, we will discuss this scenario and the considerable changes that have taken place in recent years in our ideas about the early stages of the evolution of the universe, and about its large-scale structure.

1.7 The chaotic inflation scenario

We will now illustrate the basic idea of the chaotic inflation scenario with an example drawn from the simplest theory of the scalar field φ minimally coupled to gravity, with the Lagrangian

$$L = \frac{1}{2}\partial_\mu\varphi\,\partial^\mu\varphi - V(\varphi)\,. \qquad (1.7.1)$$

We shall also assume that when $\varphi \gtrsim M_P$, the potential $V(\varphi)$ rises more slowly than (approximately) $\exp\left(\frac{6\varphi}{M_P}\right)$. In particular, this requirement is satisfied by any potential that follows a power law for $\varphi \gtrsim M_P$:

$$V(\varphi) = \frac{\lambda\,\varphi^n}{n\,M_P^{n-4}}\,, \qquad (1.7.2)$$

$n > 0,\ 0 < \lambda \ll 1$.

In order to study the evolution of a universe filled with a scalar field φ, we must somehow set the initial values of the field and its derivatives at different points in space, and also specify the topology of the space and its metric in a manner consistent with the initial conditions for φ. We might assume, for example, that from the very beginning, the field φ over all space is in the equilibrium state $\varphi = \varphi_0$ corresponding to a minimum of

$V(\varphi)$. But this would be even more unconvincing than assuming that the whole universe is perfectly uniform and isotropic from the very beginning. Actually, regardless of whether the universe was originally hot or its dynamical behavior was determined solely by the classical field φ, at a time $t \sim t_P \sim M_P^{-1}$ after the singularity (or after the quantum birth of the universe — see below) the energy density ρ (and consequently the value of $V(\varphi)$) was determined only to accuracy $O(M_P^4)$ by virtue of the Heisenberg uncertainty principle. Assuming that the field φ initially taken value $\varphi = \varphi_0$ is therefore no more plausible than assuming it taken any other value with

$$\partial_0\varphi\, \partial^0\varphi \lesssim M_P^4, \tag{1.7.3}$$

$$\partial_i\varphi\, \partial^i\varphi \lesssim M_P^4, \quad i = 1, 2, 3, \tag{1.7.4}$$

$$V(\varphi) \lesssim M_P^4, \tag{1.7.5}$$

$$R^2 \lesssim M_P^4. \tag{1.7.6}$$

The last of these inequalities is taken to mean that invariants constructed from the curvature tensor $R_{\mu\nu\alpha\beta}$ are less than corresponding powers of the Planck mass ($R_{\mu\nu\alpha\beta} R^{\mu\nu\alpha\beta} \lesssim M_P^4$, $R_\mu{}^\nu R_\nu{}^\alpha R_\alpha{}^\mu \lesssim M_P^6$, etc.). It is usually assumed that the first instant at which the foregoing conditions hold is the instant after which the region of the universe under consideration can be described as a classical space-time (in nonstandard versions of gravitation theory, the corresponding conditions may generally differ from (1.7.3)–(1.7.6)). It is precisely this instant after which one can speak of specifying the initial distribution of a classical scalar field φ in a region of classical space-time.

Since there is absolutely no *a priori* reason to expect that $\partial_\mu\varphi\, \partial^\mu\varphi \ll M_P^4$, $R^2 \ll M_P^4$, or $V(\varphi) \ll M_P^4$, it seems reasonable to suppose that the most natural initial conditions at the moment when the classical description of the universe first becomes feasible are

$$\partial_0\varphi\, \partial^0\varphi \sim M_P^4, \tag{1.7.7}$$

$$\partial_i\varphi\, \partial^i\varphi \sim M_P^4, \quad i = 1, 2, 3, \tag{1.7.8}$$

$$V(\varphi) \sim M_P^4, \tag{1.7.9}$$

$$R^2 \sim M_P^4 . \tag{1.7.10}$$

We shall return to a discussion of initial conditions in the early universe more than once in the main body of this book, but for the moment, we will attempt to understand the consequences of the assumption made above [56, 118].

Investigation of the expansion of the universe with initial conditions (1.7.7)–(1.7.10) is still an extremely complicated problem, but there is a simplifying circumstance that carries one a long way toward a solution. Specifically, we are most interested in studying the possibility that regions of the universe will form that look like part of an exponentially expanding Friedmann universe. As we have already noted in Section 1.4, the latter is a de Sitter space, with only a small part of that space, of radius H^{-1}, being accessible to a stationary observer. This observer sees himself as surrounded by a black hole situated at a distance H^{-1}, corresponding to the event horizon of the de Sitter space. It is well known that nothing entering a black hole can reemerge, nor can anything that has been captured affect physical processes outside the black hole. This assertion (with certain qualifications that need not concern us here) is known as the theorem that "a black hole has no hair" [119]. There is an analogous theorem for de Sitter space as well: all particles and other inhomogeneities within a sphere of radius H^{-1} will have left that sphere (crossed the event horizon) by a time of order H^{-1}, and will have no effect on events taking place within the horizon (de Sitter space "has no hair" [120, 121]). As a result, the local geometrical properties of an expanding universe with energy-momentum tensor $T_{\mu\nu} \approx g_{\mu\nu} V(\varphi)$ approach those of de Sitter space at an exponentially high rate; that is, the universe becomes homogeneous and isotropic, and the total size of the homogeneous and isotropic region rises exponentially [120–122].

In order for such behavior to be feasible, the size of the domain within which the expansion takes place must exceed $2H^{-1}$. When $V(\varphi) \sim M_P^4$, the horizon is as close as it can be, with $H^{-1} \sim M_P^{-1}$; that is, we are dealing with the smallest domains that can still be described in terms of classical space-time. Moreover, it is necessary that expansion be approximately exponential in order for the event horizon $H^{-1}(t)$ to recede slowly enough, and for inhomogeneities at the time of expansion to escape beyond the horizon, without engendering any back influence on the expansion taking place within the horizon. This condition will be satisfied if $\dot{H} \ll H^2$, and this is just the situation during the stage of inflation.

Thus, to assess the possibility of inflationary regions arising in a universe with initial conditions (1.7.7)–(1.7.10), it is sufficient to consider whether inflationary behavior could arise at the Planck epoch in an isolated domain of the universe, with the minimum size l that could still be treated in terms of classical space-time, $l \sim H^{-1}(\varphi) \sim M_P^{-1}$.

The significance of (1.7.9) is that the typical initial value φ_0 of the field φ in the early universe is exceedingly large. For example, in a theory with $V(\varphi) = \frac{\lambda}{4}\varphi^4$ and $\lambda \ll 1$,

$$\varphi_0(x) \sim \lambda^{-1/4} M_P \gg M_P. \qquad (1.7.11)$$

According to (1.7.4) and (1.7.11), in any region whose size is of the order of the event horizon $H^{-1}(\varphi) \sim M_P^{-1}$, the field $\varphi_0(x)$ changes by a relatively insignificant amount, $\Delta\varphi \sim M_P \ll \varphi_0$. In each such domain, as we have said, the evolution of the field proceeds independently of what is happening in the rest of the universe.

Let us consider such a region of the universe having initial size $O(M_P^{-1})$, in which $\partial_\mu\varphi\,\partial^\mu\varphi$ and the squares of the components of the curvature tensor $R_{\mu\nu\alpha\beta}$, which are responsible for the inhomogeneity and anisotropy of the universe,[5] are several times smaller than $V(\varphi) \sim M_P^4$. Since all these quantities are typically of the same order of magnitude according to (1.7.7)–(1.7.10), the probability that regions of the specified type do exist should not be much less than unity. The subsequent evolution of such regions turns out to be extremely interesting.

In fact, the relatively low degree of anisotropy and inhomogeneity of space in such regions enables one to treat each of them as being a locally Friedmann space, with a metric of the type (1.3.1), governed by Eq. (1.3.7):

---

[5] Note that the quantities $\partial_\mu\partial^\mu\varphi$ and R^2 cannot exceed $V(\varphi)$ in one small *part* of the region considered and be less than $V(\varphi)$ in another, since it is not possible to subdivide *classical* space into parts less than M_P^{-1} in size and consider the *classical* field φ separately in each of these parts, due to the large quantum fluctuations of the metric at this scale.

$$H^2 + \frac{k}{a^2} \equiv \left(\frac{\dot{a}}{a}\right)^2 + \frac{k}{a^2} = \frac{8\pi}{3M_P^2}\left(\frac{\dot{\varphi}^2}{2} + \frac{(\nabla\varphi)^2}{2} + V(\varphi)\right). \tag{1.7.12}$$

At the same time, the field φ satisfies the equation

$$\Box\varphi = \ddot{\varphi} + 3\frac{\dot{a}}{a}\dot{\varphi} - \frac{1}{a^2}\Delta\varphi = -\frac{dV}{d\varphi}, \tag{1.7.13}$$

where $\Box$ is the covariant d'Alembertian operator, and Δ is the Laplacian in three-dimensional space with the time-independent metric

$$dl^2 = \frac{dr^2}{1-kr^2} + r^2(d\theta^2 + \sin^2\theta\, d\varphi^2). \tag{1.7.14}$$

For a sufficiently uniform and slowly varying field φ $\left(\dot{\varphi}^2,\ (\nabla\varphi)^2 \ll V;\ \ddot{\varphi} \ll \frac{dV}{d\varphi}\right)$, Eqs. (1.7.12) and (1.7.13) reduce to

$$H^2 + \frac{k}{a^2} \equiv \left(\frac{\dot{a}}{a}\right)^2 + \frac{k}{a^2} = \frac{8\pi}{3M_P^2}V(\varphi), \tag{1.7.15}$$

$$3H\dot{\varphi} = -\frac{dV}{d\varphi}. \tag{1.7.16}$$

It is not hard to show that if the universe is expanding $(\dot{a} > 0)$ and, as we have said, the initial value for φ satisfies (1.7.11), then the solution of the system of equations (1.7.15) and (1.7.16) rapidly proceeds to its asymptotic limit of quasiexponential expansion (inflation), whereupon the term $\frac{k}{a^2}$ in (1.7.15) can be neglected. Such behavior is understandable, inasmuch as Eq. (1.7.15) tells us that when a^2 is large, $H^2 = \frac{8\pi V(\varphi)}{3M_P^2}$. It then follows from (1.7.16) that

$$\frac{1}{2}\dot{\varphi}^2 = \frac{M_P^2}{48\pi V}\frac{dV}{d\varphi}. \tag{1.7.17}$$

Hence, for $V(\varphi) \sim \varphi^n$, we have that

$$\frac{1}{2}\dot{\varphi}^2 = \frac{n^2 M_P^2}{48\pi\varphi^2} V(\varphi), \tag{1.7.18}$$

i.e., that $\frac{1}{2}\dot{\varphi}^2 \ll V(\varphi)$ when

$$\varphi \gg \frac{n}{4\sqrt{3\pi}} M_P. \tag{1.7.19}$$

This means that for large φ, the energy-momentum tensor $T_{\mu\nu}$ of the field φ is determined almost entirely by the quantity $g_{\mu\nu}V(\varphi)$, or in other words, $p \approx -\rho$, and the universe expands quasiexponentially. Because of the fact that when $\varphi \gg M_P$ the rates at which the field φ and potential $V(\varphi)$ vary are much less than the rate of expansion of the universe $\left(\frac{\dot{\varphi}}{\varphi} \ll H,\ \dot{H} \ll H^2\right)$, over time intervals $\Delta t \lesssim \frac{H}{\dot{H}} \gg H^{-1}$ the universe looks approximately like de Sitter space with the expansion law

$$a(t) \sim e^{Ht} \tag{1.7.20}$$

where the quantity

$$H(\varphi(t)) = \sqrt{\frac{8\pi V(\varphi)}{3M_P^2}} \tag{1.7.21}$$

decreases slowly with time [56].

Under these conditions, the behavior of the field $\varphi(t)$ (see Fig. 5) is

$$\varphi(t) = \varphi_0 \exp\left(-\sqrt{\frac{\lambda}{6\pi}} M_P t\right) \tag{1.7.22}$$

for a theory with $V(\varphi) = \frac{\lambda}{4}\varphi^4$, and

$$\varphi(t)^{2-\frac{n}{2}} = \varphi_0^{2-\frac{n}{2}} + t\left(2-\frac{n}{2}\right)\sqrt{\frac{n\lambda}{24\pi}} M_P^{3-\frac{n}{2}} \tag{1.7.23}$$

for $V(\varphi) \sim \varphi^n$ (1.7.2), with $n \neq 4$. In particular, for a theory with $V(\varphi) = \frac{m^2\varphi^2}{2}$ (i.e., for $n = 2$ and $\lambda M_P^2 = m^2$),

$$\varphi(t) = \varphi_0 - \frac{m M_P}{2\sqrt{3\pi}}\, t\,. \tag{1.7.24}$$

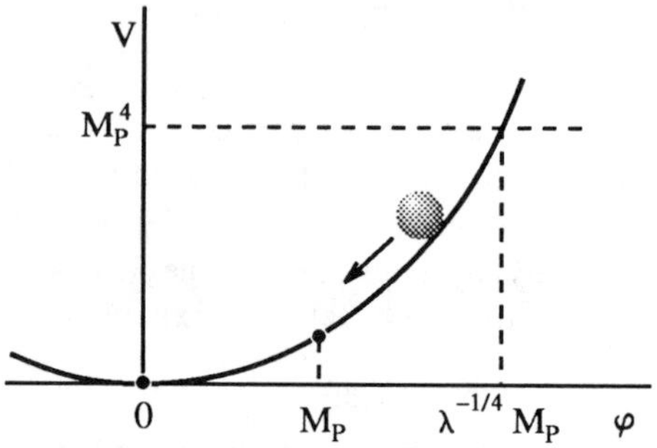

Figure 5. Evolution of a homogeneous classical scalar field φ in a theory with $V(\varphi) = \frac{\lambda}{4}\varphi^4$, neglecting quantum fluctuations of the field. When $\varphi > \lambda^{-1/4} M_P$, the energy density of the field φ is greater than the Planck density, and the evolution of the universe cannot be described classically. When $\frac{M_P}{3} \lesssim \varphi \lesssim \lambda^{-1/4} M_P$, the field φ slowly decreases, and the universe then expands quasiexponentially (inflates). When $\varphi \lesssim \frac{M_P}{3}$, the field φ oscillates rapidly about the minimum of $V(\varphi)$, and transfers its energy to the particles produced thereby (reheating of the universe).

Meanwhile, the behavior of the scale factor of the universe is given by the general equation

$$a(t) = a_0 \exp \frac{4\pi}{n M_P^2} \left(\varphi_0^2 - \varphi^2(t)\right), \tag{1.7.25}$$

which yields Eq. (1.7.20) for sufficiently small t. Making use of the estimate (1.7.18), one can easily see that this regime (the inflation regime) ends when $\varphi \lesssim \frac{n}{12} M_P$. If $\varphi_0 \gg M_P$, then (1.7.24) implies that the overall inflation factor P for the universe at that time is

$$P \sim \exp\left(\frac{4\pi}{n M_P^2}\varphi_0^2\right). \tag{1.7.26}$$

According to (1.7.26), then, the degree of inflation is small for small initial values of the field φ, and it grows exponentially with increasing φ_0. This means that most of the physical volume of the universe comes into being not by virtue of the expansion of regions which initially, and randomly, contained a small field φ (or a markedly inhomogeneous and rapidly varying field φ that failed to lead to exponential expansion of the universe), but as a result of the inflation of regions of a size exceeding the radius of the event horizon $H^{-1}(\varphi)$ which were initially filled with a sufficiently homogeneous, slowly varying, extremely large field $\varphi = \varphi_0$. The only fundamental constraint on the magnitude of the homogeneous, slowly varying field φ is $V(\varphi) \lesssim M_P^4$ (1.7.5). As we have already mentioned, the probability that domains of size $\Delta l \gtrsim H^{-1}(\varphi) \sim M_P^{-1}$ exist in the early universe with $\dot{\varphi}^2$, $(\vec{\nabla}\varphi)^2 \lesssim V(\varphi) \sim M_P^4$ should not be significantly suppressed. In conjunction with (1.7.26), this leads one to believe that most of the physical volume of the present-day universe came into being precisely as a result of the exponential expansion of regions of the aforementioned type.

If in the initial state, as we are assuming,

$$V(\varphi_0) \sim \frac{\lambda\varphi_0^n}{n M_P^{n-4}} \sim M_P^4, \tag{1.7.27}$$

the inflation factor of the corresponding region is

$$P \sim \exp\left[\frac{4\pi}{n}\left(\frac{\lambda}{n}\right)^{-\frac{2}{n}}\right]. \tag{1.7.28}$$

In particular, for a $\frac{\lambda}{4}\varphi^4$ theory

$$P \sim \exp\left(\frac{\pi}{\sqrt{\lambda}}\right), \tag{1.7.29}$$

while for an $\frac{m^2\varphi^2}{2}$ theory,

$$P \sim \exp \frac{\pi \sqrt{2}\, M_P^2}{m^2}. \tag{1.7.30}$$

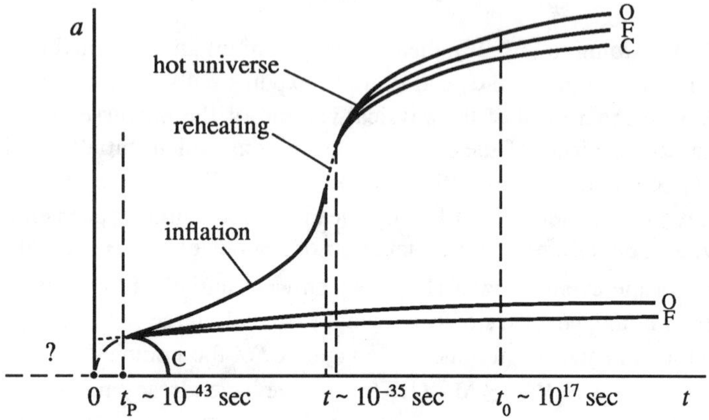

Figure 6. The lighter set of curves depicts the behavior of the size of the hot universe (or more precisely, its scale factor) for three Friedmann models: open (O), flat (F), and closed (C). The heavy curves show the evolution of an inflationary region of the universe. Because of quantum gravitational fluctuations, the classical description of the expansion of the universe cannot be valid prior to $t \sim t_P = M_P^{-1} \sim 10^{-43}$ sec after the Big Bang at $t = 0$ (or after the start of inflation in the given region). In the simplest models, inflation continues for approximately 10^{-35} sec. During that time, the inflationary region of the universe grows by a factor of from 10^{10^7} to $10^{10^{14}}$. Reheating takes place afterwards, and the subsequent evolution of the region is described by the hot universe theory.

After the field φ decreases in magnitude to a value of order M_P (1.7.18), the quantity H, which plays the role of a coefficient of friction in Eq. (1.7.13), is no longer large enough to prevent the field φ from rapidly rolling down to the minimum of the effective potential. The field φ starts its oscillations near the minimum of $V(\varphi)$, and its energy is transferred to the particles that are created as a result of these oscillations. The particles thus created collide with one another, and approach a state of thermodynamic equilibrium — in other words, the universe heats up [53, 123, 124] (see Fig. 6).

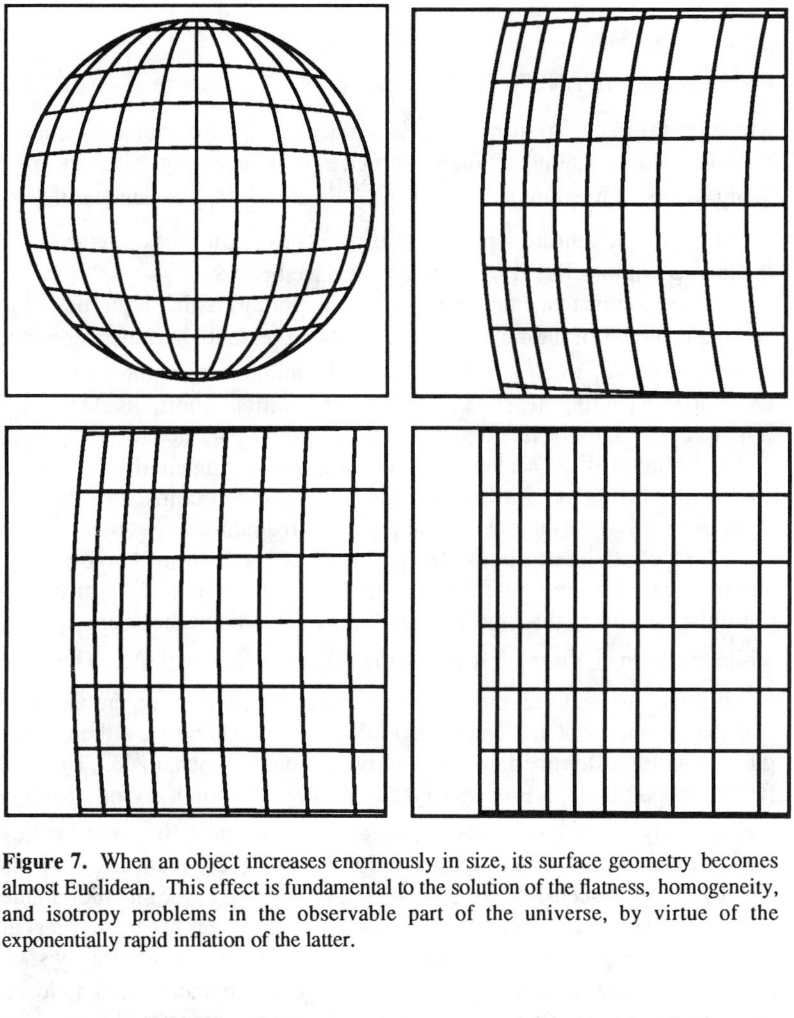

Figure 7. When an object increases enormously in size, its surface geometry becomes almost Euclidean. This effect is fundamental to the solution of the flatness, homogeneity, and isotropy problems in the observable part of the universe, by virtue of the exponentially rapid inflation of the latter.

If this reheating of the universe occurs rapidly enough (during the time $\Delta t \lesssim H^{-1}(\varphi \sim M_P)$), virtually all of the energy from the oscillating field will be transformed into thermal energy, and the temperature of the universe after reheating will be given by

$$\frac{\pi^2 N(T_R)}{30} T_R^4 \sim V(\varphi \sim \frac{n}{12} M_P). \qquad (1.7.31)$$

For example, with $N(T) \sim 10^3$ for the $V(\varphi) = \frac{\lambda}{4}\varphi^4$ theory, $T_R = c\,\lambda^{1/4} M_P$, where $c = O(10^{-1})$. In many realistic versions of the inflationary scenario, however, the temperature of the universe after reheating is found to be many orders of magnitude lower than $V^{1/4}\,(\varphi \sim \frac{n}{12} M_P)$ because of the inefficiency of the reheating process that results from the weak interaction of the field φ with itself and with other fields (see below).

One circumstance that is especially important is that both the value and the behavior of the field φ near $\varphi \sim M_P$ are essentially independent of its initial value φ_0 when $\varphi_0 \gg M_P$; that is, the initial temperature of the universe after reheating depends neither on the initial conditions during the inflationary stage nor its duration, etc. The only parameter that changes during inflation is the scale factor, which grows exponentially in accordance with (1.7.28)–(1.7.30). This is precisely the circumstance that enables us to solve the majority of the problems recounted in Section 1.5.

First of all, let us discuss the problems of the flatness, homogeneity, and isotropy of space. Note that during the quasiexponential expansion of the universe, the right-hand side of Eq. (1.7.12) decreases very slowly, while the term $\frac{k}{a^2}$ on the left-hand side falls off exponentially. Thus, the local difference between the three-dimensional geometry of the universe and the geometry of flat space also falls off exponentially, although the global topological properties of the universe remain unchanged. To solve the flatness problem, it is necessary that during inflation a region of initial size $\Delta l \sim M_P^{-1} \sim 10^{-33}$ cm grow by a factor of roughly 10^{30} (see Section 1.5). This condition is amply satisfied in most specific realizations of the chaotic inflation scenario (see below), and in contrast to the situation in the new inflationary universe scenario, inflation can begin in the present scenario at energy densities as high as one might wish, and arbitrarily soon after the universe starts to expand, i.e., prior to the moment when a closed universe starts to recollapse. After a closed universe has passed through its inflationary stage, its size (and therefore its lifetime) becomes exponentially large. The flatness problem in the chaotic inflation scenario is thereby solved, even if the universe is closed.

The solution of the flatness problem in this scenario has a simple, graphic interpretation: when a sphere inflates, its topology is unaltered, but its geometry becomes flatter (Fig. 7). The analogy is not perfect, but it is reasonably useful and instructive. It is clear, for instance, that if the Himalayas were drastically stretched horizontally while their height remained fixed, we would find a plain in place of the mountains. The same thing happens during inflation of the universe. Thus, for example, rapid inflation inhibits time-dependent changes in the amplitude of the field φ (the term $3H\dot{\varphi}$ in (1.7.13) plays the role of viscous damping), i.e., the distribution of the field φ in coordinates r, θ, ϕ is "frozen in." At the same time, the overall scale of the universe $a(t)$ grows exponentially, so that the distribution of the classical field φ *per unit physical volume* approaches spatial uniformity at an exponential rate, $\partial_i\varphi\, \partial^i\varphi \to 0$. At the same time, the energy-momentum tensor rapidly approaches $g_{\mu\nu} V(\varphi)$ (to within small corrections $\sim \dot{\varphi}^2$), the curvature tensor acquires the form

$$R_{\mu\nu\alpha\beta} = H^2(g_{\mu\nu}\, g_{\alpha\beta} - g_{\mu\beta}\, g_{\nu\alpha}), \tag{1.7.32}$$

$$R_{\mu\nu} = 3H^2 g_{\mu\nu}, \tag{1.7.33}$$

$$R = 12H^2 = \frac{32\pi}{M_P^2} V(\varphi), \tag{1.7.34}$$

and the difference between the properties of this domain of the universe and those of the homogeneous, isotropic Friedmann universe (1.3.1) becomes exponentially small (in complete accord with the "no hair" theorem for de Sitter space). After inflation, this homogeneous and isotropic domain becomes exponentially large. This explains the homogeneity and isotropy of the observable part of the universe [54–56, 120–122].

The stretching of the scales of all inhomogeneities leads to an exponential decrease in the density of monopoles, domain walls, gravitinos, and other entities produced before or during inflation. If T_R, the temperature of the universe after reheating, is not high enough to produce monopoles, domain walls, and gravitinos again, the corresponding problems disappear.

Simultaneously with the smoothing of the original inhomogeneities and ejection of monopoles and domain walls beyond the limits of the observable universe, inflation itself gives rise to specific large-scale inhomogeneities [107, 114, 125]. The theory of this phenomenon is quite complicated; it will be considered in Section 7.5. Physically, the reason for the appearance of large-scale inhomogeneities in an inflationary universe is related to the restructuring of the vacuum state resulting from the exponential expansion of the universe. It is well known that the expansion of the universe often leads to the production of elementary particles [74]. It turns out that the usual particles are produced at a very low rate during inflation, but inflation converts short-wavelength quantum fluctuations $\delta\varphi$ of the field φ into long-wavelength fluctuations. In an inflationary universe, short-wavelength fluctuations of the field φ are no different from short-wavelength fluctuations in the Minkowski space (1.1.13) (a field with momentum $k \gg \mathrm{H}$ does not "feel" the curvature of space). After the wavelength of a fluctuation $\delta\varphi$ exceeds the horizon H^{-1} in size, however, its amplitude is "frozen in" (due to the damping term $3\mathrm{H}\dot{\varphi}$ in (1.7.13)); that is, the field $\delta\varphi$ stops oscillating, but the wavelength of the field $\delta\varphi$ keeps growing exponentially. Looked at from the standpoint of conventional scalar field quantization in a Minkowski space, the appearance of such scalar field configurations may be interpreted not as the production of particles of the field φ (1.1.13), but as the creation of an inhomogeneous (quasi)classical field $\delta\varphi(x)$, where the degree to which it can be considered quasiclassical rises exponentially as the universe expands. One could say that in a certain sense an inflationary universe works like a laser, continuously generating waves of the classical field φ with wavelength $l \sim k^{-1} \sim \mathrm{H}^{-1}$. There is an important difference, however, in that the wavelength of the inhomogeneous classical field $\delta\varphi$ that is produced then grows exponentially with time. Small-scale inhomogeneities of the field φ that arise are therefore stretched to exponentially large sizes (with their amplitudes changing very slowly), while new small-scale inhomogeneities $\delta\varphi(x)$ are generated in their place.

The typical time scale in an inflationary universe is of course $\Delta t = \mathrm{H}^{-1}$. The mean amplitude of the field $\delta\varphi(x)$ with wavelength $l \sim k^{-1} \sim \mathrm{H}^{-1}$ generated over this period is [126–128]

$$|\delta\varphi(x)| \sim \frac{\mathrm{H}(\varphi)}{2\pi}. \tag{1.7.35}$$

Since $\mathrm{H}(\varphi)$ varies very slowly during inflation, the amplitude of perturbations of the field φ that are formed over a time $\Delta t = \mathrm{H}^{-1}$ will have only weak time dependence. Bearing in mind, then, that the wavelength $l \sim k^{-1}$ of fluctuations $\delta\varphi(x)$ depends exponentially on the inflation time t, it can be shown that the spectrum of inhomogeneities of the field φ formed during inflation and the spectrum of density inhomogeneities $\delta\rho$ proportional to $\delta\varphi$ are almost independent of wavelength l (momentum k) on a logarithmic scale. As we have already mentioned, inhomogeneity spectra of this type were proposed long ago by cosmologists studying galaxy formation [76, 214]. The theory of galaxy formation requires, however, that the relative amplitude of density fluctuations with such a spectrum be fairly low,

$$\frac{\delta\rho(k)}{\rho} \sim 10^{-4}\text{–}10^{-5}. \tag{1.7.36}$$

At the same time, estimates of the quantity $\frac{\delta\rho}{\rho}$ in the $V(\varphi) \sim \frac{\lambda}{4}\varphi^4$ theory yield [114, 116]

$$\frac{\delta\rho}{\rho} \sim 10^2 \sqrt{\lambda}\,, \tag{1.7.37}$$

whereupon we find that the constant λ should be extremely small,

$$\lambda \sim 10^{-12} - 10^{-14}, \tag{1.7.38}$$

exactly as in the new inflationary universe scenario. With this value of λ, the typical inflation factor for the universe is of order

$$P \sim \exp\frac{\pi}{\sqrt{\lambda}} \sim 10^{10^5}. \tag{1.7.39}$$

During inflation, a region of initial size $\Delta l \sim l_P \sim M_P^{-1} \sim 10^{-33}$ cm will grow to

$$L \sim M_P^{-1} \exp\frac{\pi}{\sqrt{\lambda}} \sim 10^{10^5}\ \text{cm}, \tag{1.7.40}$$

which is many orders of magnitude larger than the observable part of the universe, $R_p \sim 10^{28}$ cm. According to (1.7.22), the total duration of inflation will be

$$\tau \sim \frac{1}{4}\sqrt{\frac{6\pi}{\lambda}}\, M_P^{-1} \ln\frac{1}{\lambda} \sim 10^8 M_P^{-1} \sim 10^{-35}\,\text{sec}. \tag{1.7.41}$$

The estimates (1.7.39) and (1.7.40) make it clear how the horizon problem is resolved in the chaotic inflation scenario: expansion began practically simultaneously in different regions of the observable part of the universe with a size $l \lesssim 10^{28}$ cm, since they all came into being as a result of inflation of a region of the universe no bigger than 10^{-33} cm, which started simultaneously to within $\Delta t \sim t_P \sim 10^{-43}$ sec. The exponential expansion of the universe makes it causally connected at scales many orders of magnitude greater than the horizon size in a hot universe, $R_p \sim ct$.

These results may seem absolutely incredible, especially when one realizes that the entire observable part of the universe, which according to the hot universe theory has now been expanding for about 10^{10} years, is incomparably smaller than a single inflationary domain which started with the smallest possible initial size, $\Delta l \sim l_P \sim M_P^{-1} \sim 10^{-33}$ cm (1.7.40), and expanded in a matter of 10^{-37} sec. Here we must again emphasize that such a rapid increase in the size of the universe is not at variance with the conventional limitation on the speed at which a signal can propagate, $v \leq c = 1$ (see Section 1.4). On the other hand, it must also be understood that the actual numerical estimates (1.7.39) and (1.7.40) depend heavily on the model used. For example, if $\frac{\delta\rho}{\rho} \sim 10^{-5}$ in the theory with $V(\varphi) = \frac{m^2\varphi^2}{2}$, the characteristic inflation factor P becomes $10^{10^{14}}$ instead of 10^{10^7}; the inflation factor is much smaller in some other models. For our purposes, it will only be important that after inflation, the typical size of regions of the universe that we consider become many orders of magnitude larger than the observable part of the universe. Accordingly, the quantity $\frac{k}{a^2}$ in (1.3.7) will then be many orders of magnitude less than $\frac{8\pi}{3}G\rho$; i.e., the universe after inflation becomes (locally) indistinguishable from a flat universe. This implies that the density of the universe at the present time must be very close to the critical value,

$$\Omega = \frac{\rho}{\rho_c} = 1, \tag{1.7.42}$$

to within $\frac{\delta\rho}{\rho_c} \sim 10^{-3}$–$10^{-4}$, which is related to local density inhomogeneities in the observable part of the universe. This is one of the most important predictions of the inflationary universe scenario, and in principle it can be verified using astronomical observations.

Let us now turn to the problem of reheating of the universe after inflation. For $\lambda \sim 10^{-14}$ in the $\frac{\lambda}{4}\varphi^4$ theory, the temperature of the universe after reheating, according to (1.7.31), cannot typically exceed

$$T_r \sim 10^{-1}\lambda^{1/4} M_P \sim 3 \cdot 10^{14}\ \text{GeV}. \qquad (1.7.43)$$

As a rule, T_R actually turns out to be even lower. In the first place, in this theory, the oscillation frequency of the field φ near the minimum of $V(\varphi)$ is $\sqrt{\lambda}\, M_P \sim 10^{12}$ GeV at most, and in some theories it is impossible to reheat the universe to higher temperatures. Furthermore, the weakness with which φ interacts with other fields retards reheating. As a result, the oscillation amplitude of the field φ decreases as the universe expands, due to the term $3H\dot{\varphi}$ in the equation of motion for φ, and the temperature of the universe after inflation turns out to be much lower in certain theories than the value in (1.7.43).

Generally speaking, this can lead to some difficulties in treating the baryon asymmetry problem. Actually, during inflation, any initial baryon asymmetry in the universe dies out exponentially, and for such asymmetry to arise after inflation becomes not just esthetically attractive, as in the usual hot universe theory, but necessary. Moreover, the mechanism for producing the baryon asymmetry that was proposed in [36–38] and worked out in the context of grand unified theories is only effective if the temperature T is high enough that superheavy particles appear in the hot plasma, with their subsequent decay producing an excess of baryons over antibaryons. Usually, for this to happen, the temperature of the universe must be higher than 10^{15} GeV, and this is seldom achievable in the inflationary universe scenario. Fortunately, however, baryon production can also proceed at much lower temperatures after inflation, due to nonequilibrium processes that take place during reheating [123]. Moreover, a number of models have recently been suggested which allow for the onset of baryon asymmetry even if the temperature of the universe after inflation never exceeds 10^2 GeV [97–99, 129]. Thus, the inflationary universe scenario can successfully incorporate all basic results of the hot universe theory, and the resulting theory proves to be free of the main difficulties of the standard hot Big Bang cosmology.

The main stages in the evolution of an inflationary domain of the universe are shown in Fig. 8. The initial and final stages in the development of each individual inflationary domain depend on the global structure of the inflationary universe, which we will discuss in Section 1.8.

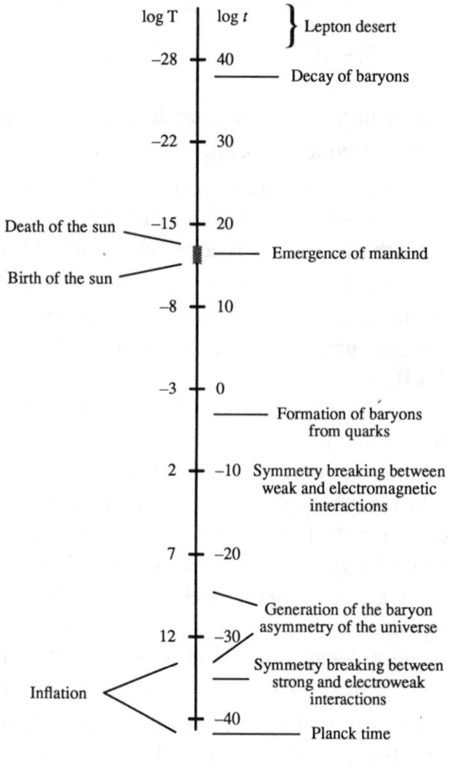

Figure 8 (facing page). The main stages in the evolution of an inflationary region of the universe. The time t is measured in seconds from the start of inflation, and the temperature T is measured in GeV (1 GeV ~ 10^{13} K). The typical lifetime of a region of the universe, between the start of inflation and the region's collapse (if it exceeds the critical density), is many orders of magnitude greater than the proton decay lifetime in the simplest grand unified theories.

1.8 The self-regenerating universe

The attentive reader probably already has noticed that in discussing the problems resolved with the aid of the inflationary universe scenario, we have silently skirted the most important one — the problem of the cosmological singularity. We have also said nothing about the global structure of the inflationary universe, having limited ourselves to statements to the effect that its local properties are very similar to those of the observable world. The study of the global structure of the universe and the problem of the cosmological singularity within the scope of the inflationary universe scenario conceals a number of surprises. Prior to the advent of this scenario, there was absolutely no reason to suppose that our universe was markedly inhomogeneous on a large scale. On the contrary, the astronomical data attested to the fact that on large scales, up to the very size of the entire observable part of the universe $R_P \sim 10^{28}$ cm, inhomogeneities $\frac{\delta\rho}{\rho}$ on the average were at most 10^{-3}. To understand the evolution of the universe, it was in large measure thought to be sufficient to investigate homogeneous (or slightly inhomogeneous) cosmological models like the Friedmann model (or anisotropic Bianchi models) [65].

Meanwhile, the results of the preceding section make it clear that the observable part of the universe is most likely just a minuscule part of the universe as a whole, and it is an impermissible extrapolation to draw any conclusions about the homogeneity of the latter based on observations of such a tiny component. On the contrary, an investigation of the global geometry of the inflationary universe shows that the universe, which is locally Friedmann, should be completely inhomogeneous on the largest scales, and its global geometry and dynamical behavior as a whole have nothing in common with the geometry and dynamics of a Friedmann universe [57, 78, 132, 133].

In order to obtain a simple derivation of this important and somewhat surprising result, let us consider more carefully the behavior of the scalar field φ for the minimal model (1.7.1) with $V(\varphi) = \frac{\lambda}{4}\varphi^4$ in the chaotic inflation scenario, taking into account long-wave fluctuations of φ that arise during inflation [57]. We have from (1.7.21) and (1.7.22) that in a typical time

$$\Delta t = H^{-1}(\varphi) = \sqrt{\frac{3}{2\pi\lambda}}\,\frac{M_P}{\varphi^2}, \qquad (1.8.1)$$

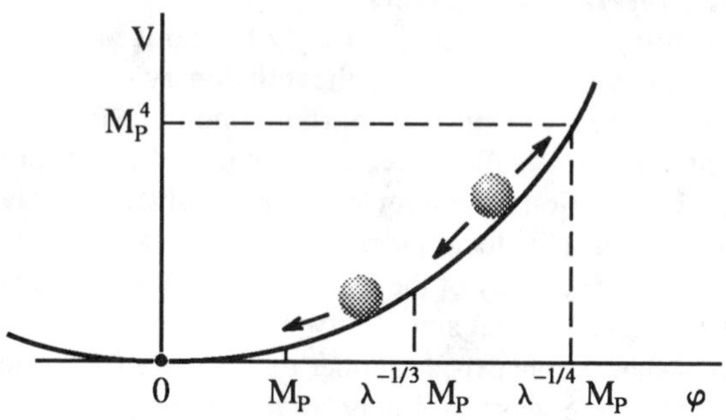

Figure 9. Evolution of the scalar field φ in the simplest field theory with the potential $V(\varphi) = \frac{\lambda}{4}\varphi^4$, with quantum fluctuations of the field φ taken into account. When $\varphi > \lambda^{-1/4} M_P (V(\varphi) \gtrsim M_P^4)$, quantum gravity fluctuations of the metric are large, and no classical description of space is possible in the simplest theories. For $\frac{M_P}{3} \lesssim \varphi \lesssim \lambda^{-1/4} M_P$, the field φ evolves relatively slowly, and the universe expands quasiexponentially. When $\lambda^{-1/6} M_P \lesssim \varphi \ll \lambda^{-1/4} M_P$, the amplitude of φ fluctuates markedly, leading to the endless birth of ever newer regions of the universe. For $\frac{M_P}{3} \lesssim \varphi \ll \lambda^{-1/6} M_P$, fluctuations of the field are of relatively low amplitude. The field φ rolls downhill, and fluctuations engender the density inhomogeneities required for the formation of galaxies. When $\varphi \lesssim \frac{M_P}{3}$, the field starts to oscillate rapidly about the point $\varphi = 0$, particle pairs are produced, and all of the energy of the oscillating field is converted into heat.

the classical homogeneous field φ decreases by

$$\Delta\varphi = \frac{M_P^2}{2\pi\varphi} . \tag{1.8.2}$$

During this same period of time, according to (1.7.36), inhomogeneities of the field φ are generated having wavelength $l \gtrsim H^{-1}$ and mean amplitude

$$|\delta\varphi(x)| \sim \frac{H(\varphi)}{2\pi} = \sqrt{\frac{\lambda}{6\pi}}\,\frac{\varphi^2}{M_P} . \tag{1.8.3}$$

It is not hard to show that when $\varphi \ll \varphi^*$, where

$$\varphi^* = \lambda^{1/6} M, \tag{1.8.4}$$

quantum fluctuations of φ have a negligible influence on its evolution, $|\delta\varphi(x)| \ll \Delta\varphi$. It is precisely at the later stages of inflation, when the field φ becomes less than $\varphi^* = \lambda^{1/6} M_P$, that small inhomogeneities $\delta\varphi$ in the field φ and small density inhomogeneities $\delta\rho$ are produced, leading to the formation of galaxies. On the other hand, when $\varphi \gg \varphi^*$, only the mean field φ is governed by Eq. (1.7.22), and the role played by fluctuations becomes extremely significant (see Fig. 9).

Consider a region of an inflationary universe of size $\Delta l \sim H^{-1}(\varphi)$ that contains the field $\varphi \gg \varphi^*$. According to the "no hair" theorem for de Sitter space, inflation in this region of space proceeds independently of what happens in other regions. In such a region, the field can be assumed to be largely homogeneous, as initial inhomogeneities of the field φ are reduced by inflation, and nascent inhomogeneities (1.8.3) that make their appearance during inflation have wavelengths $l > H^{-1}$. In the typical time $\Delta t = H^{-1}$, the region in question will have grown by a factor of e, and its volume will have grown by a factor of $e^3 \approx 20$, so that it could be subdivided approximately into e^3 regions of size $O(H^{-1})$, each once again containing an almost homogeneous field φ, which differs from the original field φ by $\delta\varphi(x) - \Delta\varphi \approx \delta\varphi(x)$. This means, however, that in something like $\frac{e^3}{2}$ regions of size $O(H^{-1})$, instead of decreasing, the field φ increases by a quantity of order $|\delta\varphi(x)| \sim \frac{H}{2\pi} \gg \Delta\varphi$ (see Fig. 10). This process repeats during the next time interval $\Delta t = H^{-1}$, and so on. It is not hard to show that the total volume of the universe occupied by the *continually growing* field φ increases approximately as

$$\exp\left[(3 - \ln 2) H t\right] \gtrsim \exp\left(3\sqrt{\lambda}\, \frac{\varphi^2}{M_P}\, t\right),$$

while the total volume occupied by the *non-decreasing* field φ grows almost as fast as $\exp\left[3H(\varphi)t\right]$.

This means that regions of space containing a field φ continually spawn brand new regions with even higher field values, and as φ grows the birth and expansion process in the new regions takes place at an ever increasing rate.

To better understand the physical meaning of this phenomenon, it is useful to examine those regions which, while rare, are still constantly appearing, where the field φ *increases continuously*, i.e., it is typically augmented by $\delta\varphi \sim \frac{H(\varphi)}{2\pi}$ in each successive time interval $\Delta t = H^{-1}(\varphi)$. The rate of growth of the field in such regions is given by

$$\frac{d\varphi}{dt} = \frac{H^2(\varphi)}{2\pi} = \frac{4\,V(\varphi)}{3M_P^2} = \frac{\lambda\varphi^4}{3\,M_P^2}, \tag{1.8.5}$$

whereupon

$$\varphi^{-3}(t) = \varphi_0^{-3} - \frac{\lambda t}{M_P^2}. \tag{1.8.6}$$

This means that in a time

$$\tau = \frac{M_P^2}{\lambda\,\varphi_0^3} \tag{1.8.7}$$

the field φ should become infinite. Actually, of course, one can only say that the field in such regions approaches the limiting value φ for which $V(\varphi) \sim M_P^4$ (i.e., $\varphi \sim \lambda^{-1/4} M_P$). At higher densities, it becomes impossible to treat such regions of space classically. Furthermore, formal consideration of inflationary regions with $V(\varphi) \gg M_P^4$ indicates that most of their field energy is concentrated not at $V(\varphi)$, but at a value related to the inhomogeneities $\delta\varphi(x)$ and proportional to $H^4 \sim \frac{V^2}{M_P^4}$. Therefore, in the overwhelming majority of regions of the universe with $V(\varphi) \gg M_P^4$, inflation is most likely cut short, and in any case we cannot describe it in terms of classical space-time.

To summarize, then, many inflationary regions with $V(\varphi) \sim M_P^4$ are created over a time $\tau \sim \frac{M_P^2}{\lambda\,\varphi_0^3}$ in a part of the universe originally filled with a field $\varphi_0 \gg \varphi^*$. Some fraction of these regions will ultimately expand to become regions with $V(\varphi) \gg M_P^4$, that is, a space-time foam, which we are presently not in a position to describe. It will be important for us, however, that the volume of the universe filled by the extremely large and non-decreasing field φ, such that $V(\varphi) \sim M_P^4$, continue to grow at the highest possible rate, as $\exp(cM_P t)$, $c = O(1)$. The net result is that in time $t \gg \tau$ (1.8.7) (in a synchronous coordinate system; see Section 10.3), most of the physical volume of an original inflationary region of the universe with $\varphi = \varphi_0 \gg \varphi^*$ should contain an exceedingly large field φ, with $V(\varphi) \sim M_P^4$.

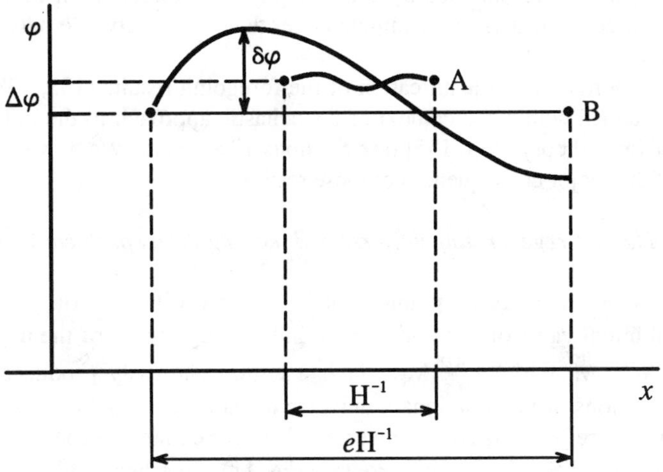

Figure 10. Evolution of a field $\varphi \gg \varphi^* = \lambda^{-1/6} M_P$ in an inflationary region of the universe of initial size $\Delta l = H^{-1}(\varphi)$. Initially (A), the field φ in this domain is relatively homogeneous, since inhomogeneities $\delta\varphi(x)$ with a wavelength $l \sim H^{-1}(\varphi)$ that result from inflation are of order $\delta\varphi \sim \frac{H}{2\pi} \ll \varphi$. After a time $\Delta t = H^{-1}$, the field has grown (B) by a factor of e. When $\varphi \gg \varphi^*$, the average decrement $\Delta\varphi$ of the field φ in this region is much less than $|\delta\varphi| \sim \frac{H}{2\pi}$. This means that in almost half the region under consideration, the field φ grows instead of shrinking. Thus, in a time $\Delta t = H^{-1}$, the volume occupied by *increasing* φ values grows by a factor of approximately $\frac{e^3}{2} \sim 10$.

This doesn't at all mean that the whole universe must be in a state with the Planck density. Fluctuations of the field φ constantly lead to the formation of regions not just with $\varphi \gg \varphi^*$, but with $\varphi \ll \varphi^*$ as well. Just such regions form gigantic homogeneous regions of the universe like our own. After inflation, the typical size of each such region exceeds

$$l \sim M_P^{-1} \exp\left[\frac{\pi(\varphi^*)^2}{M_P^2}\right] \sim M_P^{-1} \exp\left(\pi\lambda^{-1/3}\right) \sim 10^{6\cdot 10^4}\ \text{cm}. \quad (1.8.8)$$

when $\lambda \sim 10^{-14}$. This is much less than $l \sim M_P^{-1} \exp\left(\pi\lambda^{-1/2}\right) \sim 10^{10^7}$ cm (1.7.40), which we obtained by neglecting quantum fluctuations, but it is still hundreds of orders of magnitude larger than the observable part of the universe.

A more detailed justification of the foregoing results [132, 133] has been obtained within the context of a stochastic approach to the inflationary universe theory [134, 135] (see Sections 10.2–10.4). We now examine two of the major consequences of these results.

1.8.1. *The self-regenerating universe and the singularity problem*

As we have already pointed out in the preceding section, the most natural initial value of the field φ in an inflationary region of the universe is $\varphi \sim \lambda^{-1/4} M_P \gg \varphi^* \sim \lambda^{-1/6} M_P$. Such a region endlessly produces ever newer regions of the inflationary universe containing the field $\varphi \gg \varphi^*$. As a whole, therefore, this entire universe will never collapse, even if it starts out as a closed Friedmann universe (see Fig. 11). In other words, contrary to conventional expectations, even in a closed (compact) universe there will never be a global singular spacelike hypersurface — the universe as a whole will never just vanish into nothingness. Similarly, there is no sufficient reason for assuming that such a hypersurface ever existed in the past — that at some instant of time $t = 0$, the universe as a whole suddenly appeared out of nowhere.

This of course doesn't mean that there are no singularities in an inflationary universe. On the contrary, a considerable part of the physical volume of the universe is constantly in a state that is close to singular, with energy density approaching the Planck density $V(\varphi) \sim M_P^4$. What is important, however, is that different regions of the universe pass through a singular state at different times, so there is no unique end of time, after which space and time disappear. It is also quite possible that there was no unique beginning of time in the universe.

It is worth noting that the standard assertion about the occurrence of a general cosmological singularity (i.e., a global singular spacelike hypersurface in the universe, or what is the same thing, a unique beginning or end of time for the universe as a whole) is *not* a direct consequence of existing topological theorems on singularities in general relativity [69, 70], or of the behavior of general solutions of the Einstein equations near a singularity [68]. This assertion is primarily based on analyses of homogeneous cosmological models like the Friedmann or Bianchi models. Certain authors have emphasized that there might in fact not be a unique beginning and end of time in the universe as a whole if our own universe is only locally a Friedmann space but is globally inhomogeneous (a so-called quasihomogeneous universe; see [34, 136]). However, in the absence of any experimental basis for hypothesizing significant large-scale inhomogeneity of the universe, this approach to settling the problem of an overall cosmological singularity has not elicited much interest.

The present attitude toward this problem has changed considerably. Actually, the only explanation that we are aware of for the homogeneity of the observable part of the universe is the one provided by the inflationary universe scenario. But as we have just shown, this same scenario implies that on the largest scales the universe must be absolutely *in*homogeneous, with density excursions ranging from $\rho \lesssim 10^{-29}$ g/cm^3 (as in the observable part of the universe) to $\rho \sim M_P^4 \sim 10^{94}$ g/cm^3. Hence, there is presently no compelling basis for maintaining that there is a unique beginning or end of the universe as a whole. (For a more detailed discussion of this question, see also Section 10.4.)

It is not impossible in principle that the universe as a whole might have been born "out of nothing," or that it might have appeared from a unique initial singularity. Such a suggestion could be fairly reasonable if in the process a compact (closed, for example) universe of size $l = O(M_P^{-1})$ were produced. But for a noncompact universe, this hypothesis is not only hard to interpret, it is completely implausible, since there would seem to be absolutely no likelihood that all causally disconnected regions of an infinite universe could spring *simultaneously* from a singularity (see the discussion of the horizon problem in Section 1.5). Fortunately, this hypothesis turns out to be unnecessary for the scenario being developed, and in that sense it seems possible to avoid one of the main conceptual difficulties associated with the problem of the cosmological singularity [57].

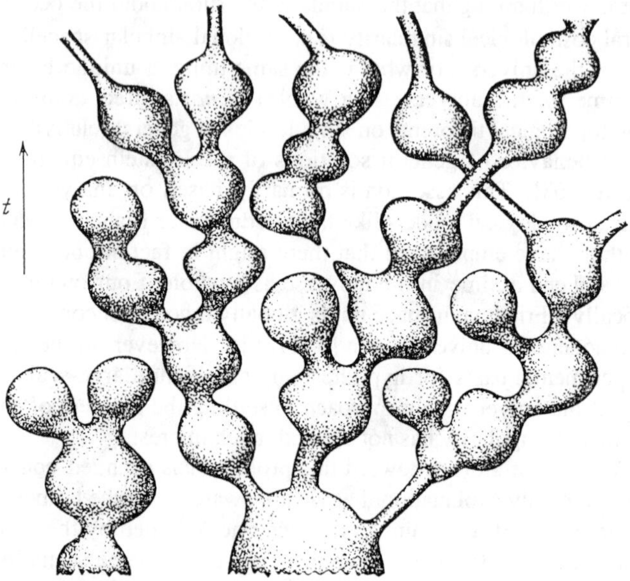

Figure 11. A rather fanciful attempt to convey some impression of the global structure of the inflationary universe. One region of the inflationary universe gives rise to a multitude of new inflationary regions; in different regions, the properties of space-time and elementary particle interactions may be utterly different. In this scenario, the evolution of the universe as a whole has no end, and may have no beginning.

1.8.2. *The problem of the uniqueness of the universe, and the Anthropic Principle*

The ceaseless creation of new regions of the inflationary universe takes place when $\lambda^{-1/6} M_P \lesssim \varphi \lesssim \lambda^{-1/4} M_P$; then $\lambda^{-1/3} M_P^4 \lesssim V(\varphi) \lesssim M_P^4$ ($10^{-5} M_P^4 \lesssim V(\varphi) \lesssim M_P^4$ for $\lambda \sim 10^{-14}$), or in other words it is not necessary to appeal for a description of this process to speculative phenomena that take place above the Planck density.

On the other hand, it is important that much of the physical volume of the universe must at all times be close to the Planck density, and expanding exponentially with a Hubble constant H of order M_P. In realistic elementary particle theories, apart from the scalar field φ responsible for inflation, there are many other types of scalar fields Φ, H, etc., with masses $m \ll M_P$. Inflation leads to the generation of long-wave fluctuations not just in the field φ, but in all scalar fields with $m \ll H \sim M_P$. As a result, the universe becomes filled with fields φ, Φ, and so forth, which vary slowly in space and take on all allowable values for which $V(\varphi, \Phi, \cdots) \lesssim M_P^4$. In those regions where inflation has ended, the scalar fields "roll down" to the nearest minimum of the effective potential $V(\varphi, \Phi, \cdots)$, and the universe breaks up into exponentially large domains (mini-universes) filled with the fields φ, Φ, etc., which in the different domains take on values corresponding to all the local minima of $V(\varphi, \Phi, \cdots)$. In Kaluza–Klein and superstring theories, quantum fluctuations can result in a local change in the type of compactification on a scale $O(H^{-1}) \sim O(M_P^{-1})$. If the region continues to inflate after this change, then by virtue of the "no hair" theorem for de Sitter space, the properties of the universe outside this region (its size and the type of compactification) cannot exert any influence on the region, and after inflation an exponentially large mini-universe with altered compactification will have been created [333].

The result is that the universe breaks up into mini-universes in which all possible types of (metastable) vacuum states and all possible types of compactification that support inflationary behavior are realized. We live in a region of the universe in which there are weak, strong, and electromagnetic interactions, and in which space-time is four-dimensional. We cannot rule out the possibility, however, that this is so not because our region is the only one or the best one, but because such regions exist, are exponentially plentiful (or more likely, infinitely plentiful), and life of our type would be impossible in any other kind of region [57, 78].

This discussion is based on the Anthropic Principle, whose validity the author himself previously cast into doubt (in Section 1.5). But now the situation is different — it is not at all necessary for someone to sit down and create one universe after another until he finally succeeds. Once the universe has come into being (or if it has existed eternally), that universe itself will create exponentially large regions (mini-universes), each having different elementary-particle and space-time properties. If good conditions for the appearance of life in a solar-system environment are then to ensue, it turns out that the same conditions necessarily appear on a scale much larger than the entire observable part of the universe. In fact, for galaxies to arise in the simple model we are considering, one must have $\lambda \sim 10^{-14}$, and as we have seen, this leads to a characteristic size $l \gtrsim 10^{6 \cdot 10^4}$ cm for the uniform region. Thus, within the context of the approach being developed, it is possible to remove the main objections to the cosmological application of the Anthropic Principle (or to be more precise, of the Weak Anthropic Principle; see Sections (10.5 and 10.7, where stronger versions of the Anthropic Principle are also discussed).

This result may have important methodological implications. Attempts to construct a theory in which the observed state of the universe and the observed laws of interaction between elementary particles are the only ones possible and are realized throughout the entire universe become unnecessary. Instead, we are faced with the problem of constructing theories that can produce large regions of the universe that resemble our own. The question of the most reasonable initial conditions near a singularity and the probability of an inflationary universe being created is supplanted by the question of what values the physical fields might take, what the properties of space are in most of the inflationary universe, and what the most likely way is to form a region of the universe of size $R_p \sim 10^{28}$ cm with observable properties and observers resembling our own.

This new statement of the problem greatly enhances our ability to construct realistic models of the inflationary universe and realistic elementary particle theories.

1.9 Summary

In this introductory chapter, we have discussed some basic features of inflationary cosmology. One should take into account, however, that many details of the inflationary universe scenario look different in the context of different theories of elementary particles. For example, it is not necessary to assume that the field φ which drives inflation is an elementary scalar field. In certain theories, the role played by this field can be assumed by the curvature scalar R, a fermion condensate $\langle \bar{\psi}\psi \rangle$ or vector meson condensate $\langle G^a_{\mu\nu} G^a_{\mu\nu} \rangle$, or even the logarithm of the radius of a compactified space. More detailed discussions of phase transitions in the unified theories of weak, strong, and electromagnetic interactions, of various versions of the inflationary universe scenario, and of different aspects of quantum inflationary cosmology are to be found in subsequent sections of this book.

2

Scalar Field, Effective Potential, and Spontaneous Symmetry Breaking

2.1 Classical and quantum scalar fields

As we have seen, classical (or quasiclassical) scalar fields play an essential role in present-day cosmological models (and also in modern elementary particle theories). We will often deal with homogeneous and inhomogeneous classical fields, and there is sometimes a question as to which fields can be considered classical, and in what sense.

Let us recall, first of all, that in accordance with the standard approach to quantization of the scalar field $\varphi(x)$, the functions $a^+(\boldsymbol{k})$ and $a^-(\boldsymbol{k})$ in (1.1.3) can be put into correspondence with the creation and annihilation operators a_k^+ and a_k^- for particles with momentum k. The commutation relations take the form [58]

$$\frac{1}{2k_0}\left[\varphi_k^-,\ \varphi_q^+\right] \equiv \left[a_k^-,\ a_q^+\right] = \delta\,(\boldsymbol{k}-\boldsymbol{q}), \tag{2.1.1}$$

where the operator a_k^- acting on the vacuum gives zero:

$$a_k^-|0\rangle = 0;\ \langle 0|a_k^+ = 0;\ \langle 0|\varphi(x)|0\rangle. \tag{2.1.2}$$

The operator a_k^+ creates a particle with momentum k,

$$a_k^+|\psi\rangle = |\psi, \boldsymbol{k}\rangle, \tag{2.1.3}$$

while the operator a_k^- annihilates it,

$$a_k^-|\psi, \boldsymbol{k}\rangle = |\psi\rangle. \qquad (2.1.4)$$

Now consider the Green's function for the scalar field φ [58],

$$\mathrm{G}(x) = \langle 0|\mathrm{T}[\varphi(x)\varphi(0)]|0\rangle = \frac{1}{(2\pi)^4}\int \frac{e^{-ikx}}{m^2 - k^2 - i\varepsilon} d^4k. \qquad (2.1.5)$$

Here T is the time-ordering operator, and ε shows how to perform integration near the singularity at $k^2 = m^2$ (from here on, we omit both symbols). Evaluation of this expression indicates that when $t = 0$ and $x \gtrsim m^{-1}$, $\mathrm{G}(x)$ falls off exponentially with increasing x; that is, the correlation between $\varphi(x)$ and $\varphi(0)$ becomes exponentially small. When $m = 0$, $\mathrm{G}(x)$ has a power-law dependence on x.

It is also useful to calculate $\mathrm{G}(0)$, which after transforming to Euclidean space (by a Wick rotation $k_0 \to -ik_4$) may be written in the form

$$\mathrm{G}(0) = \langle 0|\varphi^2|0\rangle = \frac{1}{(2\pi)^4}\int \frac{d^4k}{k^2 + m^2} = \frac{1}{(2\pi)^3}\int \frac{d^3k}{2\sqrt{\boldsymbol{k}^2 + m^2}}. \qquad (2.1.6)$$

If averaging is carried out, for example, over a state containing particles rather than over the conventional vacuum in Minkowski space, we can represent the quantity $\langle 0|\varphi^2|0\rangle \equiv \langle\varphi^2\rangle$ in the form

$$\begin{aligned}\langle\varphi^2\rangle &= \frac{1}{(2\pi)^3}\int \frac{d^3k}{2\sqrt{\boldsymbol{k}^2 + m^2}}\left(1 + 2\langle a_k^+ a_k^-\rangle\right) \\ &= \frac{1}{(2\pi)^3}\int \frac{d^3k}{\sqrt{\boldsymbol{k}^2 + m^2}}\left(\frac{1}{2} + n_k\right).\end{aligned} \qquad (2.1.7)$$

Here n_k is the number density of particles with momentum k. For instance, for a Bose gas at nonzero temperature T, one has

$$n_k = \frac{1}{\exp\left(\frac{\sqrt{\boldsymbol{k}^2 + m^2}}{\mathrm{T}}\right) - 1}. \qquad (2.1.8)$$

Another important example is a Bose condensate φ_0 of noninteracting particles of the field φ, with mass m and vanishing momentum $\boldsymbol{k}$, for which

$$n_k = (2\pi)^3 \varphi_0^2 m\,\delta(\boldsymbol{k}), \tag{2.1.9}$$

or a coherent wave of particles with momentum $\boldsymbol{p}$:

$$n_k = (2\pi)^3 \varphi_p^2 \sqrt{\boldsymbol{p}^2 + m^2}\,\delta(\boldsymbol{k} - \boldsymbol{p}). \tag{2.1.10}$$

In both cases, n_k tends to infinity at some value of $\boldsymbol{k}$. The fact that the $a_k^{\pm}$ operators of (2.1.1) do not commute can then be ignored, as $n_k \gg 1$ in (2.1.7). Therefore, the condensate φ_0 and the coherent wave φ_p can be called *classical* scalar fields. In performing calculations, it is convenient to separate the field φ into a classical field (condensate) φ_0 (φ_p) and field excitations (scalar particles), with quantum effects being associated only with the latter. This is formally equivalent to the appearance of a nonvanishing vacuum average of the original field φ, $\langle 0|\varphi|0\rangle = \varphi_0$, and reversion to the standard formalism (2.1.2) requires that we subtract the classical part φ_0 from the field φ; see (1.1.12).

The foregoing instances are not the most general. If the condensate results from dynamic effects (minimization of a relativistically invariant effective potential), the properties of its constituent particles will be altered, and the condensate itself (in contrast to (2.1.9) and (2.1.10)) can turn out to be relativistically invariant. This is precisely what happens in theories incorporating the Glashow–Weinberg–Salam model, where $\langle\varphi^2\rangle$ can be put in the form

$$\langle\varphi^2\rangle = \frac{1}{(2\pi)^3}\int \frac{d^3k}{2\sqrt{\boldsymbol{k}^2 + m^2}} + \frac{1}{(2\pi)^3}\int \frac{d^3k}{\sqrt{\boldsymbol{k}^2}}\,n_k \tag{2.1.11}$$

with $k = \sqrt{\boldsymbol{k}^2}$, and

$$n_k = (2\pi)^3 \varphi_0^2\, k\,\delta(\boldsymbol{k}). \tag{2.1.12}$$

The gist of this representation is that the constant classical scalar field φ_0 (1.1.12) is Lorentz invariant, and it can therefore only form a condensate if the particles comprising it have zero momentum and zero energy — in other words, zero mass (compare (2.1.11) and (2.1.7)).

It is not obligatory that the constant classical field be interpreted as a condensate, but this proves to be a very useful, fruitful approach to the analysis of phase transitions in gauge theories. There, the relativistically invariant form of the condensate (2.1.11), (2.1.12) leads to a number of effects which are lacking from solid-state theory with a condensate (2.1.9). We shall return to this problem in the next chapter.

Note that $n_k \gg 1$ when $\sqrt{\boldsymbol{k}^2 + m^2} \ll T$ for an ultrarelativistic Bose gas (2.1.8). We can therefore tentatively divide the field φ into a quantum part corresponding to $\sqrt{\boldsymbol{k}^2 + m^2} \gtrsim T$, and a (quasi)classical part with $\sqrt{\boldsymbol{k}^2 + m^2} \ll T$. This sort of partitioning is not very useful, however, since it is excitations with $\sqrt{\boldsymbol{k}^2 + m^2} \sim T$ that make the main contribution to most thermodynamic functions.

Much more interesting effects arise in the inflationary universe, where the main contribution to $\langle\varphi^2\rangle$, to density inhomogeneities, and to a number of other quantities comes precisely from long-wavelength modes with $k \ll H$, for which $n_k \gg 1$. The interpretation of these modes as inhomogeneous classical fields $\delta\varphi$ significantly facilitates one's understanding of a great many of the fundamental features of the inflationary universe scenario. Corresponding effects were discussed in Section 1.8, and we shall continue the discussion in Chapters 7 and 10.

Let us formulate a few more criteria that could help one decide whether the field φ is (quasi)classical. One has already been discussed, namely the presence of modes with $n_k \gg 1$. Another is the behavior of the correlation function of $G(x)$ at large x. At large x, this function usually (when there are no classical fields) falls off either exponentially or according to a power law (as x^{-2}). When there really is a condensate (2.1.9), (2.1.11) or a coherent wave (2.1.10), the correlation function no longer decreases at large x (since the condensate is everywhere the same, i.e., the values at different points are correlated). The onset of long-range order is thus another criterion for the existence of a classical field in a medium, one that has long been successfully applied in the theory of phase transitions. As will be shown in Chapter 7, the corresponding correlation function in the inflationary universe theory falls off only at exponentially large distances $x \sim H^{-1} \exp(Ht)$, $Ht \gg 1$, which enables us to speak of a classical field $\delta\varphi(x)$ being produced during inflation.

Somewhat surprisingly, the classical field φ cannot be overly nonuniform (unless it is a coherent wave with a single well-defined momentum (2.1.10)). Suppose, in fact, that in some region of space

$\nabla\varphi \sim k\varphi \gg m\varphi$. In order for this field to be distinguishable from the quantum fluctuation background, the field φ must be greater than the contribution to the rms value $\sqrt{\langle\varphi^2\rangle}$ coming from quantum fluctuations with momentum $\sim k \gg m$. Making use of (2.1.6), we obtain

$$\varphi^2 \gtrsim C k^2, \tag{2.1.13}$$

where $C = O(1)$, or

$$(\nabla\varphi)^2 \lesssim \varphi^4. \tag{2.1.14}$$

In particular, this means that the initial value of the *classical* scalar field φ cannot be arbitrary; inhomogeneities in the classical scalar field cannot exceed a certain limit.

Even more important constraints can be obtained by taking quantum gravitation into consideration. At energy densities of the order of the Planck density, fluctuations of the metric become so large that one can no longer speak of classical space-time with a classical metric $g_{\mu\nu}$ (in the same sense as one would speak of the classical field φ). This means that it is impossible to treat fields φ as being classical unless

$$\partial_\mu\varphi\, \partial^\mu\varphi \lesssim M_P^4, \quad \mu = 0, 1, 2, 3, \tag{2.1.15}$$

$$V(\varphi) \lesssim M_P^4. \tag{2.1.16}$$

We made essential use of these constraints in discussing initial conditions in the inflationary universe in Section 1.7.

2.2 Quantum corrections to the effective potential $V(\varphi)$

In Section 1.1, we investigated the theory of symmetry breaking in the simplest quantum field theoretical models, neglecting quantum corrections to the effective potential of the scalar field φ. Nevertheless, in some cases quantum corrections to $V(\varphi)$ are substantial.

According to [137, 138], quantum corrections to the classical expression for the effective potential are given by a set of all one-particle irreducible vacuum diagrams (diagrams that do not dissociate into two when a single line is cut) in a theory with the Lagrangian $L(\varphi + \varphi_0)$ without the terms linear in φ. Corresponding diagrams with one, two, or more loops for the theory (1.1.5) have been drawn in Fig. 12. In the present case, expansion in the number of loops corresponds to expansion in the small coupling constant λ. In the one-loop approximation (taking only the first diagram of Fig. 12 into account),

$$V(\varphi) = -\frac{\mu^2}{2}\varphi^2 + \frac{\lambda}{4}\varphi^4 + \frac{1}{2(2\pi)^4}\int d^4k \ln\left[k^2 + m^2(\varphi)\right]. \qquad (2.2.1)$$

Here $k^2 = k_4^2 + \boldsymbol{k}^2$ (i.e., we have carried out a Wick rotation $k_0 \to -ik_4$ and integrated over Euclidean momentum space), and the effective mass squared of the field φ is

$$m^2(\varphi) = 3\lambda\varphi^2 - \mu^2. \qquad (2.2.2)$$

As before, we have omitted the subscript 0 from the classical field φ in Eqs. (2.2.1) and (2.2.2). The integral in (2.2.1) diverges at large k. To supplement the definition given by (2.2.1), it is necessary to renormalize the wave function, mass, coupling constant, and vacuum energy [2, 8, 9]. To do so, we may add to $L(\varphi + \varphi_0)$ of (1.1.5) the counterterms $C_1\partial_\mu(\varphi+\varphi_0)\partial^\mu(\varphi+\varphi_0)$, $C_2(\varphi+\varphi_0)^2$, $C_3(\varphi+\varphi_0)^4$ and C_4.

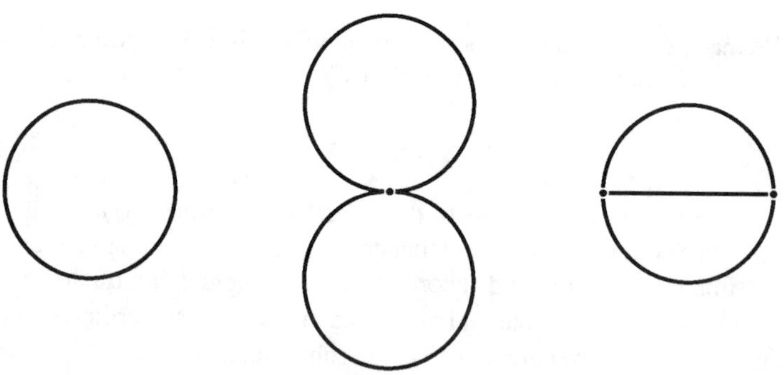

Figure 12. One-and two-loop diagrams for $V(\varphi)$ in the theory of the scalar field (1.1.5).

The meaning of (2.2.1) becomes particularly clear after integrating over k_4. The result (up to an infinite constant that is eliminated by renormalization of the vacuum energy, i.e., by the addition of C_4 to $L(\varphi + \varphi_0)$) is

$$V(\varphi) = -\frac{\mu^2}{2}\varphi^2 + \frac{\lambda}{4}\varphi^4 + \frac{1}{(2\pi)^3}\int d^3k\sqrt{k^2 + m^2(\varphi)}\,. \tag{2.2.3}$$

Thus, in the one-loop approximation, the effective potential $V(\varphi)$ is given by the sum of the classical expression for the potential energy of the field φ and a φ-dependent vacuum energy shift due to quantum fluctuations of the field φ. To determine the quantities C_i, normalization conditions must be imposed on the potential, and these, for example, can be chosen to be [139]

$$\left.\frac{dV}{d\varphi}\right|_{\varphi = \mu/\sqrt{\lambda}} = 0, \qquad \left.\frac{d^2V}{d\varphi^2}\right|_{\varphi = \mu/\sqrt{\lambda}} = 2\mu^2. \tag{2.2.4}$$

These normalization conditions are chosen to ensure that the location of the minimum of $V(\varphi)$ for $\varphi = \mu/\sqrt{\lambda}$ and the curvature of $V(\varphi)$ at the minimum (which is the same to lowest order in λ as the mass squared of the scalar field φ) remain the same as in the classical theory. Other types of normalization conditions also exist; for example, the Coleman–Weinberg conditions [137] are

$$\left.\frac{d^2V}{d\varphi^2}\right|_{\varphi = 0} = m^2, \qquad \left.\frac{d^4V}{d\varphi^4}\right|_{\varphi = M} = \lambda\,. \tag{2.2.5}$$

where M is some normalization point. All physical results obtained via the normalization conditions (2.2.4) and (2.2.5) are equivalent, after one establishes the appropriate correspondence between the parameters μ, m, M, and λ in the renormalized expressions for $V(\varphi)$ in the two cases. The conditions (2.2.4) are usually more convenient for practical purposes in work with theories that have spontaneous symmetry breaking, although (2.2.5) is sometimes the most suitable approach in certain instances involving the study of the fundamental features of the theory, since the first condition determines the mass squared of the scalar field prior to symmetry breaking. Since we are most interested in the present section in the properties of $V(\varphi)$ for certain values of $m^2(\varphi) = \frac{d^2V}{d\varphi^2}$ at the minimum of $V(\varphi)$, we will use the conditions (2.2.4). The effective potential $V(\varphi)$ then takes the form [23]

$$V(\varphi) = -\frac{\mu^2}{2}\varphi^2 + \frac{\lambda}{4}\varphi^4 + \frac{(3\lambda\varphi^2 - \mu^2)^2}{64\pi^2}\ln\left(\frac{3\lambda\varphi^2 - \mu^2}{2\mu^2}\right) + \frac{21\lambda\mu^2}{64\pi^2}\varphi^2 - \frac{27\lambda^2}{128\pi^2}\varphi^4 . \tag{2.2.6}$$

Clearly, for $\lambda \ll 1$, quantum corrections only become important for asymptotically large φ (when $\lambda \ln(\varphi/\mu) \gg 1$), where it becomes necessary to take account of all higher-order corrections. When $\lambda > 0$, it becomes extremely difficult to sum all higher-order corrections to the expression for $V(\varphi)$ at large φ. This problem can only be solved for a special class of $\lambda\varphi^4$ theories discussed in the next section.

We can make much more progress in clarifying the role of quantum corrections in theories with several different coupling constants. As an example, let us consider the Higgs model (1.1.15) in the transverse gauge $\partial_\mu A_\mu = 0$. In the one-loop approximation, the effective potential in that case is given by the diagrams in Fig. 13.

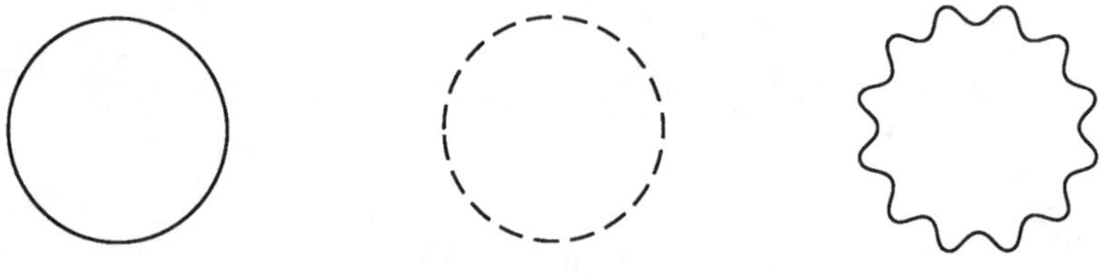

Figure 13. Diagrams for $V(\varphi)$ in the Higgs model. The solid, dashed, and wavy lines correspond to the χ_1, χ_2, and A_μ fields, respectively.

For $e^2 \ll \lambda$, the contribution of vector particles can be neglected, and the situation is analogous to the one described above. When $e^2 \gg \lambda$, we can ignore the contribution of scalar particles. In that event, the expression for $V(\varphi)$ takes the form [139]

$$V(\varphi) = -\frac{\mu^2\varphi^2}{2}\left(1 - \frac{3e^4}{16\pi^2\lambda}\right) + \frac{\lambda\varphi^4}{4}\left(1 - \frac{9e^4}{32\pi^2\lambda}\right) + \frac{3e^4\varphi^4}{64\pi^2}\ln\left(\frac{\lambda\varphi^2}{\mu^2}\right). \tag{2.2.7}$$

Clearly, then, when $\lambda < \frac{3e^4}{16\pi^2}$, the effective potential acquires an additional minimum at $\varphi = 0$, and when $\lambda < \frac{3e^4}{32\pi^2}$, this minimum becomes even deeper than the usual minimum at $\varphi = \varphi_0 = \frac{\mu}{\sqrt{\lambda}}$; see Fig. 14.

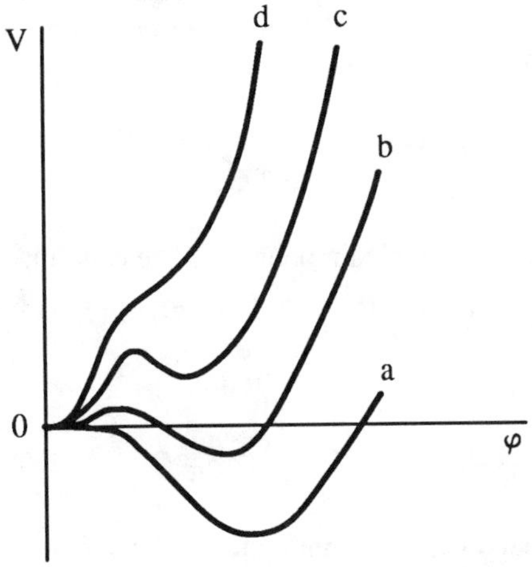

Figure 14. Effective potential in the Higgs model. a) $\lambda > \frac{3e^4}{16\pi^2}$; b) $\frac{3e^4}{16\pi^2} > \lambda > \frac{3e^4}{32\pi^2}$; c) $\frac{3e^4}{32\pi^2} > \lambda > 0$; d) $\lambda = 0$.

Hence, when $\lambda < \frac{3e^4}{16\pi^2}$, symmetry breaking in the Higgs model becomes energetically unfavorable. This effect is due not to large logarithmic factors like $\lambda \ln\frac{\varphi}{\mu} \gtrsim 1$, but to special relations between λ and e^2 ($\lambda \sim e^4$), whereby the classical terms in the expression for the effective potential (2.2.7) become of the same order as the quantum corrections to order e^2. Higher-order corrections to (2.2.7) are proportional to λ^2 and e^6, and do not lead to any substantial modification of the form of $V(\varphi)$ given by (2.2.7), over the range $\varphi \lesssim \mu/\sqrt{\lambda}$ in which we are most interested.

We remark here that $m_A^2 = e^2\varphi_0^2 = \frac{e^2\mu^2}{\lambda}$, $m_\varphi^2 = 2\lambda\varphi_0^2$ up to higher-order corrections in e^2. This means that symmetry breaking is only favorable in the Higgs model if

$$m_\varphi^2 > \frac{3e^4}{16\pi^2} m_A^2 . \tag{2.2.8}$$

The significance of this result for the Glashow–Weinberg–Salam model is that the mass of the Higgs boson in that theory (more precisely, in the standard version with one kind of Higgs boson and no superheavy fermions, and with $\sin^2\theta_W \sim 0.23$) should be more than approximately 7 GeV [139, 140],

$$m_\varphi \gtrsim 7 \text{ GeV}. \tag{2.2.9}$$

From Eq. (2.2.7), we also obtain bounds on the coupling constant between Higgs bosons, $\lambda(\varphi = \varphi_0) = \frac{1}{6}\frac{d^4V}{d\varphi^4}\Big|_{\varphi = \varphi_0}$ [139]. In fact, $\lambda > 0$, and

$$\lambda(\varphi_0) = \lambda + \frac{e^4}{2\pi^2}, \tag{2.2.10}$$

and this means that $V(\varphi)$ has a minimum at $\varphi_0 \neq 0$ if

$$\lambda(\varphi_0) > \frac{11e^4}{16\pi^2}, \tag{2.2.11}$$

and the minimum at $\varphi = \varphi_0$ is deeper than the one at $\varphi = 0$ if

$$\lambda(\varphi_0) > \frac{19e^4}{32\pi^2}. \qquad (2.2.12)$$

A bound like (2.2.12) in the Weinberg–Salam model yields

$$\lambda(\varphi_0) \gtrsim 3 \cdot 10^{-3}. \qquad (2.2.13)$$

With cosmological considerations taken into account, the corresponding bound can be improved somewhat. As we have already said in the Introduction, symmetry was restored in the early universe at $T \gtrsim 10^2$ GeV in the Glashow–Weinberg–Salam theory, and the only minimum of $V(\varphi, T)$ was the one at $\varphi = 0$. A minimum appears at $\varphi \neq 0$ only as the universe cools, and if the effective potential then continues to have a minimum at $\varphi = 0$, it is not clear *a priori* whether the field φ will be able to jump out of the local minimum at $\varphi = 0$ to a global minimum at $\varphi = \varphi_0 \sim$ 250 GeV, nor is it clear what the properties of the universe would be after such a phase transition. By making use of high-temperature tunneling theory [62], it has been shown that this transition has an exceedingly low probability of occurrence in the Glashow–Weinberg–Salam model. The phase transition can therefore only take place if the minimum of $V(\varphi)$ at $\varphi = 0$ is very shallow, $\left.\frac{d^2V}{d\varphi^2}\right|_{\varphi=0} \ll \mu^2$. This then leads to a somewhat more rigorous bound on the mass of the Higgs boson [141–144],

$$m_\varphi \gtrsim 10 \text{ GeV}. \qquad (2.2.14)$$

One particular case which is especially interesting from the standpoint of cosmology (as well as from the standpoint of elementary particle theory) is that in which $\left.\frac{d^2V}{d\varphi^2}\right|_{\varphi=0} = 0$. This is known as the Coleman–Weinberg theory [137]. The effective potential in this theory, which is based on the Higgs model (1.1.15), takes the form

$$V(\varphi) = \frac{25e^4}{128\pi^2}\left(\varphi^4 \ln\frac{\varphi}{\varphi_0} - \frac{\varphi^4}{4} + \frac{\varphi_0^4}{4}\right). \qquad (2.2.15)$$

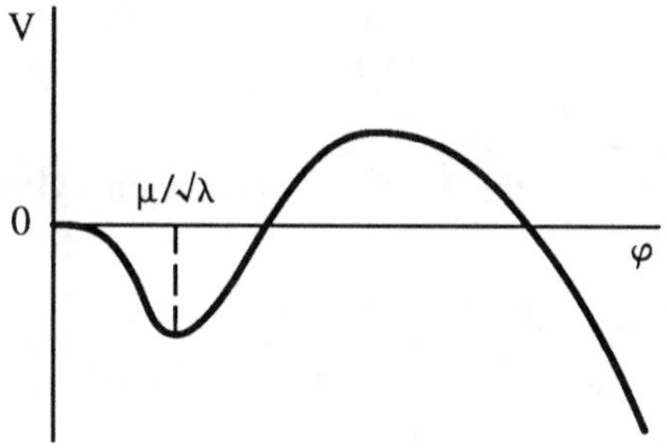

Figure 15. Effective potential in the theory (1.1.13) with $m_\psi \gg m_\varphi$.

We have added the term $\frac{25e^4}{512\pi^2}\varphi_0^4$ here in order to ensure that $V(\varphi_0) = 0$.
In the SU(5) model, the corresponding effective potential takes the form

$$V(\varphi) = \frac{25g^4}{128\pi^2}\varphi^4\left(\ln\frac{\varphi}{\varphi_0} - \frac{1}{4}\right) + \frac{9}{32\pi^2}M_X^2, \tag{2.2.16}$$

where g^2 is the SU(5) gauge coupling constant, M_X is the mass of the X boson, and φ is defined by Eq. (1.1.19). Equation (2.2.16) lay at the foundation of the first version of the new inflationary universe scenario, so we will have a number of occasions to return to it.

Whereas quantum fluctuations of vector fields stimulate the dynamical restoration of symmetry, quantum fluctuations of fermions enhance symmetry breaking. We now consider the simplified σ-model (1.1.13) as an example. At large φ, the effective potential in this theory is given by [145]

$$V(\varphi) = -\frac{\mu^2}{2}\varphi^2 + \frac{\lambda}{4}\varphi^4 + \frac{9\lambda^2 - 4h^4}{64\pi^2}\varphi^4\ln\frac{\lambda\varphi^2}{\mu^2} \tag{2.2.17}$$

Fermions evidently make a negative contribution at large φ, and when $3\lambda < 2h^2$, the effective potential $V(\varphi)$ is unbounded from below (Fig. 15).

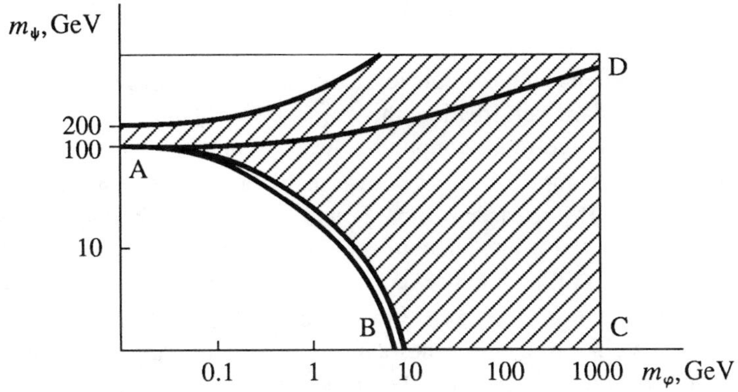

Figure 16. The hatched region corresponds to allowable mass values for the Higgs boson, m_φ, and heavy fermions, m_ψ (or more precisely, $\sum_i (m_{\psi_i}^4)^{1/4}$) when one takes account of both cosmological considerations and quantum corrections to the effective potential in the Glashow–Weinberg– Salam model. The area bounded by the curve ABCD is the region of absolute phase stability with spontaneous symmetry breaking, $\varphi = \frac{\mu}{\sqrt{\lambda}}$.

Of course when $\varphi \to \infty$, the one-loop approximation is longer applicable. However, if $\lambda \ll h^2$, there is a range of values of the field φ ($\varphi^2 \sim \frac{\mu^2}{\lambda} \exp \frac{\lambda}{h^4}$ for which $V(\varphi) < V\left(\frac{\mu}{\sqrt{\lambda}}\right)$, and the one-loop approximation still gives reliable results. Thus, in the σ-model with $\lambda \ll h^2$, or what is the same thing, with $m_\varphi \ll m_\psi$, the state $\varphi = \frac{\mu}{\sqrt{\lambda}}$ is unstable, and strong dynamical symmetry breaking takes place.

We can readily generalize this result to a wider class of theories, including the Glashow–Weinberg–Salam theory, which leads to a set of constraints on the mass of the Higgs meson and the fermion masses in this theory [139–151]; see Fig. 16. We shall take advantage of cosmological considerations in Chapter 6 to strengthen these constraints.

2.3 The 1/N expansion and the effective potential in the $\lambda\varphi^4/N$ theory

As a rule, it is not possible to study the behavior of the effective potential in standard perturbation theory as $\varphi \to \infty$, but theories that are asymptotically free in all coupling constants constitute an important exception. For example, it can be shown that in a massless $\lambda\varphi^4$ theory with negative λ, $V(\varphi)$ decreases without bound as $\varphi \to \infty$ both in the classical approximation and when quantum corrections are taken into consideration [137, 152]. It is difficult to use standard perturbation theory in λ to investigate the behavior of $V(\varphi)$ as $\varphi \to \infty$ in the $\lambda\varphi^4$ theory with $\lambda > 0$. There does exist a class of theories, however, in which one can make substantial progress toward understanding the properties of $V(\varphi)$ for both small and large φ, bringing with it a number of surprising results.

Consider the O(N) symmetric theory of the scalar field $\Phi = \{\Phi_1, \cdots, \Phi_N\}$, with the Lagrangian

$$L = \frac{1}{2}(\partial_\mu\Phi)^2 - \frac{\mu^2}{2}\Phi^2 - \frac{\lambda}{4!N}\left(\Phi^2\right)^2, \tag{2.3.1}$$

where $\Phi^2 = \sum_i \Phi_i^2$. The field Φ may have a classical part $\Phi_0 = \sqrt{N}\{\varphi, 0, \cdots, 0\}$. Let us also introduce the composite field

$$\hat{\chi} = \mu^2 + \frac{\lambda}{6N}\Phi^2 \tag{2.3.2}$$

with a classical part χ, and let us add to (2.3.1) the term

$$\Delta L = \frac{3N}{2\lambda}\left(\hat{\chi} - \mu^2 - \frac{\lambda}{6N}\Phi^2\right)^2, \tag{2.3.3}$$

so that

$$L' = L + \Delta L = L = \frac{1}{2}(\partial_\mu\Phi)^2 - \frac{3N}{\lambda}\mu^2\hat{\chi} + \frac{3N}{\lambda}\hat{\chi}^2 - \frac{1}{2}\hat{\chi}\Phi^2. \tag{2.3.4}$$

The theory described by (2.3.4) is equivalent to the theory (2.3.1), since the Lagrange equation for the field $\hat{\chi}$ in the theory (2.3.4) is exactly (2.3.2), while the Lagrange equation for the field Φ in the theory (2.3.4), taking

(2.3.2) into account, gives the Lagrange equation for the field Φ in the theory (2.3.1) [153]. In the one-loop approximation, the effective potential $V(\varphi, \chi) \equiv N\,\mathcal{V}(\varphi, \chi)$ which corresponds to the theory (2.3.4) is given by [154]

$$\mathcal{V}(\varphi, \chi) = -\frac{3}{2}\left(\frac{1}{\lambda} + \frac{1}{96\pi^2}\right)\chi\,(\chi - 2\mu^2) + \frac{1}{2}\chi\varphi^2 + \frac{\chi^2}{128\pi^2}\left(2\ln\frac{\chi}{M^2} - 1\right), \tag{2.3.5}$$

where M is the normalization parameter, and

$$(\chi - \mu^2)\left(\frac{1}{\lambda} + \frac{1}{96\pi^2}\right) = \frac{\varphi^2}{6} + \frac{\chi}{96\pi^2}\ln\frac{\chi}{M^2}. \tag{2.3.6}$$

The effective potential $V(\varphi) \equiv N\,\mathcal{V}(\varphi)$ in the original theory (2.3.1) is equal to $V(\varphi, \chi(\varphi))$. It is important to note that all of the higher-order corrections to Eqs. (2.3.5) and (2.3.6) contain higher powers of 1/N, and vanish in the limit as $N \to \infty$. In that sense, Eqs. (2.3.5) and (2.3.6) are *exact* in the limit $N \to \infty$.

We now impose the following normalization conditions on μ^2 and λ in (2.3.5) and (2.3.6):

$$\mathrm{Re}\left.\frac{d^2\mathcal{V}}{d\varphi^2}\right|_{\varphi=0} = \mu^2, \tag{2.3.7}$$

$$\mathrm{Re}\left.\frac{d^4\mathcal{V}}{d\varphi^4}\right|_{\varphi=0} = \lambda\,. \tag{2.3.8}$$

This then tells us that after renormalization, the parameter M^2 in (2.3.5) should be put equal to μ^2.

The signs of μ^2 and λ in (2.3.7) and (2.3.8) are arbitrary. For simplicity, we will consider the case in which $\mu^2 > 0$, $\lambda > 0$. The field χ is found to be a double-valued function of φ when $\varphi < \bar{\varphi}$, where

$$1 - \frac{\lambda}{96\pi^2}\ln\frac{\chi(\bar{\varphi})}{\mu^2} = 0\,. \tag{2.3.9}$$

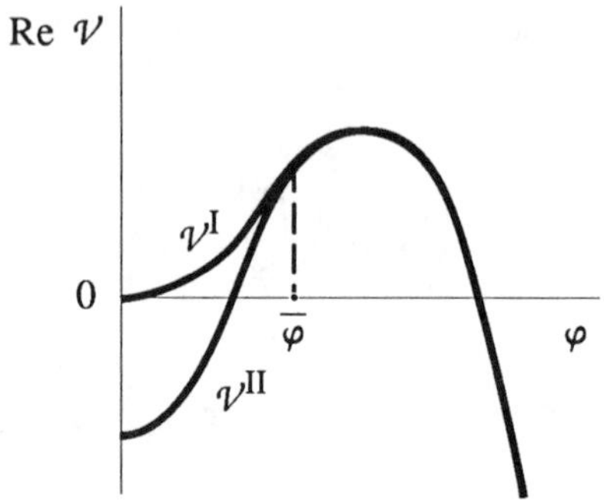

Figure 17. Effective potential in the theory (2.3.1) with $\mu^2 > 0$.

As a result, for $\varphi < \bar{\varphi}$, the effective potential $\mathcal{V}(\varphi)$ turns out to be a double-valued function of φ (with branches $\mathcal{V}^{\mathrm{I}}(\varphi)$ and $\mathcal{V}^{\mathrm{II}}(\varphi)$, $\mathcal{V}^{\mathrm{I}} > \mathcal{V}^{\mathrm{II}}$; see Fig. 17) [154]. The normalization conditions (2.3.7) and (2.3.8) hold on the upper branch of $\mathcal{V}(\varphi)$.

On the branch $\mathcal{V}^{\mathrm{II}}$, the field χ is extremely large ($\frac{\lambda}{96\pi^2} \ln \frac{\chi}{\mu^2} > 1$), and one may well ask whether Eqs. (2.3.5) and (2.3.6) are actually valid for such large χ and for any large but finite N. The answer to this question is affirmative, since on the branch $\mathcal{V}^{\mathrm{II}}$, χ is large in magnitude but finite, and is independent of N. For any arbitrarily large χ, there should therefore exist an N such that corrections $\sim O(1/N)$ to Eqs. (2.3.5) and (2.3.6) for this χ are small [155].

When $\varphi = 0$, as was shown in [153] to lowest order in 1/N, the Green's function $G_{\chi\chi}(k^2)$ of the field χ on the upper branch $\mathcal{V}^{\mathrm{I}}$ has a tachyon pole at $k^2 = -\mu^2 e^{1/\lambda}$. Using the same arguments as above, it can be shown that higher-order corrections in 1/N to $G_{\chi\chi}(k^2)$ can change the type of singularity at $k^2 < 0$, but they cannot alter the fact that $G_{\chi\chi}(k^2)$ changes sign at $k^2 < 0$. Such behavior of $G_{\chi\chi}(k^2)$ is incompatible with the Källén–Lehmann theorem, and indicates that the theory is unstable against production of the classical field χ, the reason simply being that on the branch $\mathcal{V}^{\mathrm{I}}$, the point $\varphi = 0$ is not a minimum but a saddle point of the potential $\mathcal{V}(\varphi, \chi)$, and a transition takes place to the minimum at $\varphi = 0$ on the branch $\mathcal{V}^{II}$.

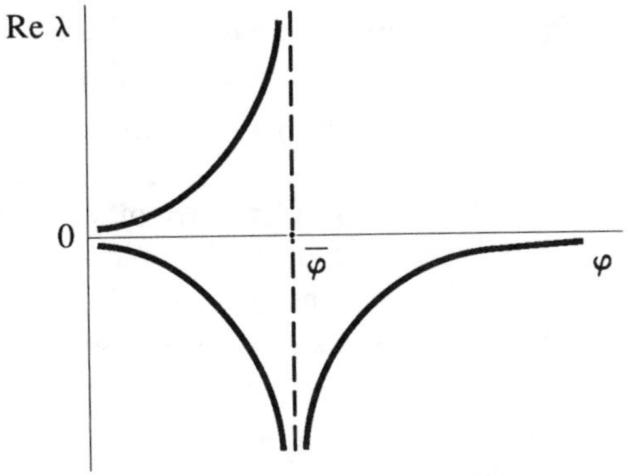

Figure 18. Effective coupling constant $\lambda(\varphi)$ in the theory (2.3.1).

However, even this point is not an absolute minimum of $\mathcal{V}(\varphi)$. In fact, according to (2.3.5) and (2.3.6),

$$\mathcal{V}(\varphi) = -4\pi^2 \frac{\varphi^4}{\ln \frac{\varphi^2}{\mu^2}} \left(1 + \frac{i\pi}{\ln \frac{\varphi^2}{\mu^2}} \right) \tag{2.3.10}$$

as $\varphi \to \infty$. This means that the potential $\mathcal{V}(\varphi)$ is not bounded from below, and the theory (2.3.1) is unstable against production of arbitrarily large fields φ [155].

A number of objections can be raised to this conclusion, the principal one being the following. Equation (2.3.10) holds when $N = \infty$, but for any finite N there might exist a field $\varphi = \varphi_N$ so large that when $\varphi > \varphi_N$ the expression (2.3.10) becomes unreliable; an absolute minimum of $\mathcal{V}(\varphi)$ might thus exist for $\varphi > \varphi_N$.

One response to this objection can be found by combining the 1/N expansion and the renormalization group equation [155]. In order to do so, we first note that the magnitude of the effective coupling constant $\lambda(\varphi) = \frac{d^4V}{d\varphi^4}$ which can be calculated using (2.3.5) and (2.3.6), behaves as shown in Fig. 18. This then leads to several consequences:

a) for large enough N, a $\frac{\lambda}{N}\varphi^4$ theory with $\lambda > 0$ is equivalent to a theory with $\lambda < 0$, representing merely another branch of the same theory;

b) contrary to the usual expectations, a $\frac{\lambda}{N}\varphi^4$ theory with $\lambda > 0$ is unstable, while a theory with $\lambda < 0$ is metastable for small φ;

c) for large enough φ, Re λ becomes negative, and tends to zero with increasing φ.

The last of these is the decisive point. We may choose such a large value $\varphi = \varphi_1$ that λ is in fact small and negative, and such a large value of $N(\varphi_1)$ that the higher-order corrections in powers of 1/N to the value of $\lambda(\varphi)$ at $\varphi \sim \varphi_1$ are also small. We can then make use of the renormalization group equation to continue the quantity $\lambda(\varphi)$ from $\varphi = \varphi_1$ to $\varphi \to \infty$, since the $\frac{\lambda}{N}\varphi^4$ theory is asymptotically free when $\lambda < 0$. We must then integrate $\lambda(\varphi)$ with respect to φ and obtain the value of $\mathcal{V}(\varphi)$. These calculations result in a value for $\mathcal{V}(\varphi)$ identical to that obtained from (2.3.10), and thereby confirm that the effective potential in this theory is actually unbounded from below for large φ [155].

This conclusion turns out to be valid regardless of the sign of μ^2 and λ at $\varphi = 0$. Interestingly enough, spontaneous symmetry breaking, which ought to occur in the theory (2.3.1) when $\mu^2 < 0$, actually takes place only on the upper (unstable) branch of $\mathcal{V}(\varphi)$; on the lower (metastable) branch, the effective mass squared of the field φ is always positive, and symmetry breaking does not occur [154].

These results are fairly surprising, and in many respects they are quite instructive. Quantum corrections are found to lead to instability even in theories where this might be least expected, such as (2.3.1) with $\mu^2 > 0$ and $\lambda > 0$. It turns out that for large N, there is no spontaneous symmetry breaking in this theory when $\mu^2 < 0$; it has also been found that in the theory (2.3.1), $\lambda < 0$ and $\lambda > 0$ actually represent two branches of a single theory. These branches coalesce at exponentially large values of φ, and in the limit of very large φ, the effective constant $\lambda(\varphi)$ becomes negative and tends to zero from below. The latter result, however, is not so very surprising, as that is just how the effective constant λ ought to behave for large fields and large momenta, according to a study based on the renormalization group equation (see [58], for example). This sort of pathological behavior of the effective coupling constant λ also lay at the basis of the so-called zero-charge problem [156, 157]. For a long time, the corresponding results were viewed as being rather unreliable, and it was considered plausible that in many realistic situations the zero-charge problem actually does not appear; for example, see [158]. On the other hand, the principal objections to the reliability of the results presented in [156, 157] do not seem to apply to derivations based on the 1/N expansion [155, 159]. More recently, the existence of the zero-charge problem in the theory $\lambda\varphi^4$ has been sufficiently well proven, both analytically [160] and numerically [161] ("triviality" of the theory $\lambda\varphi^4$).

The foregoing analysis aids in an understanding of the essence of this problem using the theory (2.3.1) as an example: according to our results, for large N, no theory of the form (2.3.1) possesses both a stable vacuum and a nonvanishing coupling constant λ.

One question that then emerges is whether this result has any bearing on realistic elementary particle theories with spontaneous symmetry breaking. first of all, then, let us analyze just how serious the shortcomings of the theory (2.3.1) actually are. At first glance, the presence of a pole at $k^2 = -\mu^2 e^{1/\lambda}$ on the upper branch of $\mathcal{V}(\varphi)$ doesn't seem so terrible, since it is usually taken to mean that the low-energy physics doesn't "feel" the structure of the theory at superhigh momenta and masses. This is actually so at large $k^2 > 0$. But the example of the theory (1.1.5) with symmetry breaking demonstrates that the presence of a tachyon pole at $k^2 = -\mu^2 < 0$ leads to more rapid development of an instability than would a large tachyon mass; see (1.1.6). The upper branch of the potential $\mathcal{V}(\varphi)$ therefore actually corresponds to an unstable vacuum state (an analogous instability also occurs in a multicomponent formulation of quantum electrodynamics at sufficiently large N [159, 162]). On the other hand, when $\lambda \ll 1$, the lifetime of the universe at the point $\varphi = 0$ on the lower branch turns out to be exponentially large, so the putative instability of the vacuum in this theory in no way implies that it cannot correctly describe our universe. One possible problem here is that at a temperature $T \gtrsim \mu e^{1/\lambda}$, the local minimum at $\varphi = 0$ on the branch $\mathcal{V}^{II}$ also vanishes [155, 163], but in the inflationary universe theory the temperature can never reach such high values.

Proceeding to a discussion of more realistic theories, it must be pointed out that when $\lambda \ll 1$, the tachyon pole on the upper branch of $\mathcal{V}(\varphi)$ is situated at $|k^2| \gg M_P^2$, and at the point $\bar{\varphi}$ where $\mathcal{V}^I$ and $\mathcal{V}^{II}$ merge, the effective potential $V(\varphi)$ exceeds the Planck energy density M_P^4. In that event, as will be shown in the following section, all of the major qualitative and quantitative results obtained neglecting quantum gravitation become unreliable. Furthermore, quantum corrections to $V(\varphi)$ associated with the presence of other matter fields can become important at lower momenta and densities. These corrections will not change the form of $V(\varphi)$ at small φ, but they can completely eliminate the instability that arises with large fields and momenta. Exactly the same thing happens with the instability in the zero-charge problem when one makes the transition to asymptotically free theories [3, 152].

The basic practical conclusion to be drawn from the last two sections is that for the most reasonable relationships between the coupling constants ($\lambda \sim e^2 \sim h^2 \ll 1$), quantum corrections to $V(\varphi)$ in theories of the weak, strong, and electromagnetic interactions become important only for exponentially strong fields, so that the classical expression for $V(\varphi)$ is often a perfectly good approximation. Quantum corrections can often lead to instability of the vacuum when the fields or momenta are exponentially large, but this difficulty can in principle be avoided by a small modification of the theory without altering the shape of the effective potential at small φ.

2.4 The effective potential and quantum gravitational effects

In our discussion of the inflationary universe scenario in Chapter 1, we often turned our attention to fields $\varphi \gg M_P$. There is some question as to whether quantum gravitational effects might substantially modify $V(\varphi)$ under such conditions, ultimately invalidating the chaotic inflation scenario. Such suspicions have been voiced by a number of authors (see [164], for example), and we must therefore dwell specifically upon this question.

Gravitational corrections $\Delta V(\varphi)$ to the potential $V(\varphi)$ are of a twofold nature. On the one hand, they are associated with the gravitational interaction between vacuum fluctuations, as in the Feynman diagrams shown in Fig. 19. The entire set of such diagrams can be summed, with the final result being [165]

$$\Delta V(\varphi) = C_1 \frac{d^2V}{d\varphi^2} \cdot \frac{V}{M_P^2} \ln \frac{\Lambda^2}{M_P^2} + C_2 \frac{V^2(\varphi)}{M_P^4} \ln \frac{\Lambda^2}{M_P^2} . \tag{2.4.1}$$

The C_i here are numerical coefficients of order unity, and Λ is the ultraviolet cutoff. These corrections clearly diverge as $\Lambda \to \infty$, and generally speaking, they fail to converge simply to a renormalized version of the original potential $V(\varphi)$. This is a manifestation of the well known difficulty associated with the nonrenormalizability of quantum gravitation. One usually assumes, however, that at momenta of order M_P, there should exist a natural cutoff due either to the nontrivial structure of the gravitational vacuum, or to the fact that when $|k^2| \gtrsim M_P^2$, gravitation becomes part of a more general theory with no divergences. If, in accordance with this assumption, Λ^2 does not exceed M_P^2 by many orders of magnitude, then

$$\Delta V = \tilde{C}_1 \frac{d^2V}{d\varphi^2} \cdot \frac{V}{M_P^2} + \tilde{C}_2 \frac{V^2}{M_P^4}, \tag{2.4.2}$$

where $\tilde{C}_i = O(1)$. It can readily be shown that when

$$m_\varphi^2 = \frac{d^2V}{d\varphi^2} \ll M_P^2, \tag{2.4.3}$$

$$V(\varphi) \ll M_P^4, \tag{2.4.4}$$

gravitational corrections to $V(\varphi)$ are negligible. In particular, for the theory $\lambda\varphi^4$, (2.4.4) is a much stronger condition than (2.4.3); it holds when

$$\varphi \ll \varphi_P = \lambda^{-1/4} M_P. \tag{2.4.5}$$

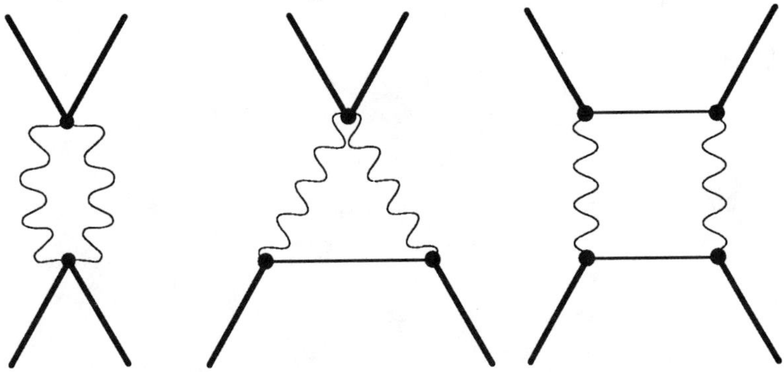

Figure 19. Typical diagrams for $V(\varphi)$ with gravitational effects taken into account. The heavy lines correspond to the external classical field φ, the lighter lines to scalar particles of φ, and the wavy lines to gravitons.

For $\lambda \sim 10^{-14}$, we obtain a very weak bound on φ from (2.4.5):

$$\varphi \ll 3000\,M_P. \tag{2.4.6}$$

Thus, in a classical space-time in which (2.4.5) holds (see Section 1.7), the indicated quantum gravitational corrections to $V(\varphi)$ are negligible.

The other type of correction to $V(\varphi)$ relates to the change in the spectrum of vacuum fluctuations in an external gravitational field. However, inasmuch as the magnitude of the field itself is proportional to $V(\varphi)$, the corresponding corrections (for $V(\varphi) \ll M_P^4$) are usually negligible. The most important exception to this is the contribution to $V(\varphi)$ from long-wavelength fluctuations of the scalar field φ that are generated at the time of inflation. But as we already noted in Section 1.8, taking this effect into account does not lead to any problems with the realization of the chaotic inflation scenario; in fact, on the contrary, it engenders a self-sustaining inflationary regime over most of the physical volume of the universe. We shall return to this question in Chapter 10.

3

Restoration of Symmetry at High Temperature

3.1 Phase transitions in the simplest models with spontaneous symmetry breaking

Having discussed the basic features of spontaneous symmetry breaking in quantum field theory, we can now turn to a consideration of symmetry behavior in systems of particles in thermodynamic equilibrium which interact in accord with unified theories of the weak, strong, and electromagnetic interactions. We will primarily examine systems of scalar particles φ with the Lagrangian (1.1.5). Such particles carry no conserved charge, nor is their number a conserved quantity. The chemical potential therefore vanishes for such particles, and their density in momentum space is

$$n_k = \frac{1}{\exp\left(\frac{k_0}{T}\right) - 1}, \qquad (3.1.1)$$

where $k_0 = \sqrt{\boldsymbol{k}^2 + m^2}$ is the energy of a particle with momentum $\boldsymbol{k}$ and mass m. All particles disappear at $T = 0$ ($n_k \to 0$), and we revert to the situation described in the previous chapter.

At finite temperature, all physically interesting quantities (thermodynamic potentials, Green's functions, etc.) in this system are given not by vacuum averages, but by Gibbs averages

$$\langle\ldots\rangle = \frac{\mathrm{Tr}\left[\exp\left(-\frac{\mathrm{H}}{\mathrm{T}}\right)\ldots\right]}{\mathrm{Tr}\left[\exp\left(-\frac{\mathrm{H}}{\mathrm{T}}\right)\right]} \tag{3.1.2}$$

where H is the system Hamiltonian. In particular, the symmetry breaking parameter (the "classical" scalar field φ) in this system is given by $\varphi(\mathrm{T}) = \langle\varphi\rangle$, rather than by $\langle 0|\varphi|0\rangle$.

In order to investigate the behavior of $\varphi(\mathrm{T})$ at $\mathrm{T} \neq 0$, let us consider the Lagrange equation for the field φ in the theory (1.1.5),

$$(\Box + \mu^2 - \lambda\varphi^2)\varphi = 0\,, \tag{3.1.3}$$

and let us take the Gibbs average of this equation, giving

$$\Box\varphi(\mathrm{T}) - [\lambda\varphi^2(\mathrm{T}) - \mu^2]\varphi(\mathrm{T}) - 3\lambda\varphi(\mathrm{T})\langle\varphi^2\rangle - \lambda\langle\varphi^3\rangle = 0\,. \tag{3.1.4}$$

Here, as in the analysis of spontaneous symmetry breaking at $\mathrm{T} = 0$, we have separated out the analog of the classical field φ by carrying out the shift $\varphi \to \varphi + \varphi(\mathrm{T})$, such that

$$\langle\varphi\rangle = 0\,. \tag{3.1.5}$$

To lowest order in λ, $\langle\varphi^3\rangle$ is equal to zero, whereas

$$\begin{aligned}\langle\varphi^2\rangle &= \frac{1}{(2\pi)^3}\int\frac{d^3k}{2\sqrt{k^2+m^2}}\left(1 + 2\langle a_k^+ a_k^-\rangle\right) \\ &= \frac{1}{(2\pi)^3}\int\frac{d^3k}{\sqrt{k^2+m^2}}\left(\frac{1}{2} + n_k\right).\end{aligned} \tag{3.1.6}$$

The first term in (3.1.6) vanishes after renormalizing the mass of the field φ in the field theory (at $\mathrm{T} = 0$). As a result,

$$\langle\varphi^2\rangle = F(T, m(\varphi)) = \frac{1}{2\pi^2}\int_0^\infty \frac{k^2\,dk}{\sqrt{k^2+m^2(\varphi)}\left(\exp\frac{\sqrt{k^2+m^2(\varphi)}}{T}-1\right)}. \tag{3.1.7}$$

Clearly, all interesting effects in this theory ($\lambda \ll 1$) take place at $T \gg m$, where we can neglect m in (3.1.7). Then

$$\langle\varphi^2\rangle = F(T, 0) = \frac{T^2}{12}, \tag{3.1.8}$$

and Eq. (3.1.4) becomes

$$\Box\varphi(T) - [\lambda\varphi^2(T) - \mu^2 + \frac{\lambda}{4}T^2]\,\varphi(T) = 0. \tag{3.1.9}$$

From (3.1.9), we obtain for the constant field $\varphi(T)$

$$\varphi(T)\,[\lambda\varphi^2(T) - \mu^2 + \frac{\lambda}{4}T^2] = 0. \tag{3.1.10}$$

At sufficiently low temperature, this equation has two solutions,

$$\begin{aligned} &1)\ \varphi(T) = 0;\\ &2)\ \varphi(T) = \sqrt{\frac{\mu^2}{\lambda} - \frac{T^2}{4}}. \end{aligned} \tag{3.1.11}$$

The second of these vanishes above a critical temperature

$$T_c = \frac{2\mu}{\sqrt{\lambda}} = 2\varphi_0. \tag{3.1.12}$$

To derive the excitation spectrum at $T \neq 0$, we must carry out the shift $\varphi \to \varphi + \delta\varphi$ in (3.1.9). When $\varphi(T) = 0$, the corresponding equation takes the form

$$\Box\delta\varphi - \left(\mu^2 + \frac{\lambda}{4}T^2\right)\delta\varphi = 0\,, \tag{3.1.13}$$

which corresponds to a mass

$$m^2 = -\mu^2 + \frac{\lambda}{4}T^2 \tag{3.1.14}$$

for the scalar field at $\varphi = 0$. This quantity is negative when $T < T_c$, and it becomes positive when $T > T_c$. For the solution of the second of Eqs. (3.1.11), $\varphi(T) = \sqrt{\frac{\mu^2}{\lambda} - \frac{T^2}{4}}$, it takes the value

$$m^2 = 3\lambda\varphi^2(T) - \mu^2 + \frac{\lambda}{4}T^2 = 2\lambda\varphi^2(T)\,. \tag{3.1.15}$$

This solution is therefore stable for $T < T_c$, and it vanishes for $T > T_c$ at the instant when the solution $\varphi = 0$ becomes stable. This then means that a phase transition with restoration of symmetry takes place at a temperature $T = T_c$ [18–24].

We illustrate the foregoing results in Fig. 20. The quantity $\varphi(T)$ clearly decreases smoothly with increasing temperature, corresponding to a second-order phase transition.

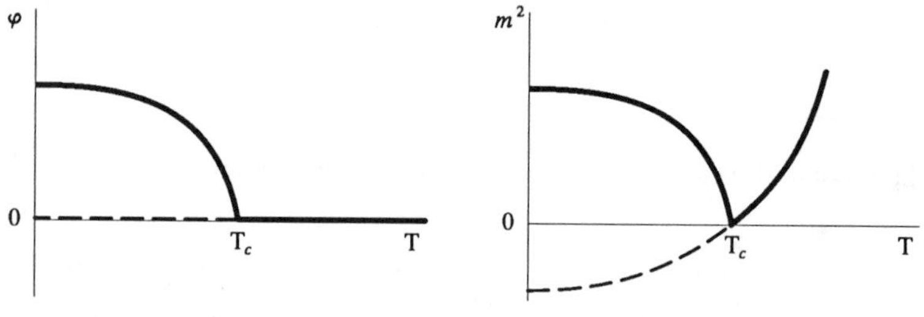

Figure 20. The quantities $\varphi(T)$ and $m^2(T)$ in the theory (1.1.5). The dashed lines correspond to the unstable phase $\varphi = 0$ at $T < T_c$.

These results can also be obtained in a different way, based on a finite-temperature generalization of the concept of the effective potential $V(\varphi)$. We will not dwell on this problem, noting simply that at its extrema, the effective potential $V(\varphi, T)$ coincides with the free energy $F(\varphi, T)$. To calculate $V(\varphi, T)$, it suffices to recall that at $T \neq 0$, quantum statistics is equivalent to Euclidean quantum field theory in a space which is periodic, with period 1/T along the "imaginary time" axis [166, 20]. To go from $V(\varphi, 0)$ to $V(\varphi, T)$ one should replace all boson momenta k_4 in the Euclidean integrals by $2\pi nT$ for bosons and $(2n+1)\pi T$ for fermions, and sum over n instead of integrating over k_4: $\int dk_4 \to 2\pi nT \sum_{n=-\infty}^{\infty}$. For example, at $T \neq 0$, Eq. (2.2.1) for $V(\varphi)$ in the theory (1.1.5) transforms into

$$\begin{aligned} V(\varphi, T) = & -\frac{\mu^2}{2}\varphi^2 + \frac{\lambda}{4}\varphi^4 \\ & + \frac{T}{2(2\pi)^3} \sum_{n=-\infty}^{\infty} \int d^3k \ln\left[(2\pi nT)^2 + k^2 + m^2(\varphi)\right], \end{aligned} \tag{3.1.16}$$

where $m^2(\varphi) = 3\lambda\varphi^2 - \mu^2$. This expression can be renormalized using the same counterterms as for $T = 0$. Equation (1.2.3) gives the result of calculating $V(\varphi, T)$ for $T \gg m$. It is straightforward to show that the equation $\frac{dV}{d\varphi} = 0$, which determines the equilibrium values of $\varphi(T)$, is the same as (3.1.10), and that the quantity $\frac{d^2V}{d\varphi^2}$, which determines the mass squared of the field φ, coincides with (3.1.14) and (3.1.15) (for equilibrium $\varphi(T)$). The description of the phase transition in terms of the behavior of $V(\varphi, T)$ is given in Section 1.2.

The methods developed above can readily be generalized to more complicated models. In the Higgs model (1.1.15), for example, in the transverse gauge $\partial_\mu A_\mu = 0$, we have

$$\left\langle \frac{\delta L}{\delta \varphi} \right\rangle = \varphi(T)\left[\mu^2 - \lambda\varphi^2(T) - 3\lambda\langle\chi_1^2\rangle - \lambda\langle\chi_2^2\rangle + e^2\langle A_\mu^2\rangle\right] = 0 \tag{3.1.17}$$

instead of Eq. (3.1.4). To start with, let us assume that $\lambda \sim e^2$. Then the phase transition takes place at $T \gg m_\chi$, m_A, as in the theory (1.1.5); hence

$$\langle\chi_1^2\rangle = \langle\chi_2^2\rangle = -\frac{1}{3}\langle A_\mu^2\rangle = \frac{T^2}{12}, \tag{3.1.18}$$

and Eq. (3.1.17) becomes

$$\varphi\left(\lambda\varphi^2 - \mu^2 + \frac{4\lambda + 3e^2}{12}T^2\right) = 0\,. \tag{3.1.19}$$

This then implies that the phase transition takes place in the Higgs model at a critical temperature

$$T_{c_1}^2 = \frac{12\mu^2}{4\lambda + 3e^2}. \tag{3.1.20}$$

According to (3.1.19), $\varphi(T)$ is a continuous function of T, that is, this is a second-order phase transition [18-20].

If we consider the case $\lambda \lesssim e^4$, however, we find that $m_A(T_{c_1}) \sim \frac{e\mu}{\lambda} \gtrsim T_{c_1}$, i.e., we no longer have $T \gg m_A$, and the contribution of vector particles to (3.1.19) at $T \sim T_{c_1}$ is strongly suppressed. We can then no longer neglect m_A compared with T when calculating $\langle A_\mu^2\rangle = -F(T, m_A)$, and all of the equations are significantly altered. The simplest way to understand this is to note that when $m < T$, the quantity $F(T, m)$ can be represented by a power series in $\frac{m}{T}$:

$$F(T, m) = \frac{T^2}{12}\left[1 - \frac{3}{\pi}\frac{m}{T} + O\left(\frac{m^2}{T^2}\right)\right]. \tag{3.1.21}$$

Bearing in mind, then, that in the lowest order of perturbation theory $m_A = e\varphi$, Eq. (3.1.19) can be rewritten as

$$\varphi\left(\lambda\varphi^2 - \mu^2 + \frac{4\lambda + 3e^2}{12}T^2 - \frac{3e^3}{4\pi}T\varphi\right) = 0\,. \tag{3.1.22}$$

In contrast to (3.1.19), this equation has three solutions rather than two in a certain temperature range $T_{c_1} < T < T_{c_2}$, corresponding to three different extrema of $V(\varphi, T)$; see Fig. 21. The solution $\varphi = 0$ is metastable when $T > T_{c_1}$. Upon heating, a phase transition from the phase $\varphi = \varphi_1$ to the phase $\varphi = 0$ begins at a temperature T_c, where

$$V(\varphi_1(T_c), T_c) = V(0, T_c). \tag{3.1.23}$$

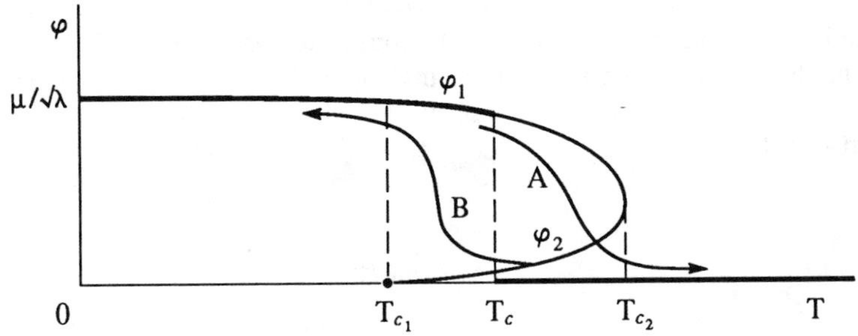

Figure 21. The function $\varphi(T)$ in the Higgs model with $\frac{3e^4}{16\pi^2} < \lambda \lesssim e^4$. The heavy curve corresponds to the stable state of the system. Arrows indicate the behavior of φ with increasing (A) and decreasing (B) temperature.

In the Higgs model with $\lambda \lesssim e^4$,

$$T_c = \left(\frac{15\lambda}{2\pi^2}\right)^{\frac{1}{4}} \mu; \tag{3.1.24}$$

see [23]. Clearly, the phase transition in the present case is a discontinuous one — a first-order phase transition (see Fig. 21).

Recall now that when $\lambda \lesssim \frac{3e^4}{16\pi^2}$, quantum corrections to $V(\varphi, T)$ lead to the existence of a local minimum of $V(\varphi)$ even at $T = 0$ (see Fig. 22), and when $\lambda < \frac{3e^4}{32\pi^2}$, this minimum becomes deeper than the usual one at

$\varphi = \frac{\mu}{\sqrt{\lambda}}$; see Section 2.2. Thus, as $\lambda \to \frac{3e^4}{32\pi^2}$, the critical temperature $T_c \to 0$. This does not mean, however, that a phase transition in such a theory becomes easy to produce in a laboratory. The point here is that a first-order phase transition occurs by virtue of the sub-barrier creation and subsequent growth of bubbles of the new phase. Bubble formation is often strongly suppressed, so it may take an extremely long time for the phase transition to occur. When the system is heated, the phase transition therefore really takes place from a superheated phase φ_1 at some temperature higher than T_c. Likewise, when the system is cooled, a first-order phase transition takes place from a supercooled phase at $T < T_c$. We will consider the theory of bubble production in Chapter 5, and the cosmological consequences of first-order phase transitions will be discussed in Chapters 6 and 7.

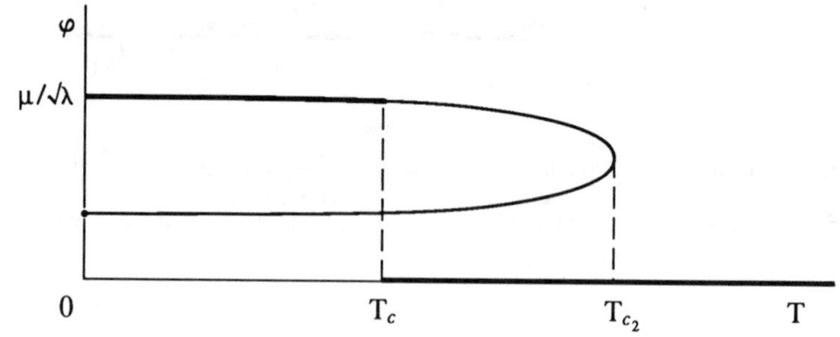

Figure 22. The function $\varphi(T)$ in the Higgs model with $\lambda < \frac{3e^4}{16\pi^2}$.

3.2 Phase transitions in realistic theories of the weak, strong, and electromagnetic interactions

As we have shown in the preceding section, when the coupling constants λ and e^2 are related in the most natural way, the phase transition in the Higgs model is second-order, but when $\lambda \sim e^4$, it becomes first-order. One can readily show that the same is true of a phase transition in the Weinberg–Salam theory.

Thus, for example, when $\lambda \sim e^2$, the counterpart of Eq. (3.1.19) in the

Weinberg–Salam theory becomes [24]

$$\varphi\left(\lambda\varphi^2-\mu^2+\left[2\lambda+\frac{e^2(1+2\cos^2\theta_W)}{\sin^2 2\theta_W}\right]\frac{T^2}{12}\right)=0\,, \qquad (3.2.1)$$

where θ_W is the Weinberg angle, $\sin^2\theta_W \sim 0.23$. Equation (3.2.1) then gives

$$T_c^2=\frac{4\mu^2}{2\lambda+\dfrac{e^2(1+2\cos^2\theta_W)}{\sin^2 2\theta_W}}=\frac{2\varphi_o^2}{1+\dfrac{e^2(1+2\cos^2\theta_W)}{2\lambda\sin^2 2\theta_W}}, \qquad (3.2.2)$$

where $\varphi_0 \sim 250$ GeV. Putting $\lambda \sim e^2 \sim 0.1$, we obtain

$$T_c \sim 200 \text{ GeV}, \qquad (3.2.3)$$

which is more than twice the mass of the $W^\pm$ and Z particles and the mass of the Higgs boson for $\lambda \sim e^2, T=0$. In the case at hand, an analysis similar to the one carried out in Section 3.1 indicates that to high accuracy, the phase transition can be called a second-order transition: the jump in the field φ at the moment of the phase transition turns out to be more than an order of magnitude less than φ_0.

On the other hand, the transitions that take place in grand unified theories at $T \gtrsim 10^{14}$ GeV, as a rule, prove to be first-order transitions with a considerable jump in the field φ at the critical point [104]. There are two reasons why this is so. First, at $T \sim 10^{14}$ GeV, the effective gauge constant $g^2 \sim 0.3$; that is, it is three times the value of e^2 at $T \sim 10^2$ GeV. Second, there are a great many particles in grand unified theories that contribute to temperature corrections to the effective potential. The net result is that the critical temperature T_{c_1} of the phase transition turns out to be approximately of the same order of magnitude as the particle masses at that temperature. As we showed in Section 3.1, this is precisely the circumstance that leads to a first-order phase transition.

As an example, consider a theory with SU(5) symmetry [91]. The effective potential in the simplest version of this theory is

$$V(\Phi)=-\frac{\mu^2}{2}\mathrm{Tr}\,\Phi^2+\frac{a}{4}\left(\mathrm{Tr}\,\Phi^2\right)^2+\frac{b}{2}\mathrm{Tr}\,\Phi^4, \qquad (3.2.4)$$

where Φ is a traceless 5×5 matrix. If one takes $b > 0$, $\lambda > 0$, where $\lambda = a + \frac{7}{15} b$, the symmetric state $\Phi = 0$ is unstable with respect to the appearance of the scalar field (1.1.19),

$$\Phi = \sqrt{\frac{2}{15}}\, \varphi \cdot \mathrm{diag}\left(1, 1, 1, -\frac{3}{2}, -\frac{3}{2}\right), \tag{3.2.5}$$

which breaks the SU(5) symmetry to SU(3) × SU(2) × U(1). At $T = 0$, the minimum of $V(\varphi)$ corresponds to $\varphi_0 = \frac{\mu}{\sqrt{\lambda}}$. Temperature corrections to $V(\varphi)$ come from 24 different kinds of Higgs bosons and 12 X and Y vector bosons. As a result, the counterpart of Eq. (3.2.1) for $\varphi(T)$ in the SU(5) theory is of the form [104]

$$\varphi\left(\mu^2 - \beta T^2 - \lambda \varphi^2 - \frac{T\varphi}{30\pi} Q(g^2, \lambda, b)\right) = 0\,, \tag{3.2.6}$$

where

$$\beta = \frac{75 g^2 + 130\, a + 94\, b}{60}\,, \tag{3.2.7}$$

$$Q = 7\lambda \sqrt{10\, b} + \frac{16}{3} b \sqrt{10\, b} + 3\sqrt{15}\, \lambda^{3/2} + 2\sqrt{15}\, \lambda\, g + \frac{75}{4}\sqrt{2}\, g^3\,. \tag{3.2.8}$$

As we have already noted, a phase transition with symmetry breaking upon cooling takes place somewhere in the temperature range between T_{c_1} and T_c, where T_{c_1} is given by

$$T_{c_1} = \frac{\mu}{\sqrt{\beta}}\,. \tag{3.2.9}$$

To estimate the size of the jump at the phase transition point, let us determine the quantity $\varphi_1(T_{c_1})$, Fig. 21. At $T = T_{c_1}$, the first two terms in (3.2.6) cancel, and we find that

$$\varphi_1(T_{c_1}) = \frac{Q\varphi_0}{30\pi\sqrt{\beta\lambda}}\,. \tag{3.2.10}$$

For the most reasonable values of the parameters $a \sim b \sim g^2 = 0.3$, Eqs. (3.2.7)(3.2.10) imply that

$$\varphi_1(T_{c1}) \sim 0.75\, \varphi_0 , \tag{3.2.11}$$

i.e., the jump in the field at the time of the phase transition is very large (of the same order of magnitude as φ_0).

In the discussion above, we studied only one "channel" of the phase transition, in which the transition goes directly from the SU(5) phase to the SU(3) × SU(2) × U(1) phase. In actuality, the phase transition usually entails the formation of an SU(4) × U(1) intermediate phase, and other phases as well [67, 42]. Each of the intermediate phase transitions is also a first-order transition. The kinetics of the phase transition in the minimal SU(5) theory will be discussed in Chapter 6.

3.3 Higher-order perturbation theory and the infrared problem in the thermodynamics of gauge fields

Our analysis of the high-temperature restoration of symmetry in the theory (1.1.5) (Section 3.1) was based on the use of the lowest-order perturbation theory in λ. One may then ask how reliable the results obtained in this manner really are.

This is not a completely trivial question. For example, apart from small terms $\sim\lambda^n T^4$, $\lambda^n T^2 m^2$, high-order corrections in λ to the expression for $V(\varphi, T)$ at $T \neq 0$ could contain terms proportional to m^{-n}. Such terms become large when m is small.

In order to analyze this question more thoroughly, let us examine the N-th order diagrams in λ for $V(\varphi, T)$ in the theory (1.1.5), for $\varphi = 0$. The contribution of these diagrams to $V(0, T)$ can be written out as an expression of the form

$$V_N(0, T) \sim (2\pi T)^{N+1} \lambda^N \int d^3p_1 \cdots d^3p_{N+1} \sum_{n_i=-\infty}^{\infty} \prod_{k=1}^{2N} \left[(2\pi r_k T)^2 + q_k^2 + m^2(T)\right]^{-1}, \tag{3.3.1}$$

where $\boldsymbol{q}_k$ is a homogeneous linear combination of the $\boldsymbol{p}_i$, and r_k is a corresponding combination of the n_i, $i = 1, \cdots, N+1, k = 1, \cdots, 2N$. When $m \to 0$, the leading term in the sum over n_i is the one for which all $n_i = 0$ ($r_k = 0$), since the factors containing the terms $(2\pi r_k T)^2$ are nonsingular as $m \to 0, \boldsymbol{q}_k \to 0$. This leading term is given by

$$\Delta V_N(0, T) \sim (2\pi T)^{N+1} \lambda^N \int d^3p_1 \cdots d^3p_{N+1} \prod_{k=1}^{2N} [q_k^2 + m^2(T)]^{-1} \sim \lambda^3 T^4 \left(\frac{\lambda T}{m(T)}\right)^{N-3} \tag{3.3.2}$$

It can be seen that dangerous terms $\sim \left(\frac{\lambda T}{m}\right)^{N-3}$ appear in the expression for $V(0, T)$ starting with perturbations of order $N = 4$, and these make it impossible to obtain reliable results from perturbation theory with $m < \lambda T$. Fortunately, however, (3.1.14) can be used to show that $m \gg \lambda T$ everywhere outside a small region near the critical temperature T_c, within which

$$|T - T_c| \lesssim \lambda T_c . \tag{3.3.3}$$

Everywhere outside this region, the results obtained in the preceding two sections are reliable.

Matters are much more difficult when it comes to dealing with phase transitions in the non-Abelian gauge theories that describe the interaction of Yang–Mills fields A_μ^a with one another and with scalar fields φ with a coupling constant g^2. At $T \neq 0$ in such theories, higher-order perturbation terms $\sim g^{2N}$ grow with increasing N (as $N \to \infty$) for $m_A \lesssim g^2 T$. Since in the classical approximation, m_A goes to zero ($m_A \sim g\varphi(T)$) at all temperatures above the critical value $T = T_c$, we are left with the question of whether high-temperature corrections lead to sufficiently high mass $m_A(T) \neq 0$, with a corresponding cutoff of infrared-divergent powers of the form $\left(\frac{g^2 T}{m_A}\right)^N$.

The authors of [168, 169] have shown that high-temperature effects give rise to a pole of the Green's function of the Yang–Mills fields $G_{\mu\nu}^{ab}(k)$ at $k_0 \sim gT$, $\boldsymbol{k} = 0$. This might be interpreted as the appearance of an infrared cutoff at a mass $m_A \sim gT$, making the terms $\left(\frac{g^2 T}{m_A}\right)^N$ small; actually, however, that would not be correct. The foregoing analysis tells us that the leading infrared divergences as $m_A \to 0$ are associated not with the behavior of the Green's functions at $\boldsymbol{k} = 0$, $k_0 \neq 0$, but with their behavior when $k_0 = 0$, $\boldsymbol{k} \to 0$ ($k_0 = 0$ corresponds to $n_i = 0$ in (3.3.1)). In this limit, the behavior of $G^{ab}_{\mu\nu}(k)$ is most easily studied in the Coulomb gauge, for which [166, 24]

$$G^{ab}_{00} = \delta^{ab}\left[k^2 + \pi_{00}(k)\right]^{-1}, \tag{3.3.4}$$

$$G^{ab}_{i0} = G^{ab}_{0j} = 0, \tag{3.3.5}$$

$$G^{ab}_{ij} = \delta^{ab}\left(\delta_{ij} - \frac{k_i k_j}{k^2}\right)G(k), \tag{3.3.6}$$

where $k = |\boldsymbol{k}|$, a and b are isotopic spin indices, $\pi_{00}(0) \sim g^2\mathrm{T}^2$ to lowest order in g^2, and $i, j = 1, 2, 3$.

Thus, there really *is* an infrared cutoff at $m_0 \sim g\mathrm{T}$ in G^{ab}_{00}, corresponding to the usual Debye screening of the electromagnetic field in hot plasma [166]. A well known result in quantum electrodynamics, however, is that a static magnetic field in plasma cannot be screened, and there is consequently no infrared cutoff in $G_{ij}(k_0 = 0,\ \boldsymbol{k} \to 0)$ to any order of perturbation theory [166]. In the Yang–Mills gas, there will likewise be no infrared cutoff at $k_0 = 0$, $\boldsymbol{k} \to 0$ with momentum $k \sim g\mathrm{T}$. There may in principle be one, however, at momentum $k \sim g^2\mathrm{T}$, inasmuch as massless Yang–Mills particles (in contrast to photons) interact with each other directly, and the same infrared divergences appear in the thermodynamics of a Yang–Mills gas as in scalar field theory at the point of a second-order phase transition. The difference is that the mass of a scalar field at a phase transition point vanishes "by definition" (the curvature of $V(\varphi)$ changes sign at the phase transition point), while in the thermodynamics of a Yang–Mills gas, the presence or absence of an infrared cutoff does not follow from any general considerations, which imply only that the expected scale of the infrared cutoff is $k \sim g^2\mathrm{T}$. One can reach the same conclusion by analyzing the most strongly infrared-divergent part of the theory [170], as well as the specific diagrams that could contribute to such a cutoff [24, 171, 172]. Unfortunately, when $k \sim g^2\mathrm{T}$, all high-order corrections to the diagrams for the polarization operator of the Yang–Mills field are of comparable size, so the infrared behavior of the Green's functions of the Yang–Mills field at $k \lesssim g^2\mathrm{T}$ thus far remains an open problem. Meanwhile, the degree of confidence that we have in our understanding of many of the fundamental features of gauge-theory thermodynamics depends on the solution to this problem. Let us consider the three main possibilities, which illustrate the significance of this problem.

1) *There is no infrared cutoff at $k \sim g^2$T in the thermodynamics of a Yang–Mills gas.* In that case, higher-order perturbations will be larger than lower-order ones for all thermodynamic quantities, making it impossible to use perturbation theory to study the thermodynamic properties of gauge theories at $T > T_c$. The only thing we might possibly be able to verify with reasonable assurance is that the energy density should be proportional to T^4 at superhigh temperature (from dimensional considerations). This would only suffice for the crudest approach to the theory of the evolution of a hot universe at $T > T_c$.

2) *There is a tachyon pole or a sign change at some momentum $k \sim g^2$T in the Green's function $G_{ij}^{ab}(k)$.* This implies instability with respect to creation of classical Yang–Mills fields. The second case is particularly interesting — the instability could result in the spontaneous crystallization of the Yang–Mills gas at superhigh temperature, which might lead to nontrivial cosmological consequences.

3) *In the best possible case (from the standpoint of perturbation theory), the theory contains a cutoff by virtue of the fact that $G^{-\frac{1}{2}}(0)$ turns out to be a positive quantity m(T) of order g^2T.* In that event, one can reliably calculate several of the lowest-order perturbation terms in g^2 for the thermodynamic potential of the Yang–Mills gas (up to $\sim g^6 T^4$) [171, 172]. In principle, the appearance of such a cutoff can lead to monopole confinement in a hot Yang–Mills plasma [173]; see Chapter 6.

Thus, the thermodynamic properties of hot dense matter described by gauge theories are still far from being well-understood, and one must not lose sight of the inherent difficulties and uncertainties. Nevertheless, many results have been reasonably reliably established. As applied to the theory of phase transitions studied in this chapter, the infrared problem in the thermodynamics of a Yang–Mills gas does not modify the results obtained for $T < T_c$. It can also be shown that $\varphi(T)$ should be much smaller than φ_0 when $T > T_c$, and that at large T it cannot exceed $O(gT)$. (If one assumes that $\varphi(T) \gg gT$, then the Yang–Mills fields acquire a mass $m_A \gg g^2 T$, perturbation theory becomes reliable, and the latter predicts that $\varphi(T) = 0$ at $T > T_c$.) One should keep these uncertainties in mind in discussing such complicated problems as production and evolution of monopoles in grand unified theories. However, for most of the effects to be discussed below, these uncertainties will not be important, and we will usually presume that $\varphi(T) = 0$ at a sufficiently high temperature $T > T_c$, in accordance with the results obtained in the preceding sections.

We must now state one last (but very important) reservation. We have assumed throughout that the field φ has sufficient time to roll down to a minimum of $V(\varphi, T)$. This natural assumption is valid if the field φ is not too large initially — violations of this "rule" are just what lead to the chaotic inflation scenario, which we discussed in Chapter 1 and will examine further in Chapter 7.

4

Phase Transitions in Cold Superdense Matter

4.1 Restoration of symmetry in theories with no neutral currents

In Chapters 2 and 3, we studied phase transitions in hot superdense matter, where the increase in density resulted from an increase in temperature. But it is also possible to examine phase transitions in *cold* superdense matter, where the density is increased by increasing the density of conserved charge or the number of particles at zero temperature T. In the first papers in which this problem was studied it was claimed that raising the density of cold matter would also result in the restoration of symmetry [25, 26]. The basic idea in those papers was that the energy of fermions interacting with a scalar field is proportional to $g\varphi\langle\bar{\psi}\psi\rangle$. When the fermion density $j_0 = \langle\bar{\psi}\gamma_0\psi\rangle$ increases, so does $\langle\bar{\psi}\psi\rangle$, and states with $\varphi \neq 0$ become energetically unfavorable.

As an example, consider the theory (1.1.13) with the Lagrangian

$$L = \frac{1}{2}(\partial_\mu\varphi)^2 + \frac{\mu^2}{2}\varphi^2 - \frac{\lambda}{4}\varphi^4 + \bar{\psi}(i\partial_\mu\gamma_\mu - h\varphi)\psi . \qquad (4.1.1)$$

It is possible for fermions with $j_0 \neq 0$ to exist if they have a chemical potential α, which (for $\alpha \gg m_\psi = h\varphi$) is related to j_0 [61] by

$$j_0 = \frac{\alpha^3}{3\pi^2} . \qquad (4.1.2)$$

To calculate the corrections to $V(\varphi)$ which appear due to the existence of the current j_0 (4.1.2), we must add $i\alpha$ to the component k_4 of the momentum of the fermions when the one-loop contribution of fermions to $V(\varphi)$ is computed [166]. As a result, the equation for the equilibrium value of the field φ in the theory (4.1.1) takes the form [25, 26]

$$\frac{dV}{d\varphi} = 0 = \varphi\left[\lambda\varphi^2 - \mu^2 + \frac{h^2}{2}\left(\frac{g j_0^2}{\pi^2}\right)^{\frac{1}{3}}\right]. \quad (4.1.3)$$

When $j_0 = 0$, we obtain $\varphi = \pm\frac{\mu}{\sqrt{\lambda}}$, as before. But it is clear that the presence of fermions with $j_0 \neq 0$ changes the effective value of μ^2, and for $j_0 > j_c$, where

$$j_c = \frac{2\pi\sqrt{2}}{3}\left(\frac{\mu}{h}\right)^3, \quad (4.1.4)$$

symmetry is restored in the theory (4.1.1).

4.2 Enhancement of symmetry breaking and the condensation of vector mesons in theories with neutral currents

Effects leading to the restoration of symmetry in the theory (4.1.1) appear only due to quantum corrections to $V(\varphi)$ for $\alpha \neq 0$. This is because the fermion current $j_\mu = \langle\bar{\psi}\gamma_\mu\psi\rangle$ in the model (4.1.1) does not interact directly with any physical fields. At the same time [27, 24], for realistic theories with neutral currents, in which the fermion current j_μ interacts with the neutral massive vector field Z_μ, an increase in the fermion density j_0 leads to an enhancement of symmetry breaking, while the effects considered in [25, 26] are but minor quantum corrections relative to the effects examined in [27, 24]. Subsequent study of this problem has shown that the effects appearing in cold superdense matter do not simply amount to enhanced symmetry breaking. At high enough density, a condensate of charged vector fields appears, and a redistribution of charge takes place among bosons and fermions [28, 29].

As an example, let us consider the effects that take place in the Glashow–Weinberg–Salam theory with $\lambda \gg e^4$ in the presence of a nonvanishing neutrino density $n_\nu = \frac{1}{2}\langle \bar{\nu}_e \gamma_0 (1-\gamma_5)\nu_e \rangle$. The conserved fermion density in this theory is

$$l = \bar{e}\gamma_0 e + \frac{1}{2}\bar{\nu}_e \gamma_0 (1-\gamma_5)\nu_e . \tag{4.2.1}$$

Clearly, given a lepton charge density $\langle l \rangle$, the most energetically favorable fermion distribution is

$$n_{e_R} = \frac{1}{2}\langle \bar{e}\gamma_0 (1+\gamma_5) e \rangle = n_{e_R}\frac{1}{2}\langle \bar{e}\gamma_0 (1-\gamma_5) e \rangle = n_\nu .$$

This would imply the appearance of a large charge density of electrons, however, which is only possible if some sort of charge-cancelling subsystem comes into being at the same time. In the Glashow– Weinberg–Salam model, this subsystem may appear in the form of a condensate of W bosons. Recall that in this theory, there are three fields A_μ^a, $a = 1, 2, 3$, and a field B_μ, from which — after symmetry breaking — the electromagnetic field

$$A_\mu = B_\mu \cos\theta_W + A_\mu^3 \sin\theta_W , \tag{4.2.2}$$

the massive neutral field

$$Z_\mu = B_\mu \sin\theta_W - A_\mu^3 \cos\theta_W , \tag{4.2.3}$$

and the charged field

$$W_\mu^\pm = \frac{1}{\sqrt{2}}(A_\mu^1 \mp A_\mu^2) \tag{4.2.4}$$

are formed. To be able to describe effects associated with nonvanishing lepton density, we must append to the Lagrangian of the theory a term αl, where α is the chemical potential corresponding to the lepton charge density. The vector field condensate that arises at sufficiently high fermion density is of the form

$$W_1^\pm = C , \tag{4.2.5}$$

$$W_0^{\pm} = W_2^{\pm} = W_3^{\pm} = A_i^3 = 0\,, \tag{4.2.6}$$

$$A_0^3 = \pm\,\frac{\varphi}{2}\,, \tag{4.2.7}$$

where C and φ are determined by the equations

$$\left\langle \frac{\delta L}{\delta A_0^3} \right\rangle = \frac{2e^2}{\sin\theta_W} C^2 A_0^3 + e\,(n_{e_L} + n_{e_R}) = 0\,, \tag{4.2.8}$$

$$\left\langle \frac{\delta L}{\delta \varphi} \right\rangle = \varphi \left(\mu^2 - \lambda\varphi^2 + \frac{e^2 Z_0^2}{\sin^2 2\theta_W} + \frac{e^2 C^2}{2\sin^2\theta_W} \right) = 0\,, \tag{4.2.9}$$

$$\left\langle \frac{\delta L}{\delta Z_0} \right\rangle = \frac{e^2\varphi^2 Z_0}{2\sin 2\theta_W} + e\,(2n_{e_R} + n_{e_L} + 2n_\nu) = 0\,, \tag{4.2.10}$$

and n_ν, n_{e_R}, and n_{e_L} are given by

$$n_\nu = \frac{1}{6\pi^2}\left(\alpha + \frac{eZ_0}{\sin 2\theta_W}\right)^3 \tag{4.2.11}$$

$$n_{e_R} = \frac{1}{6\pi^2}\left(\alpha + eZ_0\tan\theta_W + eA_0\right)^3 \tag{4.2.12}$$

$$n_{e_L} = \frac{1}{6\pi^2}\left(\alpha - eZ_0\cot\theta_W + eA_0\right)^3 \tag{4.2.13}$$

The solution $W_1^{\pm} = C \neq 0$ can only appear at high enough lepton density, $n_L = n_\nu + n_{e_R} + n_{e_L}$. To determine the critical value $n_L = n_L^c$, one should take into account that (as can be verified *a posteriori*) at $n_L \sim n_L^c$ the field Z_0 is of higher order in e^2 than are A_0^3 or C. In determining n_L^c, we can therefore put $Z_0 = 0$ in (4.2.8)–(4.2.13), making subsequent analysis quite simple.

Specifically, only the trivial solution $W_\mu^{\pm} = 0$ exists at low densities, and we deduce from (4.2.8)–(4.2.13) that $A_0^3 = A_0\sin\theta_W = -\frac{\alpha}{e}\sin\theta_W$, $n_{e_L} = n_{e_R} = 0$. Starting with $A_0^3 = \frac{\alpha}{e}\sin\theta_W = \frac{\varphi_0}{2}$, the condensate solution $W_1^{\pm} = C \neq 0$ appears. At that point

$$n_\nu = n_L^c = \frac{1}{6\pi^2}\left(\frac{e\varphi_0}{2\sin\theta_W}\right)^3 = \frac{M_W^3}{6\pi^2}, \tag{4.2.14}$$

where M_W is the mass of the W boson. With a further increase in the fermion density, this solution becomes energetically favored over the solution $C = 0$. As the reader can verify, this is because the energy required to create the classical field $W^\pm$ is small compared with the energy gain achieved by redistributing the lepton charge among the neutrinos and electrons, thereby reducing the Fermi energy of the leptons.

Our final result is that both C and Z_0 increase in magnitude with increasing fermion density. When the latter is high enough, charged $W^\pm$ boson condensates appear. This then leads to an asymptotic equalization of the partial densities of right- and left-handed leptons (baryons) of various kinds in superdense matter: $n_{\nu_e} = n_{e_R} = n_{e_L}$, $n_{\nu_\mu} = n_{\mu_R} = n_{\mu_L}$, etc. On the other hand, according to (4.2.9), the growth of both C and Z_0 will lead to growth of the field φ, i.e., to the enhancement of symmetry breaking between the weak and electromagnetic interactions.

One should note that for a particular chemical composition of cold superdense matter ($n_B = \frac{4}{3} n_L$, where n_B and n_L are the baryon and lepton densities, respectively), the fermion matter proves to be neutral both with respect to the field A_0 and the field Z_0. In that case, no W condensate is formed in the superdense matter, and at high enough density, the field φ will tend to vanish [29]. Interesting nonperturbative effects may then come to the fore [174]. For the time being, we have no idea what might cause this special regime to be realized during expansion of the universe. Strictly speaking, this reservation also pertains to the more general case $n_B \neq \frac{4}{3} n_L$ considered above, the point being that at the present time $n_L \sim n_B \ll n_\gamma$. The neutrino density is not known accurately, but in grand unified theories with nonconservation of baryon charge, it is most reasonable to expect that at present $n_L \sim n_B \ll n_\gamma$. The dominant effects, at least at very early stages in the evolution of the universe, are then those that are related not to the chemical potential α of cold fermionic matter, but to the temperature $T \gg \alpha$. It is possible in principle that effects considered in this chapter may be important in the study of certain intermediate stages in the evolution of the universe, after which there was an abrupt increase in the specific entropy $\frac{n_\gamma}{n_B}$, as induced by processes considered in [97, 98, 129], for example. A combined study of both high-temperature effects and effects related to nonvanishing lepton- and baryon-charge density, as well as an investigation of concomitant nonperturbative effects, can be found in a number of recent papers on this topic; for example, see [130, 175–178].

5

Tunneling Theory and the Decay of a Metastable Phase in a First-Order Phase Transition

5.1 General theory of the formation of bubbles of a new phase

One important and somewhat surprising property of field theories with spontaneous symmetry breaking is that the lifetime of the universe in an energetically unfavorable metastable vacuum state can be exceptionally long. This phenomenon forms the basis of the first versions of the inflationary universe scenario, according to which inflation takes place from a supercooled metastable vacuum state ("false vacuum") $\varphi = 0$ [53–55]. This same phenomenon can lead to a partitioning of the universe into enormous, exponentially long-lived regions in different metastable vacuum states, each corresponding to a different local minimum of the effective potential.

For definiteness, we shall discuss the decay of the vacuum state with $\varphi = 0$ in the theory with the Lagrangian

$$L(\varphi) = \frac{1}{2}(\partial_\mu \varphi)^2 - V(\varphi), \qquad (5.1.1)$$

where the effective potential $V(\varphi)$ has a local minimum at $\varphi = 0$ and a global minimum at $\varphi = \varphi_0$. Decay of the vacuum state with $\varphi = 0$ proceeds via tunneling, with the formation of bubbles of the field φ. A theory of bubble production at zero temperature was suggested in [179], and was substantially developed in [180, 181], where the Euclidean approach to the theory of the decay of a metastable vacuum state was proposed.

We know from elementary quantum mechanics that the tunneling of a particle through a one-dimensional potential barrier $V(x)$ can be treated as motion with imaginary energy, or to put it differently, as motion in imaginary time, i.e., in Euclidean space. To generalize this approach to the case of tunneling through the barrier $V(\varphi)$, one should consider the wave *functional* $\Psi(\varphi(x,t))$ in place of the particle wave function $\psi(x,t)$, and investigate its evolution in Euclidean space. This generalization was proposed in [180, 181]. The Euclidean approach to tunneling theory is simple and elegant, and it enables one to progress rather far in calculating the decay probability of the false vacuum. We shall therefore refer below (without proof) to the basic results obtained in [180, 181], and to their generalization to nonzero temperature [62]. In subsequent sections, these methods will be applied to the study of tunneling in several specific theories.

As in conventional quantum mechanics, determination of the tunneling probability requires first of all that we solve the classical equation of motion for the field φ in Euclidean space,

$$\Box\varphi = \frac{d^2\varphi}{dt^2} + \Delta\varphi = \frac{dV}{d\varphi}, \tag{5.1.2}$$

with boundary condition $\varphi \to 0$ as $x^2 + t^2 \to \infty$. If we then normalize $V(\varphi)$ so that $V(0) = 0$ (that is, we redefine $V(\varphi)$ by $V(\varphi) \to V(\varphi) - V(0)$), the tunneling probability per unit time and per unit volume will be given by

$$\Gamma = A\, e^{-S_4(\varphi)}, \tag{5.1.3}$$

where $S_4(\varphi)$ is the Euclidean action corresponding to the solution of Eq. (5.1.2),

$$S_4(\varphi) = \int d^4x \left[\frac{1}{2}\left(\frac{d\varphi}{dt}\right)^2 + \frac{1}{2}(\nabla\varphi)^2 + V(\varphi)\right], \tag{5.1.4}$$

and the factor A preceding the exponential is given by

$$A = \left(\frac{S_4}{2\pi}\right)^2 \left(\frac{\det'[-\Box + V''(\varphi)]}{\det[-\Box + V''(0)]}\right)^{-\frac{1}{2}}. \tag{5.1.5}$$

Here $V''(\varphi) = \dfrac{d^2V}{d\varphi^2}$, and the notation "det ′ " means that in calculating the functional determinant of the operator $[-\Box + V''(\varphi)]$, its vanishing eigenvalues, corresponding to the so-called zero modes of the operator, are to be omitted. This operator has four zero modes, corresponding to the possibility of translating the solution $\varphi(x)$ along any of the four axes in Euclidean space. Contributions of $\left(\dfrac{S_4}{2\pi}\right)^{\frac{1}{2}}$ from each of the zero modes result in the factor $\left(\dfrac{S_4}{2\pi}\right)^2$ in (5.1.5).

The derivation of Eqs. (5.1.3) and (5.1.5) may be found in [181], and is based on a calculation of the imaginary part of the magnitude of the potential $V(\varphi)$ at $\varphi = 0$. To a large extent, the equations here are analogous to the corresponding expressions in the theory of Yang–Mills instantons [182]. In essence, the solutions of Eq. (5.1.2) with the indicated boundary conditions are scalar instantons in the theory (5.1.1). We now wish to make several remarks before moving on to a generalization of these results to $T \neq 0$.

First of all, notice that in order to obtain a complete answer, it is necessary to sum over the contributions to Γ from all possible solutions of Eq. (5.1.2). Fortunately, however, it is usually sufficient to limit consideration to the simplest O(4)-symmetric solution $\varphi(\mathbf{x}^2 + t^2)$, inasmuch as these are usually the very solutions that minimize the action S_4. In that event, Eq. (5.1.2) takes on an even simpler form,

$$\frac{d^2\varphi}{dr^2} + \frac{3}{r}\frac{d\varphi}{dr} = V'(\varphi), \tag{5.1.6}$$

where $r = \sqrt{\mathbf{x}^2 + t^2}$, with boundary conditions $\varphi \to 0$ as $r \to \infty$ and $\dfrac{d\varphi}{dr} = 0$ at $r = 0$.

The high degree of symmetry inherent in the solution $\varphi(\mathbf{x}^2 + t^2)$ helps us to obtain a graphic description of the structure and evolution of a bubble of the field φ after it is created. To do so, we analytically continue the solution to conventional time, $t \to it$, or in other words $\varphi(\mathbf{x}^2 + t^2) \to \varphi(\mathbf{x}^2 - t^2)$. Since the solution $\varphi(\mathbf{x}^2 - t^2)$ depends only on the invariant quantity $\mathbf{x}^2 - t^2$, the corresponding bubble will look the same in any reference frame, and the expansion speed of a region filled with the field φ (the speed of the bubble "walls") will asymptotically approach the speed of light. The creation and growth of bubbles is an interesting mathematical problem [180], but one that we shall not discuss here, as our main objective is to study those situations in which the probability of bubble creation is negligible.

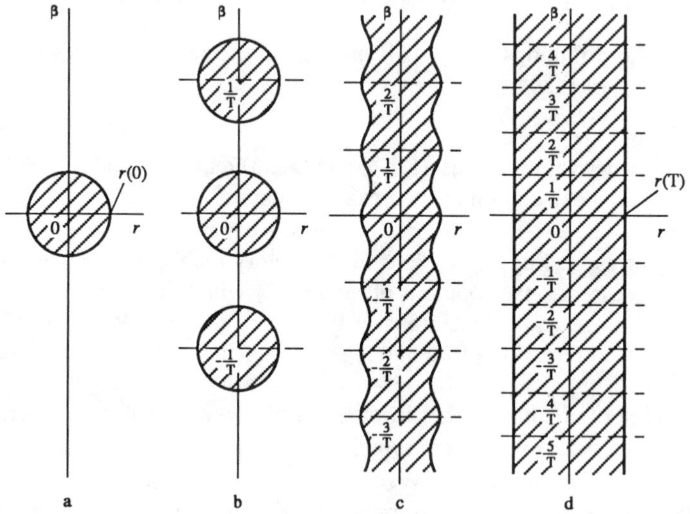

Figure 23. The form taken by the solution to Eq. (5.1.2) at various temperatures: a) $T = 0$; b) $T \ll r^{-1}(0)$; c) $T \sim r^{-1}(0)$; d) $T \gg r^{-1}(0)$. The shaded regions contain the classical field $\varphi \neq 0$. For simplicity, we have drawn bubbles for those cases in which the thickness of their walls is much less than their radii.

Unfortunately, Eq. (5.1.6) can seldom be solved analytically, so that both the solution and the associated value of the Euclidean action $S_4(\varphi)$ must often be computed numerically. In this sort of situation, determinants can only be calculated in certain special cases. It turns out, however, that in most practical problems just a rough estimate of the pre-exponential factor A will suffice. We can come up with such an estimate by noting that the factor A has dimensionality m^4, and its value is determined by three different quantities with dimensionality m, namely $\varphi(0)$, $\sqrt{V''(\varphi)}$, and r^{-1}, where r is the typical size of a bubble. In the theories that interest us most, all of these quantities lie within an order of magnitude of one another, so for a rough estimate one may assume that

$$\frac{\det'[-\Box + V''(\varphi)]}{\det[-\Box + V''(0)]} = O\left(r^{-4}, \varphi^4(0), (V'')^2\right), \tag{5.1.7}$$

where we denote by r and $V''(\varphi)$ typical mean values of these parameters for the solution $\varphi(r)$ of Eq. (5.1.6).

Next, let us proceed to the case in which $T \neq 0$ [62]. In order to generalize the preceding results to this instance, it suffices to recall that the quantum statistics of bosons (fermions) at $T \neq 0$ are formally equivalent to quantum field theory in a Euclidean space with a periodicity (antiperiodicity) of $1/T$ in the "time" β (see [166], for example). When one considers processes at fixed temperature, the quantity $V(\varphi, T)$ plays the role of the potential energy. The imaginary part of this function in an unstable vacuum can be calculated in just the same way as is done in [181] for the case $T = 0$. The only essential difference is that instead of finding an O(4)-symmetric solution of Eq. (5.1.2), one must find an O(3)-symmetric (in the spatial coordinates) solution periodic in the "time" β, with period $1/T$. As $T \to 0$, the solution of Eq. (5.1.2) that minimizes the action $S_4(\varphi)$ consists of an O(4)-symmetric bubble with some typical radius $r(0)$ (Fig. 23a). As $T \to r^{-1}(0)$, the solution becomes a series of such bubbles separated from one another by a distance $1/T$ in the "time" direction (Fig. 23b). When $T \sim r^{-1}(0)$, the bubbles start to overlap (Fig. 23c). Finally, when $T \gg r^{-1}(0)$ (which is just the case that is of most importance and interest to us), the solution becomes a cylinder whose spatial cross section is an O(3)-symmetric bubble of some new radius $r(T)$ (Fig. 23d).

When we calculate $S_4(\varphi)$ in the latter case, the integration over β reduces simply to a multiplication by $1/T$ — that is, $S_4(\varphi) = \frac{1}{T} S_3(\varphi)$, where $S_3(\varphi)$ is the three-dimensional action corresponding to the O(3)-symmetric bubble,

$$S_3(\varphi) = \int d^3x \left[\frac{1}{2}(\nabla\varphi)^2 + V(\varphi, T)\right]. \tag{5.1.8}$$

To calculate $S_3(\varphi)$, we must solve the equation

$$\frac{d^2\varphi}{dr^2} + \frac{2}{r}\frac{d\varphi}{dr} = \frac{dV(\varphi, T)}{d\varphi} = V'(\varphi, T) \tag{5.1.9}$$

with boundary conditions $\varphi \to 0$ as $r \to \infty$ and $\frac{d\varphi}{dr} = 0$ as $r \to 0$. In the high-temperature limit ($T \gg r^{-1}(0)$), the complete expression for the tunneling probability per unit time and per unit volume is obtained in a manner completely analogous to that employed in [181] to derive Eqs. (5.1.4) and (5.1.5); the result is

$$\Gamma(T) = T\left(\frac{S_3(\varphi, T)}{2\pi T}\right)^{\frac{3}{2}}\left(\frac{\det'[-\Delta + V''(\varphi, T)]}{\det[-\Delta + V''(0, T)]}\right)^{-\frac{1}{2}} \exp\left(-\frac{S_3(\varphi, T)}{T}\right). \tag{5.1.10}$$

Here as before, the notation "det ′ " means that the three zero modes of the operator $[-\Delta + V''(\varphi, T)]$ are to be omitted in calculating the determinant. These three zero modes, corresponding to translations of the solution $\varphi(x)$ in the three spatial directions, contribute the factor $\left(\frac{S_3(\varphi, T)}{2\pi T}\right)^{\frac{3}{2}}$ to (5.1.10), and the factor of T comes about when one takes into account the 1/T periodicity of Euclidean space in the direction of the "time" β.

As in the zero-temperature case, the determinants in (5.1.10) cannot usually be calculated explicitly. Moreover, it can be shown that some extra subexponential factor should appear in (5.1.10) if one takes into account that euclidean methods are only approximately applicable for a description of a non-equilibrium system where bubble formation occurs. The subexponential terms can be roughly estimated by use of dimensional arguments taking into account that typically all mass parameters related to bubble formation do not differ considerably from T:

$$\Gamma(T) \sim T^4\left(\frac{S_3(\varphi, T)}{2\pi T}\right)^{\frac{3}{2}} \exp\left(-\frac{S_3(\varphi, T)}{T}\right). \tag{5.1.11}$$

It can be seen from Eqs. (5.1.10) and (5.1.11) that the main problem to be solved in determining the probability of bubble creation is to find $S_3(\varphi, T)$ (or S_4 at $T = 0$). Furthermore, if we are to obtain reasonable estimates of the determinants, and wish to be able to study the expansion of the bubbles that are formed, we must know the form taken by the function $\varphi(r)$ and the typical size of a bubble. As we pointed out earlier, the corresponding results are obtained, as a rule, by computer solution of the equations, which seriously complicates the investigation of phase transitions in realistic theories. It is therefore of particular interest to study those instances in which the problem can be solved analytically, and we treat one such case in the next section. From here on, we shall consider not only the case $T \gg r^{-1}(0)$, but the case $T = 0$ as well, as the latter gives us information on the probability of bubble creation in the limit of a strongly supercooled metastable phase, where $T \ll r^{-1}(0)$.

5.2 The thin-wall approximation

In tunneling theory, there are two limiting cases in which the problem simplifies considerably. One of these is associated with the situation where $V(\varphi)$ at a minimum with $\varphi = \varphi_0(\tau) \neq 0$ is much larger in absolute value than the height of the potential barrier in $V(\varphi)$ between $\varphi = 0$ and $\varphi = \varphi_0$; that case will be taken up in the next chapter. Here we examine the other limit, in which $|V(\varphi_0)| = \varepsilon$ is much lower than the barrier height (see Fig. 24).

It is readily seen that as ε decreases, the volume energy involved in bubble creation ($\sim \varepsilon r^3$) becomes large compared to the surface energy ($\sim r^2$) for large r only if the bubble is big enough. When the size of a bubble greatly exceeds the bubble wall thickness (the wall being that region where derivatives $\frac{d\varphi}{dr}$ are large), one can neglect the second term in (5.1.6) and (5.1.9) compared with the first. In other words, these equations effectively reduce to one that describes tunneling in one-dimensional space-time:

$$\frac{d^2\varphi}{dr^2} = V'(\varphi, T) \,. \tag{5.2.1}$$

In the limit as $\varepsilon \to 0$, the solution of this equation takes the form

$$r = \int_{\varphi}^{\varphi_0} \frac{d\varphi}{\sqrt{2V(\varphi)}}\,, \tag{5.2.2}$$

where the functional form of $\varphi(r)$ has been sketched in Fig. 25.

Let us first consider tunneling in quantum field theory ($T = 0$). In an O(4)-symmetric bubble (5.2.2), the action S_4 is then given by

$$\begin{aligned} S_4 &= 2\pi^2 \int_0^\infty r^3\, dr \left[\frac{1}{2}\left(\frac{d\varphi}{dr}\right)^2 + V\right] \\ &= -\frac{\varepsilon}{2}\pi^2 r^4 + 2\pi^2 r^3 S_1\,, \end{aligned} \tag{5.2.3}$$

where S_1 is the surface energy of the bubble wall (surface tension), and is equal to the action in the corresponding one-dimensional problem (5.2.1):

$$\begin{aligned} S_1 &= \int_0^\infty dr \left[\frac{1}{2}\left(\frac{d\varphi}{dr}\right)^2 + V\right] \\ &= \int_0^{\varphi_0} d\varphi \sqrt{2V(\varphi)}\,, \end{aligned} \tag{5.2.4}$$

the integral in (5.2.4) being calculated in the limit as $\varepsilon \to 0$. We find the bubble radius $r(0)$ by minimizing (5.2.3):

$$r(0) = \frac{3S_1}{\varepsilon}\,, \tag{5.2.5}$$

whereupon

$$S_4 = \frac{27\pi^2 S_1^4}{2\varepsilon^3}\,. \tag{5.2.6}$$

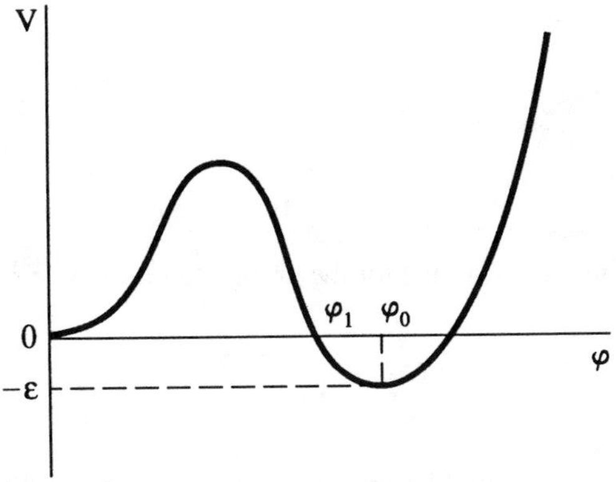

Figure 24. Effective potential $V(\varphi)$ in the case of slight supercooling of the phase $\varphi = 0$ (i.e., the quantity $\varepsilon = V(0, T) - V(\varphi_0, T)$ is small).

Notice that to order of magnitude, the bubble wall thickness is simply $(V''(0))^{-1/2}$. Taking (5.2.5) into account, therefore, the condition for the present approximation (the so-called thin-wall approximation) to be valid is

$$\frac{3S_1}{\varepsilon} \gg (V''(0))^{-1/2}. \tag{5.2.7}$$

The foregoing results were derived by Coleman [180]. We can now readily generalize these results to the high-temperature case, $T \gg r^{-1}(0)$. To do so, we merely point out that

$$\begin{aligned} S_3 &= 4\pi \int_0^\infty r^2 dr \left[\frac{1}{2}\left(\frac{d\varphi}{dr}\right)^2 + V(\varphi, T)\right] \\ &= -\frac{4\pi}{3} r^3 \varepsilon + 4\pi r^2 S_1(T), \end{aligned} \tag{5.2.8}$$

so that

$$r(T) = \frac{2S_1}{\varepsilon} \tag{5.2.9}$$

and

$$S_3 = \frac{16\pi S_1^3}{3\varepsilon^2}. \tag{5.2.10}$$

The expression thus obtained for the probability of bubble formation,

$$\Gamma \sim \exp\left(-\frac{16\pi S_1^3}{3\varepsilon^2 T}\right), \tag{5.2.11}$$

is consistent with the well-known expression found in textbooks [61]. The only difference (but an important one) is that we have the closed expression (5.2.4) for the surface tension S_1, where instead of $V(\varphi)$ one should use $V(\varphi, T)$. In many cases of interest, the function $V(\varphi, T)$ plotted in Fig. 24 can be approximated by the expression

$$V(\varphi) = \frac{M^2}{2}\varphi^2 - \frac{\delta}{3}\varphi^3 + \frac{\lambda}{4}\varphi^4. \tag{5.2.12}$$

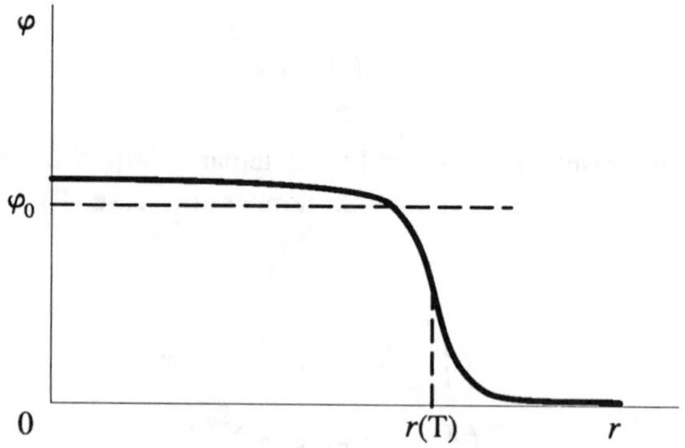

Figure 25. Typical form of the solution of Eqs. (5.1.6) and (5.1.9) when $\varepsilon \to 0$.

Let us investigate bubble formation in this theory in more detail, since for the potential (5.2.12) one can evaluate the integral in (5.2.4) exactly, and it thereby becomes possible to obtain analytic expressions for S_1, S_3, S_4, and $r(T)$.

In fact, it can readily be demonstrated that for values of the parameters M, δ, and λ such that the minima at $\varphi = 0$ and $\varphi = \varphi_0$ are of equal depth ($\varepsilon \to 0$), Eq. (5.2.12) becomes

$$V(\varphi) = \frac{\lambda}{4}\varphi^2(\varphi - \varphi_0)^2, \tag{5.2.13}$$

and in that event φ_0 is

$$\varphi_0 = \frac{2\delta}{\lambda}, \tag{5.2.14}$$

while M, δ, and λ are related by

$$2\delta^2 = 9\,M^2\lambda\,. \tag{5.2.15}$$

From (5.2.8) and (5.2.13)–(5.2.15), it follows that

$$S_1 = \sqrt{\frac{\lambda}{2}}\,\frac{\varphi_0^3}{6} = 2^{3/2}\,3^{-4}\,\delta^3\lambda^{-5/2}, \tag{5.2.16}$$

whereupon for $T = 0$ one obtains

$$S_4 = \frac{\pi^2 2^5 \delta^{12}}{3^{13}\lambda^{10}\varepsilon^3}, \qquad r(0) = \frac{2^{3/2}\delta^3}{3^3\lambda^{5/2}\varepsilon}, \tag{5.2.17}$$

while for $T \gg r^{-1}(0)$,

$$S_3 = \frac{2^{17/2}\pi\delta^9}{3^{13}\lambda^{15/2}\varepsilon^2}, \qquad r(T) = \frac{2^{5/2}\delta^3}{3^4\lambda^{5/2}\varepsilon}. \tag{5.2.18}$$

We now turn to the specific case of phase transitions in gauge theories at high temperature. Here a typical expression for $V(\varphi, T)$ is

$$V(\varphi, T) = \frac{\beta(T^2 - T_{c_1}^2)}{2}\varphi^2 - \frac{\alpha}{3}T\varphi^3 + \frac{\lambda}{4}\varphi^4, \tag{5.2.19}$$

where T_{c_1} is the temperature above which the symmetric phase $\varphi = 0$ is metastable, and β and α are numerical coefficients (compare (3.1.21), (3.1.22)). The temperature T_c at which the values of $V(\varphi, T)$ for the phases with $\varphi = 0$ and $\varphi = \varphi_0(T)$ are equal is given by

$$T_c^2 = T_{c_1}^2\left(1 - \frac{2\alpha^2}{9\beta\lambda}\right)^{-1}. \tag{5.2.20}$$

One can readily determine the quantity ε as a function of the departure of the temperature T from its equilibrium value:

$$\varepsilon = \frac{4T_c T_{c_1}^2 \alpha^2 \beta}{9\lambda^2}\Delta T, \tag{5.2.21}$$

where $\Delta T = T_c - T$. It is then straightforward, using Eqs. (5.2.14)–(5.2.20), to derive expressions for the quantities of interest, namely S_3 and $r(T)$. These may be written out for the most frequently encountered situation, with $x = \dfrac{T_c - T}{T_c} \ll 1$:

$$S_4 = \frac{S_3}{T} = \frac{2^{9/2}\pi\alpha^5}{3^9\beta^2\lambda^{7/2}}\frac{1}{x^2}, \tag{5.2.22}$$

$$r = \sqrt{\frac{2}{\lambda}}\frac{\alpha}{9\beta T_c}\frac{1}{x}. \tag{5.2.23}$$

Thus, the thin-wall approximation makes it possible to progress rather far towards an understanding of the formation of bubbles of a new phase. Unfortunately, however, this method can only be applied to relatively slow phase transitions, or to be more precise, those for which

$$S_4 = \frac{S_3}{T} \gtrsim 10\,\alpha\lambda^{-3/2}. \tag{5.2.24}$$

This restriction is not satisfied in many cases of interest, forcing us to seek ways of proceeding beyond the scope of the thin-wall approximation.

5.3 Beyond the thin-wall approximation

We have already remarked that there is one more instance in which the theory of bubble creation may be considerably simplified. Specifically, if the minimum of $V(\varphi)$ at the point φ_0 is deep enough, the maximum value of the field $\varphi(r)$ corresponding to the solution of Eqs. (5.1.6) and (5.1.9) becomes of order φ_1, where $V(\varphi_1) = V(0)$, $\varphi_1 \ll \varphi_0$. In solving (5.1.6) and (5.1.9), one can then neglect the details of the behavior of $V(\varphi)$ for $\varphi \gg \varphi_1$, and when $\varphi \lesssim \varphi_1$, it is often possible to approximate the potential $V(\varphi)$ with one of two basic types of functions:

$$V^1(\varphi) = \frac{M^2}{2}\varphi^2 - \frac{\lambda}{4}\varphi^4 , \tag{5.3.1}$$

$$V^2(\varphi) = \frac{M^2}{2}\varphi^2 - \frac{\delta}{3}\varphi^3 . \tag{5.3.2}$$

At zero temperature and with $M = 0$, Eq. (5.1.6) for the theory (5.3.1) can be solved exactly [182],

$$\varphi = \sqrt{\frac{8}{\lambda}}\,\frac{\rho}{r^2 + \rho^2} , \tag{5.3.3}$$

where ρ is an arbitrary parameter with dimensionality of length (the arbitrariness in the choice of ρ is a consequence of the absence of any mass parameter in the theory (5.3.1) for $M = 0$). For all ρ, the action corresponding to the solutions of (5.3.3) is

$$S_4 = \frac{8\pi^2}{3\lambda} . \tag{5.3.4}$$

To find the total probability of bubble formation, one must integrate (with a certain weight) the contributions from solutions (instantons) for all values of ρ, as in the theory of Yang–Mills instantons [183].

At $T = 0$ and arbitrary $M \neq 0$, Eq. (5.1.6) in the theory (5.3.1) has no exact instanton solutions of the type we have studied [184], for the same reason that there are no instantons in the theory of massive Yang–Mills fields. On the other hand, for $\rho \ll M^{-1}$, the solution of (5.3.3) is essentially insensitive to the presence of a mass M in the theory (5.3.1). Therefore, for $T = 0$ and $M \neq 0$, the theory (5.3.1) admits of "almost exact solutions" of Eq. (5.1.6) that are identical to (5.3.3) when $\rho \ll M^{-1}$. This means that an entire class of trajectories (5.3.3) exists in Euclidean space that describes formation of a bubble of the field $\varphi \neq 0$. To high accuracy, the action corresponding to each of these trajectories in the theory (5.3.1) with $M \neq 0$ is the same as (5.3.4) when $\rho \ll M^{-1}$, and it tends to the minimum (5.3.4) as $\rho \to 0$. As a result, tunneling does exist at $T = 0$ in the theory (5.3.1), and to describe it, one must integrate over ρ the contributions to Γ from all "solutions" (5.3.3) with $\rho \ll M^{-1}$, as is done in the theory of instantons when the Yang–Mills field acquires mass [183]. In this somewhat inexact sense, we will speak of solutions of Eq. (5.1.6) in the theory (5.3.1) with $M \neq 0$ and with the action (5.3.4) (a similar situation is investigated in [185]).

In all other cases under consideration (at high temperature in the theory (5.3.1), and at both high and low temperature in the theory (5.3.2)), exact solutions do exist. The form taken by the solutions is depicted in Fig. 26. At $T = 0$, the action S_4 corresponding to the solution $\varphi(r)$ in the theory (5.3.2) is

$$S_4(\varphi) \sim 205\,\frac{M^2}{\delta^2}. \tag{5.3.5}$$

In the high-temperature limit, the action $S_4(\varphi) = \frac{S_3}{T}$ for the solutions corresponding to the theories (5.3.1) and (5.3.2) is

$$S_4(\varphi) \sim \frac{19\,M}{\lambda\,T} \tag{5.3.6}$$

and

$$S_4(\varphi) \sim 44\,\frac{M^3}{\delta^2 T} \tag{5.3.7}$$

respectively.

Note that the results obtained above do not just refer to the limiting cases T = 0 and T » M. An analysis of this problem shows that Eqs. (5.3.4) and (5.3.5) continue to hold down to temperatures T ≤ 0.7 M (T ≤ 0.2 M), and at higher temperatures one can make use of the results (5.3.6) and (5.3.7) [62].

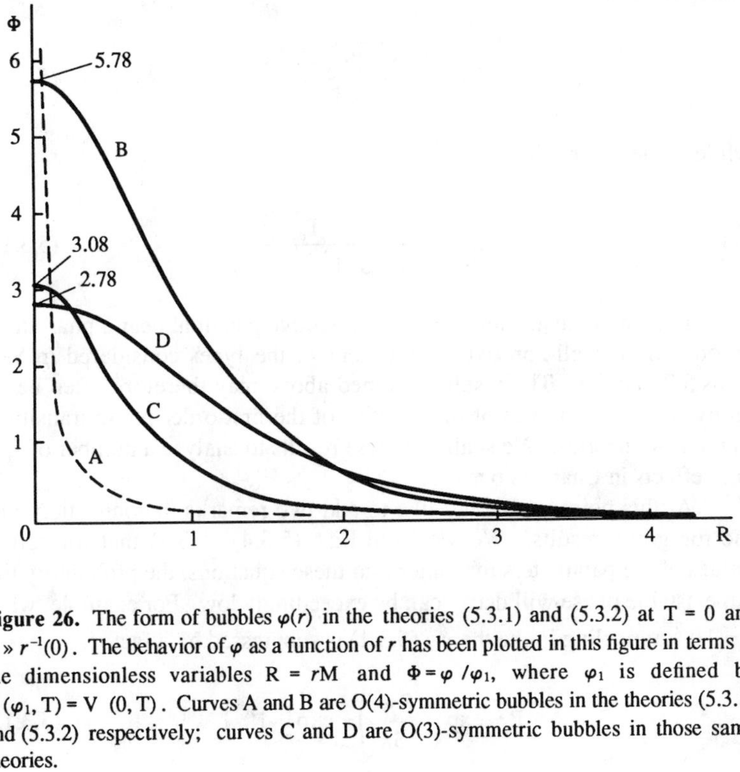

Figure 26. The form of bubbles $\varphi(r)$ in the theories (5.3.1) and (5.3.2) at T = 0 and T » $r^{-1}(0)$. The behavior of φ as a function of r has been plotted in this figure in terms of the dimensionless variables R = rM and $\Phi = \varphi / \varphi_1$, where φ_1 is defined by $V(\varphi_1, T) = V(0, T)$. Curves A and B are O(4)-symmetric bubbles in the theories (5.3.1) and (5.3.2) respectively; curves C and D are O(3)-symmetric bubbles in those same theories.

To conclude this chapter, let us consider briefly the most typical case, in which the potentials V^1 and V^2 are of the form

$$V^1(\varphi, T) = \frac{\beta(T^2 - T_{c1}^2)}{2}\varphi^2 - \frac{\lambda}{4}\varphi^4, \tag{5.3.8}$$

$$V^2(\varphi, T) = \frac{\beta(T^2 - T_{c1}^2)}{2}\varphi^2 - \frac{\alpha}{3}T\varphi^3. \tag{5.3.9}$$

From the previous results, it follows that at high enough temperature in the theory (5.3.8),

$$S_4 = \frac{19\sqrt{\beta(T^2 - T_{c1}^2)}}{\lambda T}, \tag{5.3.10}$$

while in the theory (5.3.9),

$$S_4 = \frac{44\left[\beta(T^2 - T_{c1}^2)\right]^{\frac{3}{2}}}{\alpha^2 T^3}. \tag{5.3.11}$$

In many realistic situations, the effective potential near a phase transition point is well-approximated by one of the types considered in Sections 5.2 and 5.3. The results obtained above may therefore often be directly applied to studies of the kinetics of the first-order phase transitions in realistic theories. We shall use these results to analyze a number of specific effects in Chapters 6 and 7.

At this point, we would like to add two remarks in connection with the foregoing results. We see from Eqs. (5.3.4)–(5.3.7) that for certain values of the parameters that enter into these equations, the probability that a metastable phase will decay can be exceedingly low. For example, when $\lambda \sim 10^{-2}$, tunneling in the theory (5.3.1) is suppressed by a factor

$$P \sim \exp\left(-\frac{8\pi^2}{3\lambda}\right) \sim \exp\left(-10^3\right). \tag{5.3.12}$$

This explains why in realistic theories, metastable vacuum states can turn out to be almost indistinguishable from the stable one. In particular, one has essentially no reason to think that the vacuum state in which we now reside is the one corresponding to the absolute minimum of energy. One might try, in principle, to carry out an experiment to test the stability of our vacuum by attempting to create a nucleus of a new phase (e.g., via heavy ion collisions), but both the technological feasibility and, understandably, the advisability of such an experiment are highly dubious.[1]

The second remark bears on the range of applicability of the foregoing results. These were obtained by neglecting effects associated with the expansion of the universe, an approximation that is perfectly adequate if the curvature $V''(\varphi)$ of the effective potential is much greater than the curvature tensor $R_{\mu\nu\alpha\beta}$. But in the inflationary universe scenario, $V''(\varphi) \ll R = 12\,H^2$ during inflation. Tunneling during inflation must therefore be studied as a separate issue. We shall return to this question in Chapter 7.

---

[1] One might argue, of course, that such an experiment would be quite enlightening, regardless of the results. If the experiment were to confirm the stability of our vacuum state, it would make us all very proud. On the other hand, if a bubble of a more energetically advantageous vacuum state were produced, the observable part of the universe would gradually be transformed into a better vacuum state, and no observers would remain to be dissatisfied with the experimental results.

6

Phase Transitions in a Hot Universe

6.1 Phase transitions with symmetry breaking between the weak, strong, and electromagnetic interactions

We have already pointed out in Chapter 1 that according to the standard hot universe theory, the expansion of the universe started from a state of enormously high density, at a temperature T much higher than the critical temperature of a phase transition with symmetry restoration between the strong and electroweak interactions in grand unified theories. Therefore, the symmetry between these interactions should have been restored in the very early stages of the evolution of the universe.

As the temperature decreases to $T \sim T_{c_1} \sim 10^{14}$–$10^{15}$ GeV (see Eq. (3.2.9)), a phase transition (or several) takes place, generating a classical scalar field $\Phi \sim 10^{15}$ GeV, which breaks the symmetry between the strong and electroweak interactions. When the temperature drops to $T_{c_2} \sim 200$ GeV (see Eq. (3.2.3)), the symmetry between the weak and electromagnetic interactions breaks. Finally, at $T \sim 10^2$ MeV, there should be a phase transition (or two separate transitions) which breaks the chiral invariance of the theory of strong interactions and leads to the coalescence of quarks into hadrons (confinement).

Here we must voice some reservations. The Glashow–Weinberg–Salam theory of electroweak interactions has withstood experimental tests quite well, but the situation with grand unified theories is not nearly so satisfactory. Prior to the 1980's, there seemed to be little doubt of the existence of grand unification at energies $E \sim 10^{15}$ GeV, with the most likely candidate for the role of a unified theory being minimal SU(5).

Subsequently, unified theories became more and more complicated, starting with N = 1 supergravity, then the Kaluza–Klein theory, and finally superstring theory. As the theories have changed, so has our picture of the evolution of the universe at high temperatures. But all versions of this picture have at least one thing in common: without an inflationary stage, they all lead to consequences in direct conflict with existing cosmological data. In the present section, in order to expose the sources of these problems and point out some possibilities for overcoming them, we will study the kinetics of phase transitions in the minimal SU(5) theory.

In that theory, the potential of the field Φ responsible for symmetry breaking between the strong and electroweak interactions takes the form (see Section 3.2)

$$V(\Phi) = -\frac{\mu^2}{2}\,\mathrm{Tr}\,\Phi^2 + \frac{a}{4}\left(\mathrm{Tr}\,\Phi^2\right)^2 + \frac{b}{2}\,\mathrm{Tr}\,\Phi^4. \tag{6.1.1}$$

At $T \gg \mu$, the main modification to $V(\Phi)$ consists of a change of sign of the effective parameter μ^2,

$$\mu^2(T) = \mu^2 - \beta T^2, \tag{6.1.2}$$

see (3.2.6). This leads to the restoration of symmetry at high temperatures. According to (3.2.6), however, at $T \lesssim \mu$ the modification of the effective potential does not reduce to a change in μ^2; the effective potential $V(\Phi, T)$ can acquire additional local minima that correspond not just to SU(3) × SU(2) × U(1) symmetry breaking (see Chapter 1), but also to symmetry breaking described by the groups SU(4) × U(1), SU(3) × $(SU(1))^2$, or $(SU(2))^2$ × $(SU(1))^2$ [167]. This, plus the fact that phase transitions in grand unified theories are first-order transitions, greatly complicates investigation of the kinetics of the transition from the SU(5) phase to the SU(3) × SU(2) × U(1) phase. Here we present the main results from this investigation [187].

First of all, recall that according to [167], the effective potential $V(\varphi, T)$ of the minimal SU(5) theory takes the form

$$V(\varphi, T) = -\frac{N\pi T^4}{90} - \frac{\mu^2(T)}{2}\varphi^2 - \alpha_i T\varphi^3 + \gamma_i\varphi^4 \tag{6.1.3}$$

for each of the four types of symmetry breaking mentioned above, where $\varphi^2 = \mathrm{Tr}\,\Phi^2$, and α_i and γ_i, $i = 1, 2, 3, 4$ are certain constants calculated in [167]. This effective potential is the same as the potential (5.2.12), so that all of the results we obtained in the thin-wall approximation concerning tunneling from a state $\varphi \neq 0$, with formation of bubbles of a field $\varphi \neq 0$, also apply to the theory (6.1.3). On the other hand, in those cases where the thin-wall approximation does not work, the field φ within a bubble is small, the last term in (6.1.3) can be discarded, and the potential is the same as (5.3.2), for which we also studied tunneling in Chapter 5.

Our plan of attack is thus as follows. We must understand how the quantity $V(\varphi, T)$ in (6.1.3) depends on time in an expanding universe, calculate the rate of production of each of the four types of bubbles enumerated above, determine the moment at which the bubbles thus formed occupy the whole universe, explore what happens to bubbles formed in earlier stages, and find the typical volume occupied by regions that are filled with the various phases at the end of the whole process.

Since we have already developed the theory of bubble formation, the solution of the foregoing problem should not be particularly difficult. Nevertheless, it does turn out to be a fairly tedious task, since the numerical calculations must be rerun for every new choice of parameters a and b in (6.1.1). Below we present and discuss the main results that we have obtained for the most natural case, $a \sim b \sim 0.1$ [187].

For this case, the phase transition takes place from a supercooled state in which the temperature of the universe approaches T_{c_1}; starting with this temperature, the symmetric phase φ becomes absolutely unstable. The jump in the field φ at the phase transition point is then large (of order φ_0). In that sense, the phase transition is a "strong" first-order transition.

The phase transition proceeds with the simultaneous production of all four types of phases listed above, the overwhelming majority of the bubbles containing the SU(4) × U(1) phase, and not the energetically more favorable SU(3) × SU(2) × U(1) phase, which initially occupies only a few percent of the whole volume. SU(3) × SU(2) × U(1) bubbles eventually start expanding within the SU(4) × U(1) phase, "devouring" both it and the bubbles of the other two phases. At such time as the SU(3) × SU(2) × U(1) bubbles coalesce, they have a typical size of

$$r \sim 10\, T_{c_1}^{-1} . \qquad (6.1.4)$$

Prior to the formation of a homogeneous SU(3) × SU(2) × U(1) phase, the kinetics of processes during the intermediate phase is very complex, depending on the values of a, b, and g^2. The duration of this intermediate stage, as well as that of the stage preceding the end of the phase transition, can only be significant in theories with certain specific relations between the coupling constants.

Despite the large jump in the field φ at the phase transition point, the amount of energy liberated in the phase transition process is relatively minor, as a rule, so that given the most reasonable values of the coupling constants, a symmetry-breaking transition from a supercooled SU(5)-symmetric phase will not result in a discontinuous rise in temperature, nor will it produce a marked increase in the total entropy of the expanding universe.

As the temperature drops further to $T_{c_2} \sim 10^2$ GeV, the phase transition SU(3) × SU(2) × U(1) → SU(3) × U(1) takes place, and with it the symmetry between the weak and electromagnetic interactions is broken. At the time of this transition, the temperature is many orders of magnitude lower than the mass of the superheavy bosons with $M_X \sim 10^{14}$ GeV that appear after the first phase transition. Lighter particles in this theory are described by the Glashow–Weinberg–Salam theory, so the phase transition at $T_{c_2} \sim 10^2$ GeV proceeds exactly as in the latter; see Chapter 3.

Generally speaking, the foregoing pattern of phase transitions is only relevant to the simplest grand unified theories with the most natural relation between coupling constants. In more complicated theories, phase transitions may occur with many more steps; for example, see [42, 167]. A somewhat unusual picture also emerges for certain special relations among the parameters of a theory, for which the effective potential of scalar fields contains a local minimum or a relatively flat region at small φ.

By way of example, let us consider the Glashow–Weinberg–Salam model with

$$\lambda(\varphi_0) = \frac{1}{6}\frac{d^4V}{d\varphi^4}\bigg|_{\varphi=\varphi_0} < \frac{11e^4}{16\pi^2}\cdot\frac{2\cos^4\theta_W + 1}{\sin^2 2\theta_W} \approx 3\cdot 10^{-3}, \tag{6.1.5}$$

$$m_\varphi^2(\varphi_0) = \frac{d^2V}{d\varphi^2}\bigg|_{\varphi=\varphi_0} < \frac{e^4\varphi_0^2}{16\pi^2}\cdot\frac{2\cos^4\theta_W + 1}{\sin^2 2\theta_W} \approx (10\text{ GeV})^2, \tag{6.1.6}$$

where $\sin^2\theta_W \sim 0.23$, $\varphi_0 \sim 250$ GeV. For these values of $\lambda(\varphi_0)$ and m_φ^2, the effective potential $V(\varphi)$ has a local minimum at $\varphi = 0$ even at zero temperature [139–141]; see Section 2.2.

In that case, symmetry was restored in the early universe as usual, with $\varphi = 0$. As the universe cooled, a minimum of $V(\varphi)$ then appeared at $\varphi \sim \varphi_0$, becoming deeper than the one at $\varphi = 0$ shortly thereafter. Nevertheless, the universe remained in the state $\varphi = 0$ until such time as bubbles of a new phase with $\varphi \neq 0$ formed and filled the entire universe. The formation of bubbles of a new phase in the Glashow–Weinberg–Salam theory was studied in [141, 142]. It turns out that if m_φ is even one percent less than the limiting value $m_\varphi \sim 10$ GeV (6.1.6), the probability of bubble formation with $\varphi \neq 0$ becomes exceedingly small.

The reason for this is not far to seek if we hark back to the results of the previous chapter. Consider the limiting case

$$m_\varphi^2 = \frac{e^4\varphi_0^2}{16\pi^2}\cdot\frac{2\cos^4\theta_W + 1}{\sin^2 2\theta_W}. \tag{6.1.7}$$

The curvature of $V(\varphi)$ at $\varphi = 0$, $T = 0$ then tends to zero (the Coleman–Weinberg model [137]; see Section 2.2). At $T \neq 0$, the mass of the scalar field in the vicinity of $\varphi = 0$ is, according to (3.2.1),

$$m_\varphi \sim \frac{eT}{\sin 2\theta_W}\sqrt{1 + 2\cos^2\theta_W} \tag{6.1.8}$$

(recall that in the present case $\lambda \sim e^4 \ll e^2$). In this model, at small φ, the potential $V(\varphi)$ is approximately

$$V(\varphi) = V(0) + \frac{3e^4\varphi^4}{32\pi^2}\left(\frac{2\cos^4\theta_W + 1}{\sin^2 2\theta_W}\right)\ln\frac{\varphi}{\varphi_0} + \frac{m_\varphi^2\varphi^2}{2}. \tag{6.1.9}$$

Now $\ln\frac{\varphi}{\varphi_0}$ is a fairly slowly varying function of φ, so to determine the probability P of tunneling out of the local minimum at $\varphi = 0$, we can make use of Eq. (5.3.6) [144]:

$$P \sim \exp\left(-\frac{19 m_\varphi(T)}{\lambda T}\right) \sim \exp\left(-\frac{19 \sin 2\theta_W}{\frac{3e^3}{8\pi^2}\sqrt{1+\cos^4\theta_W}\,\ln\frac{\varphi}{\varphi_0}}\right)$$
$$\sim \exp\left(-\frac{15000}{\ln\frac{\varphi}{\varphi_0}}\right). \tag{6.1.10}$$

The typical value of the field φ appearing in (6.1.10) corresponds to a local maximum of $V(\varphi)$ in (6.1.9) located at $\varphi \sim 10\,T$; that is,

$$P \sim \exp\left(-\frac{15000}{\ln\frac{T}{\varphi_0}}\right). \tag{6.1.11}$$

Hence, we find that in the theory under consideration, the phase transition in which bubbles of the field φ are formed can only take place if the temperature of the universe is exponentially low. A similar phenomenon in the Coleman–Weinberg SU(5) theory lays at the basis of the new inflationary universe scenario (see Chapter 8). But in the Glashow–Weinberg–Salam theory with

$$\left.\frac{d^2V}{d\varphi^2}\right|_{\varphi=0} = 0,$$

supercooling is actually not so strong as might be construed from (6.1.11): the phase transition occurs at $T \sim 10^2$ MeV on account of effects associated with strong interactions [144]. When it takes place, the specific entropy of the universe $\frac{n_\gamma}{n_B}$ should rise approximately 10^5–10^6-fold [144], due to liberation of the energy stored in the metastable vacuum $\varphi = 0$. Even if the effective potential $V(\varphi)$ has only a very shallow minimum at $\varphi = 0$, the increase in the specific entropy of the universe may become unacceptably large [143, 144]. Furthermore, the lifetime of the universe in a metastable vacuum state with $V''(0) \gtrsim (10^2\ \text{MeV})^2$ will be greater than the age of the observable part of the universe, $t \sim 10^{10}$ yr [141, 142]. Bubbles formed as a result of such a phase transition would make the universe strongly anisotropic and inhomogeneous. The universe would be homogeneous only inside each of the bubbles, which would be devoid of matter of any kind. This leads to the strong constraint (2.2.14) on the mass of the Higgs boson in the Glashow–Weinberg-Salam theory without superheavy fermions[1]:

$$m_\varphi \gtrsim 10 \text{ GeV}. \qquad (6.1.12)$$

As we showed in Chapter 2, the absolute minimum of $V(\varphi)$ in a theory with superheavy fermions may turn out not to be at $\varphi = \varphi_0 = \frac{\mu}{\sqrt{\lambda}}$, but at $\varphi \gg \varphi_0$, which constrains the allowable fermion masses in the theory [146–151]. When cosmological effects are taken into account, the corresponding bounds are softened somewhat, since the universe will not always succeed in going from a state $\varphi = \varphi_0$ to an energetically more favorable one [188]. The complete set of bounds on the masses of fermions and the Higgs boson, including the cosmological constraints, is shown in Fig. 16 (Chapter 2, page 81). Notice, however, that in studying tunneling, the authors of [141–151, 188] did not discuss the possibility of tunneling induced by collisions of cosmic rays with matter. If such processes could substantially increase the probability of decay of a metastable vacuum [189] (see, however, [360]), then the region above the curve AD in Fig. 16 would turn out to be forbidden, and the most stringent constraint on m_φ would be set by Eq. (2.2.9). This problem requires more detailed investigation.

6.2 Domain walls, strings, and monopoles

In the preceding section, we pointed out that a phase transition with SU(5) symmetry breaking takes place with the formation of bubbles containing several different phases, and only subsequently does all space fill with matter in a single energetically most favorable phase. For this to happen, at least two conditions must be satisfied: only one energetically most favorable phase may exist, and the typical size r of the bubbles must not exceed t, where t is the time at which the entire universe should have made the transition to a single phase. In the hot universe theory (in con-

[1] To avoid misunderstanding, we should emphasize that these bounds refer only to the simplest version of the Glashow–Weinberg–Salam theory, with a single type of scalar field φ.

trast to the inflationary universe theory), bubbles typically do not grow to be very large — $r \sim m^{-1}$ or $r \sim T^{-1}$ (see Fig. 26, page 127) — so the second requirement is usually met. But there are a great many theories in which the effective potential has several minima of the same (or almost the same) depth. The simplest example is the theory (1.1.5), which has minima at $\varphi = \frac{\mu}{\sqrt{\lambda}}$ and $\varphi = -\frac{\mu}{\sqrt{\lambda}}$ of equal depth. When a phase transition occurs at some time $t = t_c$ during the expansion of the universe, symmetry breaking takes place independently in different causally disconnected regions of size $O(t_c)$. As a result, the universe is partitioned into approximately equal numbers of regions filled with the fields $\varphi = \frac{\mu}{\sqrt{\lambda}}$ and $\varphi = -\frac{\mu}{\sqrt{\lambda}}$. These regions are separated from one another by domain walls of thickness $O(\mu^{-1})$, with the field changing from $\varphi = \frac{\mu}{\sqrt{\lambda}}$ to $\varphi = -\frac{\mu}{\sqrt{\lambda}}$ from one side of the wall to the other.

Actually, as a rule, the regions in which symmetry breaking takes place independently initially have sizes of order T_c^{-1}; that is, they have dimensions much smaller than the horizon $t \sim 10^{-2} \frac{M_P}{T_c^2}$ at the time when the phase transition starts. One example of this is the formation of regions with different phases at the time of the phase transition in the SU(5) model; see (6.1.4).

Regions that are filled with different phases at the same energy density also tend to "eat" each other, as the presence of domain walls is energetically unfavorable. But this mutual consumption proceeds independently in regions separated by distances of order t, where t is the age of the universe. As we have already noted in Section 1.5, at time $t \sim 10^5$ yr, the presently observable part of the universe consisted of approximately 10^6 causally disconnected regions, or, in other words, of 10^6 domains separated by superheavy domain walls. Since in the last $\sim 10^5$ years the observable part of the universe has been transparent to photons, the existence of such domains would lead to considerable anisotropy in the primordial background radiation. The astronomical observations, however, indicate that the background radiation is isotropic to within $\frac{\Delta T}{T} \sim 3 \cdot 10^{-5}$. This is the essence of the domain wall problem in the hot universe theory [41]. These results would force us to renounce theories with discrete symmetry breaking, such as the theory (1.1.5), theories with spontaneously broken CP invariance, the minimal SU(5) theory, in which the potential $V(\Phi)$ (6.1.1) is invariant under reflection $\Phi \to -\Phi$, etc. Most theories of the axion field θ encounter similar difficulties: the potential $V(\theta)$ in many versions of the axion theory has several minima of the same depth [49]. In some theories, this difficulty can be overcome (for example, by adding a term $c\,\mathrm{Tr}\,\Phi^3$ to $V(\Phi)$ (6.1.1)), but usually the problems are insurmountable without changing the theory fundamentally (or reverting to the inflationary universe scenario).

Besides domain walls, phase transitions can give rise to other nontrivial entities as well. Consider, for example, a model of a complex scalar field χ with the Lagrangian

$$L = \partial_\mu \chi^* \partial_\mu \chi + m^2 \chi^* \chi - \lambda (\chi^* \chi)^2 . \tag{6.2.1}$$

This is the Higgs model of (1.1.15) prior to the inclusion of the vector fields A_μ. In order to study symmetry breaking in this theory, it is convenient to change variables:

$$\chi(x) \to \frac{1}{\sqrt{2}} \varphi(x) \exp\left(\frac{i\zeta(x)}{\varphi_0}\right). \tag{6.2.2}$$

The effective potential $V(\chi, \chi^*)$ has a minimum at $\varphi(x) = \varphi_0 = \dfrac{\mu}{\sqrt{\lambda}}$ irrespective of the value of the constant part of the phase ζ_0. $V(\chi, \chi^*)$ is thus shaped like the bottom of a basin, with a maximum in the middle (at $\chi(x) = 0$), and rather than being characterized simply by the scalar φ_0, symmetry breaking is characterized by the vector $\varphi(x) \exp\left(\dfrac{i\zeta(x)}{\varphi_0}\right)$ in the (χ, χ^*) isotopic space.

The existence of fields with different phases $\zeta(x)$ in different regions of space is energetically unfavorable. But just as in the case of domain walls, the value of the phase — that is, the direction of the vector $\varphi(x) \exp\left(\dfrac{i\zeta(x)}{\varphi_0}\right)$ — cannot be correlated over distances greater than the size of the horizon, $\sim t$. Moreover, immediately after the phase transition, the direction of this vector at different points x cannot be correlated over distances much greater than $O(T_c^{-1})$.

Let us consider some two-dimensional surface in our three-dimensional space and study the possible configurations of the field φ there. Among these configurations, there is one such that upon traversing some closed contour in x-space, the vector $\varphi(x) \exp\left(\frac{i\zeta(x)}{\varphi_0}\right)$ executes a complete rotation in (χ, χ^*) isotopic space (i.e., the function $\frac{\zeta(x)}{\varphi_0}$ changes by 2π); see Fig. 27. The appearance of such an initial field distribution for φ as a result of a phase transition is in no way forbidden. Now let us gradually constrict this contour, remaining all the while in a region with $\varphi(x) \neq 0$. Since the field $\chi(x)$ is continuous and differentiable, the vector $\chi(x)$ should also execute a complete rotation in traveling along the shrinking contour. If we could shrink the contour to a point at which $\varphi(x) \neq 0$ in this manner, the field $\chi(x)$ would no longer be differentiable there; that is, the equations of motion would not hold at that point. This implies that somewhere within the original contour there must be a point at which $\varphi(x) = 0$. Let us suppose for the sake of simplicity that there is just one. Now change the section of space under consideration, appropriately moving our contour in space so that as before it does not pass through any region with $\varphi(x) = 0$. By continuity, then, in circling the contour, the vector $\chi(x)$ will also rotate by 2π.

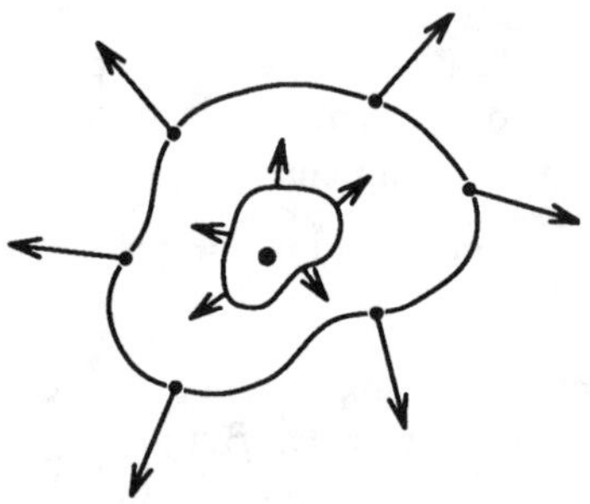

Figure 27. Distribution of the field $\chi = \varphi\ (x)\, e^{i\theta(x)}$ in isotopic space over a path surrounding a string $\varphi(x) = 0$.

Thus, there will be a point within each such contour at which $\varphi(x) = 0$. This implies that somewhere in space there exists a curve — either closed or infinite — upon which $\varphi(x) = 0$. The existence of such a curve is energetically unfavorable, since $\varphi \ll \varphi_0$ nearby and the gradient of the field φ is also nonzero. However, topological considerations indicate that such a curve, once having been produced during a phase transition, cannot break; only if it is closed can it shrink to a point and disappear. The curve $\varphi(x) = 0$ owes its topological stability to the fact that as one goes around this curve, the vector $\chi(x)$ executes either no full rotations, or one, two, or three, but there is no continuous transformation between the corresponding distributions of the field χ (in traveling along the closed contour, and returning to the same point x, the vector $\chi(x)$ cannot make 0.99 full rotations in (χ, χ^*) space). Such curves, together with their surrounding regions of inhomogeneous field $\chi(x)$, are called strings.

Similar configurations of the field χ can also arise in the Higgs model itself. In that case, however, everywhere except on the curve $\varphi(x) = 0$ one can carry out a gauge transformation like (1.1.16) and transform away the field $\zeta(x)$. However, this leads to the appearance of a field $A_\mu(x) \neq 0$ within the string, which contains a quantum of magnetic flux $\mathbf{H} = \nabla \times \mathbf{A}$. Such strings are entirely analogous to Abrikosov filaments in the theory of superconductivity [190]. Just as before, it is impossible to break such a string, in the present case by virtue of the conservation of magnetic flux. In order to distinguish such strings from those devoid of gauge fields, the latter are sometimes called global strings (their existence being related to global symmetry breaking).

Inasmuch as the directions of the isotopic vectors $\chi(x)$ are practically uncorrelated in every region of size $O(T_c^{-1})$ immediately after the phase transition, strings initially look like Brownian trajectories with "straight" segments whose characteristic length is $O(T_c^{-1})$. Gradually straightening out, these strings then accelerate as a result of their tension, and start to move at close to the speed of light. The end result is that small closed strings (with sizes less than $O(t)$) start to collapse, intersect, radiate their energy in the form of gravitational waves, and finally disappear. Very long strings, with sizes of the order of the distance to the horizon $\sim t$, become almost straight. If, as appears possible, intersecting strings have a non-negligible probability of coalescing, thereby forming small closed strings, the number of long, straight strings remaining within the horizon ought to decrease to a value of order unity.

Let α be the energy density of a string of unit length. In theories with coupling constants of order unity, $\alpha \sim \varphi_0^2$. The mass of a string inside the horizon is of order $\delta M \sim \alpha t \sim \varphi_0^2 t$, while according to (1.3.21) the total mass of matter inside the horizon is $M \sim 10\, M_P^2 t$. This means that due to the evolution of strings, the universe will eventually contain density inhomogeneities [192, 192, 81]

$$\frac{\delta\rho}{\rho} \sim \frac{\delta M}{M} \sim 10\,\frac{\alpha}{M_P^2} \sim 10\,\frac{\varphi_0^2}{M_P^2}. \tag{6.2.3}$$

For $\alpha \sim 10^{-6} M_P^2$, $\varphi_0 \sim 10^{16}$ GeV, we obtain $\frac{\delta\rho}{\rho} \sim 10^{-5}$, as required for galaxy formation.

In deriving this estimate, we have assumed that small closed strings rapidly (in a time of order t) radiate their energy and vanish. Actually, this will only happen if the value of α is large enough. More refined estimates [193] lead to values of α similar to those obtained above,

$$\alpha \sim 2 \cdot 10^{-6} M_P^2 .$$

Notice that the typical mass scale and the value of φ_0 that appear here are close to those associated with symmetry breaking in grand unified theories. Such strings can exist in some grand unified theories. Unfortunately, it is far from simple to arrange for such heavy strings to be created after inflation, since the temperature of the universe after inflation is typically much lower than $\varphi_0 \sim 10^{16}$ GeV, and the phase transition which leads to heavy string formation typically does not occur. Some possibilities for the formation of heavy strings in the inflationary universe scenario will be discussed in the next chapter.

Now let us look at another important class of topologically stable objects which might be formed at the time of phase transitions. To this end, we analyze symmetry breaking in the O(3)-symmetric model of the scalar field φ^a, $a = 1, 2, 3$:

$$L = \frac{1}{2}(\partial_\mu \varphi^a)^2 + \frac{\mu^2}{2}(\varphi^a)^2 - \frac{\lambda}{4}[(\varphi^a)^2]^2 . \tag{6.2.4}$$

Symmetry breaking occurs in this model as a result of the appearance of the scalar field φ^a, with absolute value φ_0 equal to $\frac{\mu}{\sqrt{\lambda}}$, but with arbitrary direction in isotopic space $(\varphi^1, \varphi^2, \varphi^3)$. At the time of the phase transition, domains can appear such that the vector φ^a can look either "out of" or "into" each domain (in isotopic space) at all points on its surface. One example is the so-called "hedgehog" distribution shown in Fig. 28,

$$\varphi^a(\mathbf{x}) = \varphi_0\, f(r)\frac{x^a}{r}, \tag{6.2.5}$$

where $\varphi_0 = \frac{\mu}{\sqrt{\lambda}}$, $r = \sqrt{\mathbf{x}^2}$, and $f(r)$ is some function that tends to ± 1 for $r \gg \mu^{-1}$, and tends to zero as $r \to 0$ (the latter condition derives from the continuity of to the function $\varphi^a(\mathbf{x})$). Such a distribution is a solution of equations of motion in the theory (6.2.4) (for a specific choice of function $f(r)$ with the indicated properties), and this solution turns out to be topologically stable for the same reason as do the global strings considered above.

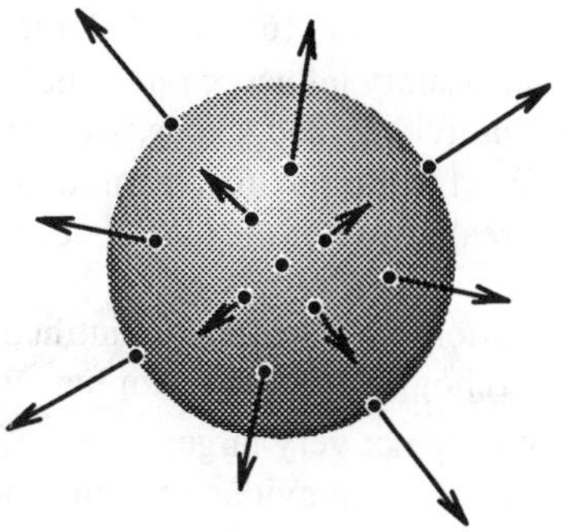

Figure 28. Distribution of the field φ^a (6.2.5) about the center of a hedgehog.

At large r, the main contribution to the hedgehog energy comes from gradient terms arising from the change in direction of the unit vector $\frac{x^a}{r}$ at different points,

$$\rho \sim \frac{1}{2}(\partial_i \varphi)^2 = \frac{3}{2}\frac{\varphi_0^2}{r^2}, \tag{6.2.6}$$

whereupon that part of the hedgehog energy contained within a sphere of radius r centered at $x = 0$ is

$$\mathrm{E}(r) = 6\pi\varphi_0^2 r. \tag{6.2.7}$$

In infinite space, the total hedgehog energy thus goes to infinity (as r). That is why the hedgehog solution (6.2.5), discovered more than ten years ago in the same paper as monopoles [83], failed until fairly recently to elicit much interest in and of itself.

When phase transitions take place in an expanding universe, however, hedgehogs can most certainly be created. The theory of hedgehog formation is similar to the theory of string creation, and in fact the first estimates of the number of monopoles created during a phase transition [40] were based implicitly on an analysis of hedgehog production. An investigation of this problem shows that rather than being created singly, hedgehogs are typically created in hedgehog-antihedgehog pairs (corresponding to $f(r) = \pm 1$ for $r \gg m^{-1}$ in (6.25)). At large distances, such a pair exerts a mutually compensatory influence on the field φ, and instead of the infinite energy of the individual hedgehogs, we obtain the energy of the pair, which is proportional to their mutual separation r (6.2.7). This is the simplest example of a realization of the idea of confinement that we know of.

Subsequent evolution of a hedgehog-antihedgehog molecule depends strongly on hedgehog interactions with matter. In a hot universe, such a molecule is initially not very large, $r \lesssim 10^2\,\mathrm{T}_c^{-1}$. Actually, though, according to the results of the previous section, domains filled with the homogeneous field φ have characteristic sizes of order $10\,\mathrm{T}_c^{-1}$ (see (6.1.4)). Simple combinatoric arguments indicate that in a region containing 10^2–10^3 such domains with uncorrelated values of φ^a, one will surely find at least one hedgehog, thereby yielding the preceding estimate.

If the fields φ^a interact weakly with matter, hedgehogs and antihedgehogs quickly approach one another, start executing oscillatory motion, radiate Goldstone bosons and gravitational waves, approach still closer, and finally annihilate, radiating their energy in the same way as do closed (global) strings. But if hedgehog motion is strongly damped by matter, the annihilation process can take much longer. We shall return to the discussion of possible cosmological effects associated with hedgehogs when we consider the production of density inhomogeneities in the inflationary universe scenario.

If we supplement the theory (6.2.4) with O(3)-symmetric Yang–Mills fields with a coupling constant e, the resultant theory will also have a solution of the equations of motion like (6.2.5) for the field φ^a, but classical Yang–Mills fields will show up as well. By a gauge transformation of the fields φ^a and A^a_μ, we can "comb out" the hedgehog, i.e., send the fields φ^a off in one direction (for example, $\varphi^a \sim x^3\delta^a_3$) everywhere except along some infinitely thin filament emanating from the point $x = 0$. Far from the point $x = 0$, then, the vector fields $A^{1,2}_\mu$ acquire a mass $m_A = e\varphi_0$, while the vector field A^3_μ remains massless. The most important feature of the resulting configuration of the fields φ^a and A^a_μ is then the presence of a magnetic field $\mathbf{H} = \nabla \times \mathbf{A}^3$ which falls off far from the center,

$$\mathrm{H} = \frac{1}{e}\frac{\boldsymbol{x}}{r^3}. \tag{6.2.8}$$

Hence, this theory gives rise to particles analogous to the Dirac monopole ('t Hooft–Polyakov monopoles) with magnetic charge

$$g = \frac{4\pi}{e}, \tag{6.2.9}$$

and these particles have an extremely high mass,

$$\mathrm{M} = c\left(\frac{\lambda}{e^2}\right)\frac{4\pi m_\mathrm{A}}{e^2} = \frac{c m_\mathrm{A}}{\alpha}, \tag{6.2.10}$$

where $\alpha = \frac{e^2}{4\pi}$, and $c\left(\frac{\lambda}{e^2}\right)$ is a quantity approximately equal to unity: $(c(0) = 1, c(0.5) = 1.42, c(10) = 1.44)$.

In contrast to hedgehogs (6.2.5), 't Hooft–Polyakov monopoles ought to exist in all grand unified theories, in which the weak, strong, and electromagnetic interactions prior to symmetry breaking are described by a single theory with a simple symmetry group (SU(5), O(10), E_6, ...). Just as for hedgehogs, monopoles are produced during phase transitions, separated from each other by distance of order $10^2 T_c^{-1}$. Their initial density n_M at the phase transition epoch was thereby 10^{-6} times the photon density n_γ. Zeldovich and Khlopov's study of the monopole-antimonopole annihilation rate [40] has shown that annihilation proceeds very slowly, so that at present we should find $\frac{n_M}{n_\gamma} \sim 10^{-9}$–$10^{-10}$, i.e., $n_M \sim n_B$, where n_B is the baryon (proton and neutron) density. The present density ρ_B of baryon matter in the universe differs from the critical density by no more than one or two orders of magnitude, $\rho_B \sim 10^{-29}$ g/cm^3. In grand unified theories, according to (6.2.10), monopoles should have a mass of $10^2 M_X \sim 10^{16}$–10^{17} GeV; that is, 10^{16}–10^{17} times the mass of the proton. But that would mean, if we believe the estimate $n_M \sim n_B$, that the density of matter in the universe exceeds the critical value by 16 orders of magnitude. Such a universe would already have collapsed long ago!

Even more stringent limits are placed on the allowable present-day density by the existence of the galactic magnetic field [194], and by theoretical estimates of pulsar luminosity [195] due to monopole catalysis of proton decay [196]. These constraints lead one to conclude that at present, most likely $\frac{n_M}{n_B} \lesssim 10^{-25}$–$10^{-30}$. Such an enormous disparity between the observational constraints on the density of monopoles in the universe and the theoretical predictions have led us to the brink of a crisis: modern elementary particle theory is in direct conflict with the hot universe theory. If we are to get rid of this contradiction, we have three options:

a) renounce grand unified theories;

b) find conditions under which monopole annihilation proceeds much more efficiently;

c) renounce the standard hot universe theory.

At the end of the 1970's, the first choice literally amounted to blasphemy. Later on, after the advent of more complicated theories based on supergravity and superstring theory, the general attitude toward grand unified theories began to change. But for the most part, rather than helping to solve the primordial monopole problem, the new theories engender fresh conflicts with the hot universe theory that are just as serious; see Section 1.5.

The second possibility has so far not been carried through to completion. The basic conclusions of the theory of monopole annihilation proposed in [40] have since been confirmed by many authors. On the other hand, it has been argued [173] that nonperturbative effects in a high-temperature Yang–Mills gas can lead to monopole confinement, accelerating the annihilation process considerably.

The basic idea here is that far from a monopole, its field is effectively Abelian, $\mathbf{H}^a = \nabla \times \mathbf{A}^a \cdot \delta_3^a$. Such a field satisfies Gauss' theorem identically, $\nabla \cdot \mathbf{H} = 0$, so its flux is conserved. However, if the Yang–Mills fields in a hot plasma acquire an effective magnetic mass $m_A \sim e^2 T$ (see Section 3.3), then the monopole magnetic field will be able to penetrate the medium only out to a distance m_A^{-1}. The only way to make this condition compatible with the magnetic field version of Gauss' theorem is to invoke filaments of thickness $\Delta l \sim m_A^{-1}$ emanating from the monopoles and incorporating their entire magnetic field. But this is exactly how the magnetic field of a monopole embedded in a superconductor behaves (and for the same reason): Abrikosov filaments (strings) of the magnetic field come into being between monopoles and antimonopoles [190]; see Fig. 29. Since the energy of each such string is proportional to its length, monopoles in a superconductor ought to be found in a confinement phase [197]. If the analogous phenomenon comes into play in the hot Yang–Mills gas, then the monopoles there should be bound to antimonopoles by strings of thickness $\Delta l \sim (e^2 T)^{-1}$. Monopole-antimonopole pairs will therefore annihilate much more rapidly than when they are bound solely by conventional attractive Coulomb forces.

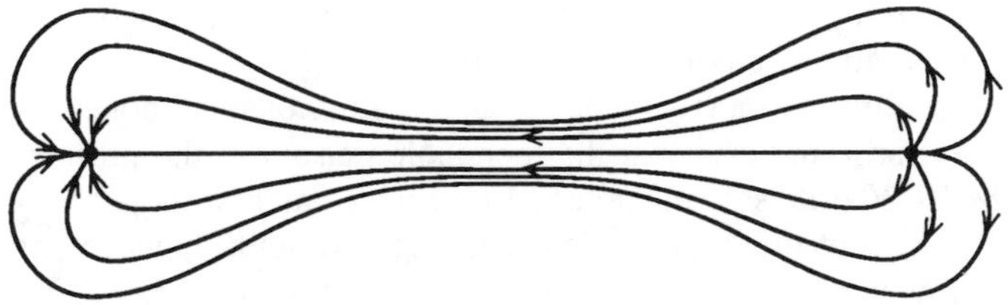

Figure 29. Magnetic field configuration for a monopole–antimonopole pair embedded in a superconductor.

Unfortunately, we still have too imperfect an understanding of the thermodynamics of the Yang–Mills gas to be able to confirm or refute the existence of monopole confinement in a hot plasma. Nonperturbative analysis of this problem using Monte Carlo lattice simulations [198, 199] is not particularly informative, since the use of the lattice gives rise to fictitious light monopoles whose mass is of the order of the reciprocal lattice spacing a^{-1}. These fictitious monopoles act to screen out the mutual interaction of 't Hooft–Polyakov monopoles during the Monte Carlo simulations; they are difficult to get rid of with presently available computing capabilities.

Besides the monopole confinement mechanism discussed above, there is another that is even simpler [200]. Specifically, it is well known that in addition to not being able to penetrate a superconductor, a magnetic field cannot penetrate the bulk of a perfect conductor either (if the field was not present in the conductor from the very start), the reason being that induced currents cancel the external magnetic field. The conductivity of the Yang–Mills plasma gas is extremely high, and that is why when monopoles appear during a phase transition, their magnetic field does not make its appearance in the medium right away. The entire magnetic field must first be concentrated into some string joining a monopole and antimonopole, as in Fig. 29 (due to conservation of total magnetic flux, induced currents cannot cancel the entire magnetic flux, which passes along the filament). The string will gradually thicken, and the field will adopt its usual Coulomb configuration. If, however, the rate at which the thickness of the string grows is small compared with the rate at which the monopoles separate from one another due to the expansion of the universe, the field distribution will remain one-dimensional for a long time; in other words, a confinement regime will prevail once again. Our estimates show that such a regime is actually possible in grand unified theories.

A preliminary analysis of the annihilation of monopoles in a confinement phase indicate that the monopole density at the present epoch may be 10–20 orders of magnitude lower than was first thought. A complete solution of this problem is exceedingly difficult, however, and it is not clear whether monopole confinement will provide a way to reconcile theoretical estimates of their density with the most stringent experimental limits, namely those based on the existence of galactic magnetic fields and the observed lack of strong X-ray emission from pulsars.

The theory of the interaction of monopoles with matter may yet harbor even more surprises. But even if a way were found to solve the primordial monopole problem within the framework of the standard hot universe theory, it would be hard to overstate the value of the contribution made by analysis of this problem to the development of contemporary cosmology. It is precisely the numerous attempts to resolve this problem that have led to wide-ranging discussions of the internal inconsistencies of the hot universe theory, and to a recognition of the need to reexamine its foundations. These attempts served as an impetus for the development of the inflationary universe scenario, and for the appearance of new concepts relating both to the initial stages of the evolution of the observable part of the universe and the global structure of the universe as a whole. We now turn to a description of these concepts.

7

General Principles of Inflationary Cosmology

7.1 Introduction

In Chapter 1, we discussed the general structure of the inflationary universe scenario. Recent developments have gone in three main directions:

a) studies of the basic features of the scenario and the revelation of its potential capacity for a more accurate description of the observable part of the universe. These studies deal basically with problems related to the production of density inhomogeneities at the time of inflation, the reheating of the universe, and the generation of the post-inflation baryon asymmetry, along with those predictions of the scenario that might be tested by analysis of the available observational evidence;

b) the construction of realistic versions of the inflationary universe scenario based on modern elementary particle theories;

c) studies of the global properties of space and time within the framework of quantum cosmology, making use of the inflationary universe scenario.

The first of these avenues of research will be discussed in the present chapter. The second will form the subject of Chapters 8 and 9, and the third, Chapter 10.

7.2 The inflationary universe and de Sitter space

As we have already noted in Chapter 1, the main feature of the inflationary stage of evolution of the universe is the slow variation (compared with the rate of expansion of the universe) of the energy density ρ. In the limiting case ρ = const, the Einstein equation (1.3.7) for a homogeneous universe has the de Sitter space (1.6.1)–(1.6.3) as its solution.

It is easy to see that when $\mathrm{H}\, t \gg 1$, the distinction between an open, closed, and flat de Sitter space tends to vanish. Much less obvious is the fact that all three of the solutions (1.6.1)–(1.6.3) actually describe the very same de Sitter space.

To facilitate an intuitive interpretation of a curved four-dimensional space, it is often convenient to imagine it to be a curved four-dimensional hypersurface embedded in a higher-dimensional space. De Sitter space is most easily represented as a hyperboloid

$$z_0^2 - z_1^2 - z_2^2 - z_3^2 - z_4^2 = -\mathrm{H}^{-2} \tag{7.2.1}$$

in the five-dimensional Minkowski space ($z_0, z_1, \cdots, z_4$). In order to represent de Sitter space as a flat Friedmann universe (1.3.2), (1.6.2), it suffices to consider a coordinate system t, x_i on the hyperboloid (7.2.1) defined by the relations

$$\begin{aligned} z_0 &= \mathrm{H}^{-1} \sinh \mathrm{H}t + \frac{1}{2}\mathrm{H}\, e^{\mathrm{H}t}\boldsymbol{x}^2 , \\ z_4 &= \mathrm{H}^{-1} \cosh \mathrm{H}t - \frac{1}{2}\mathrm{H}\, e^{\mathrm{H}t}\boldsymbol{x}^2 , \\ z_i &= e^{\mathrm{H}t} x_i , \quad i = 1, 2, 3 \ . \end{aligned} \tag{7.2.2}$$

This coordinate system spans the half of the hyperboloid with $z_0 + z_4 > 0$ (see Fig. 30), and its metric takes the form

$$ds^2 = dt^2 - e^{2\mathrm{H}t} d\boldsymbol{x}^2 . \tag{7.2.3}$$

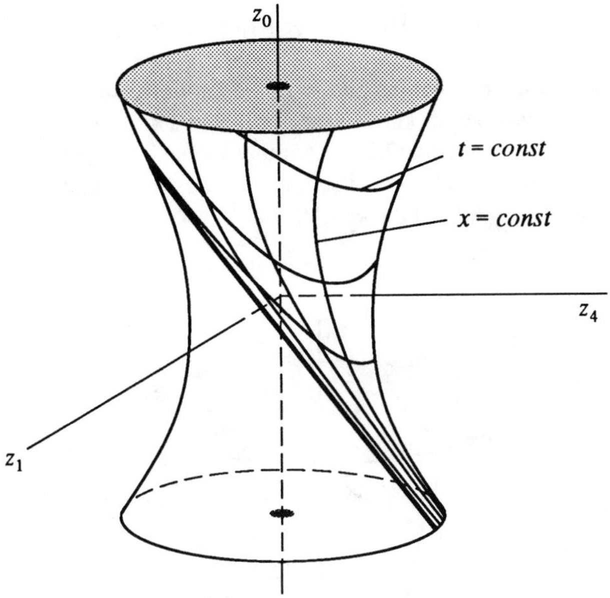

Figure 30. De Sitter space represented as a hyperboloid in five-dimensional space-time (with two dimensions omitted). In the coordinates (7.2.2), three-dimensional space at t = const is flat, expanding exponentially with increasing t — see (7.2.3). The coordinates (7.2.2) span only half the hyperboloid.

De Sitter space looks like a closed Friedmann universe in the coordinate system $(t, \chi, \theta, \varphi)$ defined by

$$\begin{aligned} z_0 &= H^{-1} \sinh Ht\,, \\ z_1 &= H^{-1} \cosh Ht \cos \chi\,, \\ z_2 &= H^{-1} \cosh Ht \sin \chi \cos \theta\,, \\ z_3 &= H^{-1} \cosh Ht \sin \chi \sin \theta \cos \varphi\,, \\ z_4 &= H^{-1} \cosh Ht \sin \chi \sin \theta \sin \varphi\,. \end{aligned} \tag{7.2.4}$$

The metric then becomes

$$ds^2 = dt^2 - H^{-2}\cosh^2 Ht\left[d\chi^2 + \sin^2\chi\,(d\theta^2 + \sin^2\theta\, d\varphi^2)\right]. \quad (7.2.5)$$

It is important to note that in contrast to the flat-universe metric (7.2.3) and the metric for an open de Sitter space (which we will not write out here), the closed-universe metric (7.2.5) describes the entire hyperboloid. In the terminology of general relativity, one can say that the closed de Sitter space, as distinct from the flat or open one, is geodesically complete (see Fig. 31).

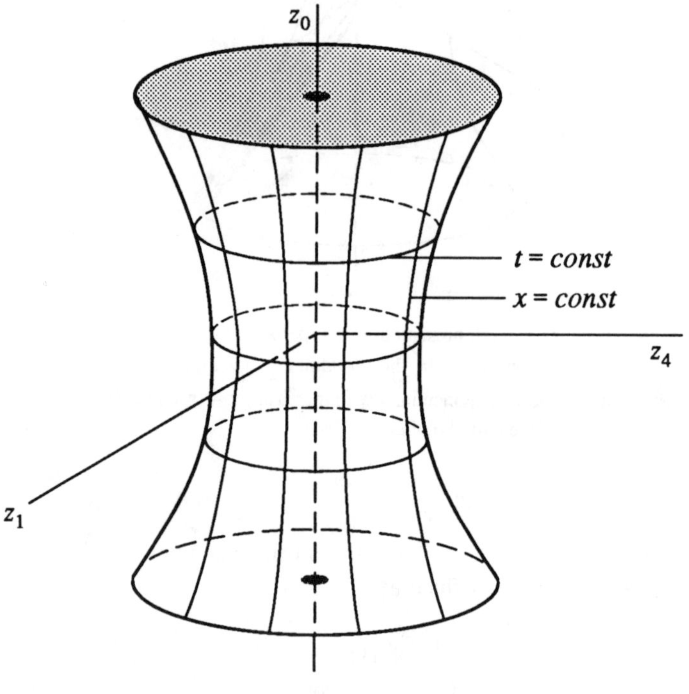

Figure 31. De Sitter space, represented as a closed Friedmann universe with coordinates (7.2.4), (7.2.5). These coordinates span the entire hyperboloid.

To gain some understanding of this situation, it is useful here to draw an analogy with what happens near a black hole. In particular, the Schwarzschild metric does not provide a description of events near the gravitational radius r_g of the black hole, but there do exist coordinate systems that enable one to describe what occurs within the black hole. In the present instance, the analog of the Schwarzschild metric is the metric for a flat (or open) de Sitter space. An even more complete analog is given by the static coordinates (r, t, θ, φ) :

$$\begin{aligned} z_0 &= \sqrt{H^{-2} - r^2}\, \sinh Ht\,, \\ z_1 &= \sqrt{H^{-2} - r^2}\, \cosh Ht\,, \\ z_2 &= r \sin\theta \cos\varphi\,, \\ z_3 &= r \sin\theta \sin\varphi\,, \\ z_4 &= r \cos\theta\,,\ 0 \le r \le H^{-1}. \end{aligned} \tag{7.2.6}$$

These coordinates span that part of the de Sitter space with $z_0 + z_1 > 0$, and the metric takes the form

$$ds^2 = (1 - r^2H^2)\, dt^2 - (1 - r^2H^2)^{-1}\, dr^2 - r^2\,(d\theta^2 + \sin^2\theta\, d\varphi^2)\,, \tag{7.2.7}$$

resembling the form of the Schwarzschild metric

$$ds^2 = (1 - r_g r^{-1})\, dt^2 - (1 - r_g r^{-1})^{-1}\, dr^2 - r^2\,(d\theta^2 + \sin^2\theta\, d\varphi^2)\,, \tag{7.2.8}$$

where $r_g = \frac{2M}{M_P^2}$, and M is the mass of the black hole. Equations (7.2.7) and (7.2.8) demonstrate that de Sitter space in static coordinates comprises a region of radius H^{-1} that looks as if it were *surrounded* by a black hole. This result was provided with a physical interpretation in Chapter 1 (see Eq. (1.4.14)) by introducing the concept of the event horizon. The analogy between the properties of de Sitter space and those of a black hole is a very important one for an understanding of many of the features of the inflationary universe scenario, and it therefore merits further discussion.

It is well known that any perturbations of the black hole (7.2.8) are rapidly damped out, and the only observable characteristic that remains is its mass (and its electric charge and angular momentum if it is rotating). No information about physical processes occurring inside a black hole leaves its surface (that is, the horizon located at $r = r_g$). This set of statements (along with some qualifications and additions), is often known in the literature as the theorem that *a black hole has no hair*; for example, see [119].

The generalization of this "theorem" to de Sitter space [120, 121] reads that any perturbation of the latter will be "forgotten" at an exponentially high rate; that is, after a time $t \gg H^{-1}$, the universe will become locally indistinguishable from a completely homogeneous and isotropic de Sitter space. On the other hand, because of the existence of an event horizon, all physical processes in a given region of de Sitter space are independent of anything that happens at a distance greater than H^{-1} from that region.

The physical meaning of the first part of the theorem is especially transparent in the coordinate system (7.2.3) (or (7.2.5) when $t \gg H^{-1}$): any perturbation of de Sitter space that is entrained by the general cosmological expansion will be exponentially stretched. Accordingly, spatial gradients of the metric, which characterize the local inhomogeneity and anisotropy of the universe, are exponentially damped. This general statement, which has been verified for a wide class of specific models [122], forms the basis for the solution of the homogeneity and isotropy problems in the inflationary universe [54–56].

The second part of the theorem means that if the initial size of an inflationary region exceeds the distance to the horizon ($r > H^{-1}$), then no events outside that region can hinder its inflation, since no information about those events can ever reach it. The indifference of inflationary regions to what goes on about them might be characterized as a sort of relatively harmless egoism: the growth of inflationary regions takes place basically by virtue of their inherent resources, rather than those of neighboring regions of the universe. This kind of process (chaotic inflation) eventually leads to a universe with very complex structure on enormous scales, but within any inflationary region, the universe looks locally uniform to high accuracy. This circumstance plays an important role in any discussion of the initial conditions required for the onset of inflationary behavior (see Sections 1.7 and 9.1) or investigation of the global structure of the universe (Sections 1.8 and 10.2).

We shall return in the next section to the analogy between physical processes inside a black hole and those in the inflationary universe, but here we should like to make one more remark apropos of de Sitter space and its relation to the inflationary universe theory.

Many classic textbooks on general relativity theory treat de Sitter space as nothing but the static space (7.2.7). As we have already pointed out, however, the space described by the metric (7.2.7) is geodesically incomplete; that is, there exist geodesics that carry one out of the space (7.2.7). In much the same way that an observer falling into a black hole does not notice anything exceptional as he makes the final plunge through the Schwarzschild sphere $r = r_g$, so an observer in de Sitter space who is located at some initial point $r = r_0 < H^{-1}$ emerges from the region described by the coordinates (7.2.7) after a definite proper time interval (as measured by his own clocks). (While this is going on, a stationary observer located at $r = \infty$ in the metric (7.2.8) or at $t = 0$ in the metric (7.2.7) will never expect to see his friend disappear beyond the horizon, but he will receive less and less information.) At the same time, the geodesically complete space (7.2.5) is non-static.

In the absence of observers, matter, or even test particles, this lack of stationarity is a "thing unto itself," since the invariant characteristics of de Sitter space itself that are associated with the curvature tensor are time-independent. Thus, for example, the scalar curvature of de Sitter space is

$$R = 12\,H^2 = \text{const}\,. \tag{7.2.9}$$

Therefore, if the inflationary universe were simply an empty de Sitter space, it would be difficult to speak of its expansion. It would always be possible to find a coordinate system in which de Sitter space looked, for example, as if it were contracting, or as if it had a size $\sim H^{-1}$ (Eqs. (7.2.5), (7.2.7)). But in the inflationary universe, the de Sitter invariance is either spontaneously broken (due to the decay of the initial de Sitter vacuum), or is broken on account of an initial disparity between the actual universe and de Sitter space. In particular, the energy-momentum tensor $T_{\mu\nu}$ in the chaotic inflation scenario, even though it is close to $V(\varphi)g_{\mu\nu}$, is never exactly equal to the latter, and in the last stages of inflation, the relative magnitude of the field kinetic energy $\frac{1}{2}\dot{\varphi}^2$ becomes large compared to $V(\varphi)$, and the difference between $T_{\mu\nu}$ and $V(\varphi)g_{\mu\nu}$ becomes significant. The distinction between static de Sitter space and the inflationary universe becomes especially clear at the quantum level, when one analyzes density inhomogeneities $\frac{\delta\rho}{\rho}$ that arise at the time of inflation. As we will show in

Section 7.5, by the end of inflation, these inhomogeneities grow to $\frac{\delta\rho}{\rho} \sim \frac{H^2}{\dot{\varphi}}$. Thus, if the field φ were constant and the inflating universe were indistinguishable from de Sitter space, then after inflation ended our universe would be highly inhomogeneous. In other words, a correct treatment of the inflationary universe requires that we not only take its similarities to de Sitter space into account, but its differences as well, especially in the latest stages of inflation, when the structure of the observable part of the universe was formed.

7.3 Quantum fluctuations in the inflationary universe

The analogy between a black hole and de Sitter space is also useful in studying quantum effects in the inflationary universe. It is well known, for example, that black holes evaporate, emitting radiation at the Hawking temperature $T_H = \frac{M_P^2}{8\pi M} = \frac{1}{4\pi r_g}$, where M is the mass of the black hole [119]. A similar phenomenon exists in de Sitter space, where an observer will feel as if he is in a thermal bath at a temperature $T_H = \frac{H}{2\pi}$. Formally, we can see this making the substitution $t \to i\tau$ in Eq. (7.2.5) in order to make the transition to the Euclidean formulation of quantum field theory in de Sitter space. The metric then becomes that of a four-sphere S^4,

$$-ds^2 = d\tau^2 + H^{-2}\cos^2 H\tau \left[d\chi^2 + \sin^2\chi\,(d\theta^2 + \sin^2\theta\, d\varphi^2)\right]. \quad (7.3.1)$$

Bose fields on the sphere are periodic in τ with period $\frac{2\pi}{H}$, which is equivalent to considering quantum statistics at a temperature $T_H = \frac{H}{2\pi}$ [201]. Physically, the appearance of a temperature T_H in de Sitter space (as is also the case for a black hole) is related to the necessity of averaging over states beyond the event horizon [119, 120]. However, the "temperature" of de Sitter space is highly unusual, in that the Euclidean sphere S^4 is periodic *in all four directions*, so the vacuum fluctuation spectrum turns out to be quite unlike the usual spectrum of thermal fluctuations.

Averages like $\langle\varphi(x)\,\varphi(y)\rangle$ and $\langle\varphi(x)^2\rangle$ will play a particularly important role in our investigation. In Minkowski space at a finite temperature T

$$\langle \varphi(x)^2 \rangle = \frac{T}{(2\pi)^3} \sum_{n=-\infty}^{\infty} \int \frac{d^3k}{(2\pi n T)^2 + k^2 + m^2}, \tag{7.3.2}$$

which reduces to Eq. (3.1.7) for $\langle \varphi^2 \rangle$ after summing over n. In S^4 -space, *all* integrations are replaced by summations over n_i, $i = 1, 2, 3, 4$, and the temperature is replaced by the quantity $\frac{H}{2\pi}$. A term with $n_i = 0$ is especially important in summing over n_i, since it makes the leading contribution to $\langle \varphi^2 \rangle$ as $m^2 \to 0$. It is readily shown that this contribution will be proportional to $\frac{H^4}{m^2}$; for $m^2 \ll H^2$, the corresponding calculation gives

$$\langle \varphi^2 \rangle = \frac{3H^4}{8\pi^2 m^2} \tag{7.3.3}$$

(a result first obtained by a different method [202, 126–128]). The pathological behavior of $\langle \varphi^2 \rangle$ as $m^2 \to 0$ is noteworthy. Formally, it occurs because now instead of one summation, we have four, and the corresponding infrared divergences of scalar field theory in de Sitter space are found to be three orders of magnitude stronger than in quantum statistics.[1] It will be very important to understand the physical basis of such a strange result.

To this end, one quantizes the massless scalar field φ in de Sitter space in the coordinates (7.2.3) in much the same way as in Minkowski space [202, 126–128]. The scalar field operator $\varphi(x)$ can be represented in the form

$$\varphi(\mathbf{x}, t) = (2\pi)^{-3/2} \int d^3p \left[a_p^+ \psi_p(t)\, e^{ipx} + a_p^- \psi_p^*(t)\, e^{-ipx} \right], \tag{7.3.4}$$

where according to (1.7.13), $\psi_p(t)$ satisfies the equation

$$\ddot{\psi}_p(t) + 3\,H\dot{\psi}_p(t) + p^2 e^{-2Ht}\psi_p(t) = 0 \ . \tag{7.3.5}$$

---

[1] Note that in vector or spinor field theory, sums over n_i contain no terms that are singular in the limit $m \to 0$.

In Minkowski space, the function $\frac{1}{\sqrt{2p}} e^{-ipt}$ takes on the role of $\psi_p(t)$, where $p = \sqrt{\mathbf{p}^2}$; see (1.1.3). In de Sitter space (7.2.3), the general solution of (7.3.5) takes the form

$$\psi_p(t) = \frac{\sqrt{\pi}}{2} \mathrm{H}\,\eta^{3/2} \left[\mathrm{C}_1(p)\, \mathrm{H}^{(1)}_{3/2}(p\eta) + \mathrm{C}_2(p)\, \mathrm{H}^{(2)}_{3/2}(p\eta) \right], \qquad (7.3.6)$$

where $\eta = -\mathrm{H}^{-1} e^{-\mathrm{H}t}$ is the conformal time, and the $\mathrm{H}^{(i)}_{3/2}$ are Hankel functions:

$$\mathrm{H}^{(2)}_{3/2}(x) = \left[\mathrm{H}^{(1)}_{3/2}(x) \right]^* = -\sqrt{\frac{2}{\pi x}}\, e^{-ix}\left(1 + \frac{1}{ix}\right). \qquad (7.3.7)$$

Quantization in de Sitter space and Minkowski space should be identical in the high-frequency limit, i.e., $\mathrm{C}_1(p) \to 0$, $\mathrm{C}_2(p) \to -1$ as $p \to \infty$. In particular, this condition is satisfied[2] for $\mathrm{C}_1 \equiv 0$, $\mathrm{C}_2 \equiv -1$. In that case,

$$\psi_p(t) = \frac{i\mathrm{H}}{p\sqrt{2p}} \left(1 + \frac{p}{i\mathrm{H}} e^{-\mathrm{H}t}\right) \exp\left(\frac{ip}{\mathrm{H}} e^{-\mathrm{H}t}\right). \qquad (7.3.8)$$

Notice that at sufficiently large t (when $pe^{-\mathrm{H}t} < \mathrm{H}$), $\psi_p(t)$ ceases to oscillate, and becomes equal to $\frac{i\mathrm{H}}{p\sqrt{2p}}$.

The quantity $\langle \varphi^2 \rangle$ may be simply expressed in terms of ψ_p:

$$\langle \varphi^2 \rangle = \frac{1}{(2\pi)^3} \int |\psi_p|^2 d^3p = \frac{1}{(2\pi)^3} \int \left(\frac{e^{-2\mathrm{H}t}}{2p} + \frac{\mathrm{H}^2}{2p^3} \right) d^3p. \qquad (7.3.9)$$

The physical meaning of this result becomes clear when one transforms from the conformal momentum p, which is time-independent, to the conventional physical momentum $k = p\,e^{-\mathrm{H}t}$, which decreases as the universe expands:

---

[2] It is important that if the inflationary stage is long enough, all physical results are independent of the specific choice of functions $\mathrm{C}_1(p)$ and $\mathrm{C}_2(p)$ if $\mathrm{C}_1(p) \to 0$, $\mathrm{C}_2(p) \to -1$ as $p \to \infty$.

$$\langle\varphi^2\rangle = \frac{1}{(2\pi)^3}\int\frac{d^3k}{k}\left(\frac{1}{2}+\frac{H^2}{2k^2}\right). \tag{7.3.10}$$

The first term is the usual contribution from vacuum fluctuations in Minkowski space (for $H = 0$; see (2.1.6), (2.1.7)). This contribution can be eliminated by renormalization, as in the theory of phase transitions (see (3.1.6)). The second term, however, is directly related to inflation. Looked at from the standpoint of quantization in Minkowski space, this term arises because of the fact that de Sitter space, apart from the usual quantum fluctuations that are present when $H = 0$, also contains φ-particles with occupation numbers

$$n_k = \frac{H^2}{2k^2}. \tag{7.3.11}$$

It can be seen from (7.3.10) that the contribution to $\langle\varphi^2\rangle$ from long-wave fluctuations of the φ field diverges, and that is why the value of $\langle\varphi^2\rangle$ in Eq. (7.3.3) becomes infinite as $m^2 \to 0$.

However, the value of $\langle\varphi^2\rangle$ for a massless field φ is infinite only in de Sitter space that exists forever, and not in the inflationary universe, which expands exponentially (or quasiexponentially) starting at some time $t = 0$ (for example, when the density of the universe becomes smaller than the Planck density). Indeed, the spectrum of vacuum fluctuations (7.3.10) differs from the fluctuation spectrum in Minkowski space only when $k \lesssim H$. If the fluctuation spectrum before inflation has a cutoff at $k \lesssim k_0 \sim T$ resulting from high-temperature effects [127], or at $k \lesssim k_0 \sim H$ due to an inflationary region of the universe having an initial size $O(H^{-1})$, then the spectrum will change at the time of inflation, due to exponential growth in the wavelength of vacuum fluctuations. The spectrum (7.3.10) will gradually be established, but only at momenta $k \gtrsim k_0 e^{-Ht}$. There will then be a cutoff in the integral (7.3.9). Restricting our attention to contributions made by long-wave fluctuations with $k \lesssim H$, which are the only ones that will subsequently be important for us, and assuming that $k_0 = O(H)$, we obtain

$$\langle\varphi^2\rangle \sim \frac{H^2}{2(2\pi)^3}\int_{He^{-Ht}}^{H}\frac{d^3k}{k^3} = \frac{H^2}{4\pi^2}\int_{-Ht}^{0} d\ln\frac{k}{H}$$
$$\equiv \frac{H^2}{4\pi^2}\int_{0}^{Ht} d\ln\frac{p}{H} = \frac{H^3}{4\pi^2}\,t\,. \qquad (7.3.12)$$

As $t \to \infty$, $\langle\varphi^2\rangle$ tends to infinity in accordance with (7.3.3). A similar result is obtained for a massive scalar field φ. In that case, long-wave fluctuations with $m^2 \ll H^2$ behave as

$$\langle\varphi^2\rangle = \frac{3H^4}{8\pi^2 m^2}\left[1 - \exp\left(-\frac{2m^2}{3H}\,t\right)\right]. \qquad (7.3.13)$$

When $t \lesssim \frac{3H}{m^2}$, $\langle\varphi^2\rangle$ grows linearly, just as in the case of the massless field (7.3.12), and it then tends to its asymptotic value (7.3.3).

Let us now try to provide an intuitive physical interpretation of these results. First, note that the main contribution to $\langle\varphi^2\rangle$ (7.3.12) comes from integrating over exponentially small k (with $k \sim H\exp(-Ht)$). The corresponding occupation numbers n_k (7.3.11) are then exponentially large. For large $l = |\boldsymbol{x} - \boldsymbol{y}|\,e^{Ht}$, the correlation function $\langle\varphi(x)\,\varphi(y)\rangle$ for the massless field φ is [203]

$$\langle\varphi(\boldsymbol{x},t)\,\varphi(\boldsymbol{y},t)\rangle \sim \langle\varphi^2(\boldsymbol{x},t)\rangle\left(1 - \frac{1}{Ht}\ln Hl\right). \qquad (7.3.14)$$

This means that the magnitudes of the fields $\varphi(x)$ and $\varphi(y)$ will be highly correlated out to exponentially large separations $l \sim H^{-1}\exp(Ht)$. By all these criteria, long-wave quantum fluctuations of the field φ with $k \ll H^{-1}$ behave like a weakly inhomogeneous (quasi)classical field φ generated during the inflationary stage; see the discussion of this point in Section 2.1.

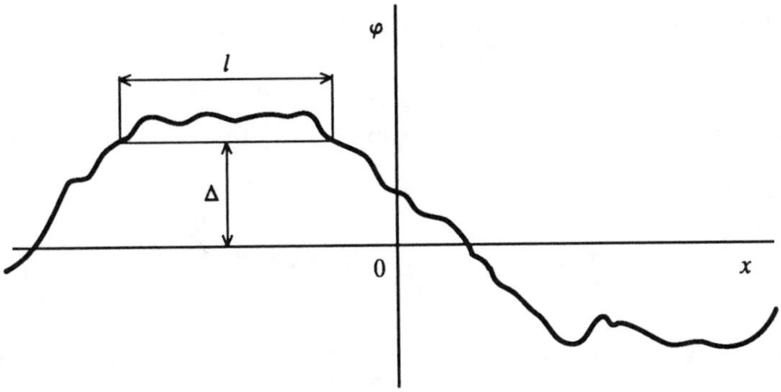

Figure 32. Distribution of the quasiclassical field φ generated at the time of inflation. For a massless field, dispersion Δ is equal to $\frac{H}{2\pi}\sqrt{Ht}$, and a typical correlation length l is equal to $H^{-1}\exp(Ht)$. For a massive field with $m \ll H$, an equilibrium distribution is established in a time $\Delta t \lesssim \frac{H}{m^2}$, having $\Delta \sim \frac{H^2}{m}$ and $l \sim H^{-1}\exp\left(\frac{3H^2}{2m^2}\right)$.

Analogous results also hold for a massive field with $m^2 \ll H^2$. There, the principal contribution to $\langle\varphi^2\rangle$ comes from modes with $k \sim H\exp\left(-\frac{3H^2}{2m^2}\right)$, and the correlation length is of order $H^{-1}\exp\left(\frac{3H^2}{2m^2}\right)$; see Fig. 32.

An important remark is in order here. In constructing theories of particle creation in an expanding universe, elementary particle theorists have had to come to grips with the fact that distinguishing real particles from vacuum fluctuations in the general theory of relativity is a rather ambiguous problem [74]. What we have encountered here in our example is a similar situation. Specifically, from the standpoint of quantization in the coordinate system (7.2.3), long-wave fluctuations with $He^{-Ht} \lesssim k \lesssim H$ correspond to momenta $H \lesssim p \lesssim H\,e^{Ht}$. The corresponding occupation numbers in p-space show no exponential rise with time whatsoever. The correlation between $\varphi(x)$ and $\varphi(y)$ at large $|\mathbf{x} - \mathbf{y}|$ is negligible. Thus, from the standpoint of quantization in de Sitter space (7.2.3), we are dealing with quantum fluctuations. But from the standpoint of occupation numbers at *physical* momenta $k = p\exp(-Ht)$ and the correlation at large *physical* separations $l = |\mathbf{x} - \mathbf{y}|\, e^{\mathrm{H}t}$, we are dealing with a quasiclassical weakly inhomogeneous field φ.

The difference in question is quite evident when we compare the functions $\psi_p(t)$ (7.3.8) and $\psi_k(t) = \psi_p \exp\left(\frac{3}{2}\mathrm{H}\, t\right)$, the square of which also gives the spectrum (7.3.10) in terms of the physical momentum k:

$$\psi_k(t) = -\frac{1}{\sqrt{2}}\, e^{-\frac{ik}{\mathrm{H}}}\left(1 + \frac{\mathrm{H}}{ik}\right) \tag{7.3.15}$$

When $k \gg \mathrm{H}$, we are dealing with a field that oscillates at constant amplitude $\frac{1}{\sqrt{2}}$. But in the course of time, when the magnitude of $k \sim p\, e^{-\mathrm{H}t}$ (p = constant) falls below H, oscillations will cease, and the amplitude of the field distribution for $\psi_k(t)$, which has had its phase frozen in, begins to grow exponentially,

$$\psi_k(t) = \frac{i\mathrm{H}}{\sqrt{2}\,k} \sim \frac{i\mathrm{H}}{\sqrt{2}\,p}\, e^{\mathrm{H}t}, \tag{7.3.16}$$

which is just the reason for the appearance of exponentially large occupation numbers.

We have already encountered this phenomenon in discussing the problem of Bose condensation and symmetry breaking in field theory. This is exactly the way in which the exponential instability involving the creation of the classical Higgs field develops; see Eq. (1.1.6). The difference here is that for symmetry breaking in Minkowski space, it is the mode with vanishing momentum k that grows most rapidly. In the inflationary universe, the momentum k at any of the modes falls off exponentially. This leads to an almost identical growth of modes with different initial momenta k, as a result of which the classical field φ becomes inhomogeneous, although this inhomogeneity becomes significant only at exponentially large distances $l \sim \mathrm{H}^{-1} \exp(\mathrm{H}\, t)$; see (7.3.14). Another important difference between the phenomenon at hand and spontaneous symmetry breaking in Minkowski space is that the production of a classical field φ in de Sitter space is an induced phenomenon. The growth of long-wave perturbations of the field φ occurs even when it is energetically unfavorable, as for instance when $m^2 > 0$ (but only when $m^2 \ll \mathrm{H}^2$).

The process for generating a classical scalar field $\varphi(x)$ in the inflationary universe can be interpreted to be the result of the Brownian motion of the field φ induced by the conversion of quantum fluctuations of that field into a quasiclassical field $\varphi(x)$. For any given mode with fixed p, this conversion occurs whenever the physical momentum $k \sim p e^{-Ht}$ becomes comparable to H. A "freezing" of the amplitude of the field $\psi_p(t)$ then occurs; see (7.3.8). Due to a phase mismatch e^{ipx}, waves with different momenta contribute to the classical field $\varphi(x)$ with different signs, and this also shows up in Eq. (7.3.9), which characterizes the variance in the random distribution of the field that arises at the time of inflation. As in the standard diffusion problem for a particle undergoing Brownian motion, the mean squared particle distance from the origin is directly proportional to the duration of the process (7.3.12).

At any given point, the diffusion of the field φ can conveniently be described by the probability distribution $P_c(\varphi, t)$ to find the field φ at that point and at a given instant of time t. The subscript c here serves to indicate the fact that this distribution, as can readily be shown, also corresponds to the fraction of the original coordinate volume d^3x (7.2.3) filled by the field φ at time t. The evolution of the distribution of the massless field φ in the inflationary universe can be found by solving the diffusion equation [204, 134, 135]:

$$\frac{\partial P_c(\varphi, t)}{\partial t} = D\frac{\partial^2 P_c(\varphi, t)}{\partial \varphi^2} \tag{7.3.17}$$

To determine the diffusion coefficient D in (7.3.17), we take advantage of the fact that

$$\langle\varphi^2\rangle \equiv \int \varphi^2 P_c(\varphi, t)\; d\varphi = \frac{H^3}{4\pi^2}\,t\,.$$

Differentiating this relation with respect to t and using (7.3.17), we obtain

$$D = \frac{H^3}{8\pi^2}$$

It is readily shown that the solution of (7.3.17) with initial condition $P_c(\varphi, 0) = \delta\ (\varphi)$ is a Gaussian distribution

$$P_c(\varphi, t) = \sqrt{\frac{2\pi}{H^3 t}}\exp\left(-\frac{2\pi^2\varphi^2}{H^3 t}\right), \tag{7.3.18}$$

with dispersion squared $\Delta^2 = \langle\varphi^2\rangle = \frac{H^3 t}{4\pi^2}$ (7.3.12).

When we consider the production of a massive classical scalar field with mass $|m^2| \ll H^2$, the diffusion coefficient D, which is related to the rate at which quantum fluctuations with $k > H$ are transferred to the range $k < H$, remains unchanged, since the contribution to $\langle\varphi^2\rangle$ from modes with $k \sim H$ does not depend on m for $|m^2| \ll H^2$. For the same reason, $\langle\varphi^2\rangle$ of (7.3.13) grows in just the same way as for a massless field, as given by (7.3.12). But subsequently, the long-wave classical field φ, which appears during the first stages of the process, begins to decrease as a result of the slow roll down toward the point $\varphi = 0$, in accordance with the classical equation of motion

$$\ddot{\varphi} + 3H\dot{\varphi} = -\frac{dV}{d\varphi} = -m^2\varphi. \tag{7.3.19}$$

This finally leads to stabilization of the quantity $\langle\varphi^2\rangle$ at its limiting value $\frac{3H^4}{8\pi^2 m^2}$ (7.3.13). To describe this process, we must write the diffusion equation in a more general form [205]:

$$\frac{\partial P_c}{\partial t} = D\frac{\partial^2 P_c}{\partial\varphi^2} + b\frac{\partial}{\partial\varphi}\left(P_c\frac{dV}{d\varphi}\right), \tag{7.3.20}$$

where as before $D = \frac{H^3}{8\pi^2}$ and b is the mobility coefficient, defined by the equation $\dot{\varphi} = -b\frac{dV}{d\varphi}$. Using (7.3.19) for the slowly varying field φ ($\ddot{\varphi} \ll 3H\dot{\varphi}$), we obtain

$$\frac{\partial P_c}{\partial t} = \frac{H^3}{8\pi^2}\frac{\partial^2 P_c}{\partial\varphi^2} + \frac{1}{3H}\frac{\partial}{\partial\varphi}\left(P_c\frac{dV}{d\varphi}\right). \tag{7.3.21}$$

This equation was first derived by Starobinsky [134] by another, more rigorous method; a more detailed derivation can be found in [186, 135,

132, 206]. Solution of this equation for the case $V(\varphi) = V(\varphi) = \frac{m^2}{2}\varphi^2 + V(0)$ actually leads to the distribution $P_c(\varphi, t)$ with dispersion determined by (7.3.13). Solutions that are valid for a more general class of potentials $V(\varphi)$ will be discussed in the next section in connection with the problem of tunneling in the inflationary universe.

To conclude, we note that in deriving Eq. (7.3.21), it has been assumed that H is independent of the field φ. More generally, Eq. (7.3.21) can be written in the form

$$\frac{\partial P_c}{\partial t} = \frac{\partial^2}{\partial \varphi^2}\left(\frac{H^3 P_c}{8\pi^2}\right) + \frac{\partial}{\partial \varphi}\left(\frac{P_c}{3H}\frac{dV}{d\varphi}\right). \tag{7.3.22}$$

Strictly speaking, this equation also holds only when variations in the field φ are small enough that the back reaction of inhomogeneities of the field on the metric is not too large. Nevertheless, with the help of this equation, one can obtain important information on the global structure of the universe; see Chapter 10.

7.4 Tunneling in the inflationary universe

The first versions of the inflationary universe scenario were based on the theory of decay of a supercooled vacuum state $\varphi = 0$ due to tunneling with creation of bubbles of the field φ at the time of inflation [53–55]. The theory of such processes in Minkowski space, which is discussed in Chapter 4, turns out to be inapplicable to the most interesting situations, where the curvature of the effective potential near its local minimum is small compared to H^2. Coleman and De Luccia [207] have developed a Euclidean theory of tunneling in de Sitter space, but the general applicability of this theory to the study of tunneling during inflation was confirmed only very recently [208]. One of the main problems was that according to [207] both the scalar field φ inside a bubble and the metric $g_{\mu\nu}(x)$ experience a quantum jump. However, in certain situations, there is a barrier only in the direction of change of the field φ. The analog of this problem is that of the motion of a particle in the (x, y)-plane in a potential $V(x, y)$ having the form of a barrier only in the x-direction. A particle encountering the barrier in this situation tunnels through in the x-direction, but it may continue to move undisturbed along its classical trajectory in the y-direction. To investigate tunnelling under these circumstances, in general one cannot simply transform to imaginary time (imaginary energy); instead, one must undertake an honest solution of the Schrödinger equation for the wave function $\Psi(x, y)$, allowing for the fact that some of the components of the particle momentum may have an imaginary part.

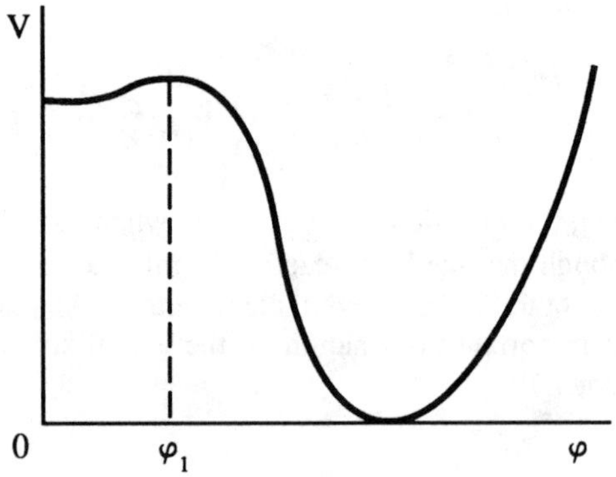

Figure 33. The potential $V(\varphi)$ used by Hawking and Moss to study tunneling.

Another problem was an ambiguity of interpretation of the results of Euclidean approach to tunneling. Let us consider e.g. a theory with the effective potential with a small local minimum a $\varphi = 0$, such that $m^2 = \left.\frac{d^2V}{d\varphi^2}\right|_{\varphi=0} \ll H^2$; tunneling in such a theory was studied by Hawking and Moss [121]. Their expression for the probability of tunneling from the point $\varphi = 0$ through a barrier with a maximum at the point φ_1 (Fig. 33) looks like

$$P \sim A \exp\left[-\frac{3M_P^4}{8\pi}\left(\frac{1}{V(0)} - \frac{1}{V(\varphi_1)}\right)\right], \tag{7.4.1}$$

where A is some multiplicative factor with dimensionality m^4. Hawking and Moss assumed in deriving this equation that by virtue of the "no hair" theorem for de Sitter space (see Section 7.2), tunneling in an exponentially expanding de Sitter space (7.2.3) should occur in just the same way it occurs in a closed space (7.2.5). Tunneling in the latter case is most likely at the throat of the hyperboloid (i.e., at $t = 0$, $a = H^{-1}$), while according to [207], a description of that tunneling requires that we calculate the appropriate action in the Euclidean version of the space (7.2.5), that is, on a sphere S^4 of radius $H^{-1}(\varphi)$. Since our concern is with tunneling in which $H(\varphi)$ increases (i.e., a decreases), which is classically forbidden, the preceding argument against a Euclidean approach to this case does not apply. A calculation of the action S on the sphere leads to the quantity

$$S_E(\varphi) = -\frac{3M_P^4}{8\pi V(\varphi)}. \tag{7.4.2}$$

Adhering to the ideology developed in the work of Coleman and De Luccia, Hawking and Moss asserted that the probability of tunneling is proportional to $\exp\left(S_E(0) - S_E(\varphi)\right)$. This also leads to Eq. (7.4.1), but the contribution to the action from the bubble walls was not taken into account — in other words, they treated purely homogeneous tunneling suddenly taking place over all space [121]. This result was later "confirmed" by numerous authors; however, simultaneous tunneling throughout an entire exponentially large universe seems quite unlikely.

In order to study this problem in more detail, a Hamiltonian approach to the theory of tunneling at the time of inflation was developed, and has succeeded in showing that the probability of homogeneous tunneling over an entire inflationary universe is actually vanishingly small [186]. Hawking and Moss themselves later remarked that their result should be interpreted not as the probability of homogeneous tunneling throughout the entire universe, but as the probability of tunneling which looks homogeneous only on some scale $l \gtrsim H^{-1}$ [209]. They argued that bubble walls and other inhomogeneities should have no effect on tunneling, due to the "no hair" theorem for de Sitter space (see Section 7.2).

The validity of that argument, and in fact the overall applicability of the Euclidean approach to this problem, was open to doubt. Only much later was it learned that when $m^2 \ll H^2$, the contribution to the Euclidean action from gradients of the field φ is small [186] (in contrast to the situation in a Minkowski space, where this contribution is of the same order as that of the potential energy of the field φ), and that tunneling in this instance is effectively one-dimensional (basically occurring because of a change in the scalar field). Equation (7.4.1) was thereby partially justified. But a genuine understanding of the physical essence of this phenomenon was not achieved until an approach to the theory of tunneling based on the diffusion equation (7.3.21) was developed [134, 135].

The fundamental idea is that for tunneling to occur, it suffices to have a bubble with a field exceeding φ_1 and radius

$$r > H^{-1}(\varphi_1) = \sqrt{\frac{3M_P^2}{8\pi V(\varphi_1)}} .$$

Further evolution of the field φ inside this bubble does not depend on what goes on outside it; in other words, the field will start to roll down to the absolute minimum of $V(\varphi)$ with $\varphi > \varphi_1$. It only remains to evaluate the probability that a region of this type will form — but that is exactly the problem we studied in the previous section!

As we have already stated, the distribution $P_c(\varphi, t)$ actually characterizes that fraction of the original coordinate volume d^3x (7.2.3) which at time t contains the field φ; the latter is homogeneous on a scale $l \gtrsim H^{-1}$. The problem of tunneling at the time of inflation thereby reduces to the solution of the diffusion equation (7.3.21) with initial condition $P_c(\varphi, 0) = \delta(\varphi)$.

At this point, we must distinguish between two possible regimes.

1) In the initial stage of this process, the dispersion $\sqrt{\langle\varphi^2\rangle}$ grows as $\frac{H}{2\pi}\sqrt{Ht}$ (7.3.12). If at that stage it becomes larger in magnitude than φ_1, which characterizes the position of the local maximum of $V(\varphi)$, the process will proceed as if there were no barrier at all [127]. In that event, diffusion will end when the field φ encounters a steep slope in $V(\varphi)$, where the rate of diffusive growth of the field drops below the rate of classical rolling. Under typical conditions, the diffusion stage lasts for a time

$$t \sim \frac{4\pi^2\varphi_1^2}{H^3}, \tag{7.4.3}$$

and the typical shape of the regions within which the field φ exceeds some given value (say φ_1) is far from that of a spherical bubble.

2) If the dispersion stops growing when $\sqrt{\langle\varphi^2\rangle} \ll \varphi_1$, the distribution $P_c(\varphi, t)$ will become quasistationary. It should then be possible to find it by putting $\frac{\partial P_c(\varphi, t)}{\partial t} = 0$ in Eq. (7.3.21), or in the more general equation (7.3.22). To clarify the physical meaning of solutions of these equations, it is convenient to rewrite Eq. (7.3.22) in the form

$$\frac{\partial P_c}{\partial t} = -\frac{\partial j_c}{\partial \varphi}, \tag{7.4.4}$$

$$-j_c = \frac{1}{3}\sqrt{\frac{3M_P^2}{8\pi V}}\left[\frac{8V^2}{3M_P^4}\frac{\partial P_c}{\partial \varphi} + P_c\frac{dV}{d\varphi}\left(1 + \frac{4V}{M_P^4}\right)\right]. \tag{7.4.5}$$

Here we have introduced the probability current $j_c(\varphi, t)$ in $(\varphi,\ t)$-space [205], so that Eq. (7.4.4) takes on the standard form of the continuity equation for the probability density $P_c(\varphi, t)$. Consideration of the standard condition $\frac{\partial P_c}{\partial t} = 0$ amounts to examination of the case in which the probability current is constant for all φ from $-\infty$ to ∞. As a rule, there are no reasonable initial conditions under which an unattenuated, nonvanishing diffusion current $j_c = \text{const} \neq 0$ arises between $\varphi = -\infty$ and $\varphi = +\infty$ (see [135], however). Furthermore, the diffusion process itself is usually feasible only within certain limited ranges of variation of the field φ (namely, where $\frac{d^2V}{d\varphi^2} \ll H^2$ and $V(\varphi) \ll M_P^4$). Outside these zones, the first (diffusion) term in (7.4.5) does not appear, and if the potential $V(\varphi)$ is an even function of φ, Eq. (7.4.5) implies that P_c must be an odd function of φ, which is impossible, inasmuch as $P_c(\varphi, t) \geq 0$. For all of these reasons, we will consider only the case $j_c = 0$ (in this regard, see also Chapter 10).

When $j_c = 0$, and also when $V(\varphi) \ll M_P^4$, Eq. (7.4.5) reverts to a very simple form,

$$\frac{\partial \ln P_c}{\partial \varphi} = -\frac{3M_P^4}{8V^2(\varphi)}\frac{dV}{d\varphi}, \tag{7.4.6}$$

whereupon

$$P_c = N\exp\left(\frac{3M_P^4}{8V(\varphi)}\right), \tag{7.4.7}$$

where N is a normalizing factor such that $\int P_c\, d\varphi = 1$. In the present instance, where the rms deviation of the field is much less than the width of the potential well ($\sqrt{\langle\varphi^2\rangle} \ll \varphi_1$), the function $\exp\left(\frac{3M_P^4}{8V(\varphi)}\right)$ has a clear-cut maximum at $\varphi = 0$, and therefore, to within an unimportant subexponential factor,

$$P_c(\varphi) = \exp\left(-\frac{3M_P^4}{8}\left(\frac{1}{V(0)} - \frac{1}{V(\varphi)}\right)\right). \qquad (7.4.8)$$

According to (7.4.8), the probability that the field at a given point (or more precisely, in a neighborhood of size $l \gtrsim H^{-1}$ at a given point) will equal φ_1 is given precisely by the exponential term in the Hawking–Moss equation (7.4.1). This is not just a coincidence, since the mean diffusion time from $\varphi = 0$ to $\varphi = \varphi_1$, that is, the mean time that tunneling goes on at a given point, is in fact proportional to $P_c(\varphi_1)$. The corresponding result in the theory of Brownian motion is well known [210]; for the present case, it was derived in [134, 135]. Its physical meaning is most easily understood if we consider motion along a Brownian trajectory at (approximately) constant speed (as happens here when $H(\varphi) \sim$ const). The value of $P_c(\varphi)$ indicates the number density of points on this trajectory at which the value of the field is equal to φ. This means that the mean time τ to move from the point $\varphi = 0$ to the point $\varphi = \varphi_1$ along a Brownian trajectory is proportional to $[P_c(\varphi)]^{-1}$, and consequently the tunneling (diffusion) probability P per unit time τ is proportional to $P_c(\varphi)$.

Strictly speaking, the tunneling process is not stationary, but if the time required for relaxation to the quasistationary state is much less than the time for tunneling, then Eq. (7.4.8) will provide a good representation of the distribution $P_c(\varphi)$ This condition is satisfied if

$$\frac{3M_P^4}{8}\left(\frac{1}{V(0)} - \frac{1}{V(\varphi_1)}\right) \gg 1 . \qquad (7.4.9)$$

One can readily show that this is equivalent to requiring that $\sqrt{\langle\varphi^2\rangle} \ll \varphi_1$. In the present context, the probability of forming large nonspherical regions of a field $\varphi > \varphi_1$ that are bigger than $H^{-1}(\varphi_1)$ is much lower than that of forming spherical bubbles of the field φ.

As an example, consider the theory with effective potential

$$V(\varphi) = V(0) + \frac{m^2}{2}\varphi^2 - \frac{\lambda}{4}\varphi^4 . \qquad (7.4.10)$$

For this theory, $\varphi_1 = \frac{m}{\sqrt{\lambda}}$, and Eq. (7.4.1) for $V(\varphi_1) - V(0) \ll V(0)$ becomes

$$P_c \sim \exp\left[-\frac{3M_P^4 m^4}{32\lambda V^2(0)}\right] = \exp\left[-\frac{2}{3\lambda}\left(\frac{m}{H}\right)^4\right], \qquad (7.4.11)$$

while (7.4.9), together with the condition that $m^2 \ll H^2$, may be cast in the form

$$\sqrt{\lambda} < \frac{m^2}{H^2} \ll 1 . \qquad (7.4.12)$$

A more detailed study of the solutions of Eq. (7.3.22) makes it possible to obtain expressions for the mean duration of tunneling which hold both for $\sqrt{\langle\varphi^2\rangle} \gg \varphi_1$ and $\sqrt{\langle\varphi^2\rangle} \ll \varphi_1$ [135]. Most important for us here has been the elucidation of the general properties of phase transitions at the time of inflation, a subject discussed in more detail elsewhere [186]. One of the most surprising features of such phase transitions is the possibility of diffusion from one local minimum of $V(\varphi)$ to another with an *increase* in energy density [211]. This effect and related ones are extremely important for an understanding of the global structure of the universe. We shall return to this question in Chapter 10.

Thus, with a stochastic approach, one can justify the Hawking–Moss equation (7.4.1) [121] and confirm their interpretation of this equation [209]. On the other hand, this same approach has provided a means of appreciating the limits of applicability of Eq. (7.4.1). The "derivation" of this result given in [121] imposed no constraints on the form of the potential $V(\varphi)$, and it was not clear why tunneling should occur through the nearest maximum of $V(\varphi)$, rather than directly to its next minimum. The answer to this last question is obvious within the context of our present approach, and it is also clear that Eq. (7.4.1) itself is only valid if the curvature of $V(\varphi)$ is much less than H^2 over the whole domain of variation of φ from 0 to φ_1.

Another important observation to be made in studying the theory of tunneling in the inflationary universe relates to the properties of the walls of bubbles of a new phase. In Minkowski space, the total energy of a bubble of a new phase that is created from vacuum is exactly zero. As the bubble grows, so does its negative energy, which is proportional to its volume $\sim \frac{4}{3}\pi r^3 \varepsilon$, and which is related to the energy gain ε realized in the transition to a new phase. At the same time (and the same rate), the positive bubble-wall energy grows as $4\pi r^2 \sigma(t)$, where σ is the surface energy density of the bubble. These two terms cancel, which is only possible because the surface energy is also proportional to r, the reason being that the speed of a wall approaches the speed of light, while its thickness decreases. Thus, even if the thin-wall approximation were inappropriate in a description of the bubble creation process, it could be usable in a description of its subsequent evolution [212, 213]. Formally, this occurs because any bubble of the field φ created from vacuum by O(4)-symmetric tunneling can be described by some function of the form

$$\varphi = \varphi(\mathbf{r}^2 - t^2)\,; \tag{7.4.13}$$

see [180]. If this bubble has a characteristic initial size of r_0 at $t = 0$, then the field at large t will reach a value $\varphi(0)$ at a distance

$$\Delta r = \frac{r_0^2}{2r} \sim \frac{r_0^2}{2t} \tag{7.4.14}$$

from the bubble boundary (i.e., from the sphere on which $\varphi(\mathbf{r}^2 - t^2) = \varphi(r_0^2) \sim 0$). With time, then, the wall thickness quickly decreases.

In the inflationary universe, everything is completely different. The total energy of the field φ within a bubble does not vanish, and is not conserved as the universe expands. This is attributable to the same gravitational forces that drive the exponential growth of the total energy of the scalar field at the time of inflation ($E \sim V(\varphi)\, a^3(t) \sim V(\varphi)\, e^{3Ht}$). Tunneling results from the formation of perturbations $\delta\varphi(x)$ with wavelengths $l \gtrsim H^{-1}$. All gradients of these perturbations are very small, and do not affect their evolution. This is also the reason why in the final analysis the Hawking–Moss equation, neglecting the contribution of boundary terms in the Euclidean action, is found to be correct. In the study of bubbles engendered by the foregoing mechanism, therefore, the thin-wall approximation is often inapplicable at any stage of bubble evolution. But if the regions that are formed contain matter in many different phase states, then the domain walls that appear between these regions in the late stages of inflation, or after inflation has terminated, can actually become thin. The powerful methods developed in [212, 213] can be utilized to investigate the structure of the universe in the vicinity of such regions.

7.5 Quantum fluctuations and the generation of adiabatic density perturbations

We now continue our study of perturbations of a scalar field with exponentially large wavelength that come into being during an inflationary stage. From the standpoint of quantization in the coordinate system (7.2.3), the wavelengths of these fluctuations do not grow (p = const in Eq. (7.3.4)), and they differ but little from conventional vacuum fluctuations. In particular, one can calculate the corrections to the energy-momentum tensor $g_{\mu\nu}V(\varphi)$ that are associated with these fluctuations; in the stationary state (φ = const), these are [202, 203]

$$\Delta T_{\mu\nu} = \frac{1}{4}\langle\varphi^2\rangle m^2 g_{\mu\nu} = \frac{3H^4}{32\pi^2} g_{\mu\nu} \tag{7.5.1}$$

regardless of the mass of the field φ (for $m^2 \ll H^2$). These corrections have a relativistically invariant form (despite the presence of the Hawking "temperature" $T_H = \frac{H}{2\pi}$ in de Sitter space).

But as we have already pointed out, from the point of view of a stationary observer armed with measuring rods that do not stretch during inflation of the universe, fluctuations of a scalar field that have wavelengths greater than the distance to the horizon ($k^{-1} \gtrsim H^{-1}$) look like a classical field $\delta\varphi$ that is weakly inhomogeneous on scales $l \gtrsim H^{-1}$. These fluctuations give rise to density inhomogeneities on an exponentially large scale. During inflation, the magnitude of these inhomogeneities is

$$\delta\rho \sim V'\,\delta\varphi\,, \tag{7.5.2}$$

where $V' = \frac{dV}{d\varphi}$. In the last stages of inflation, an ever increasing fraction of the field energy is tied up in the kinetic energy of the field φ, rather than in $V(\varphi)$. This energy then transforms into heat, and energy density inhomogeneities $\delta\rho$ lead to temperature inhomogeneities δT. The original density inhomogeneities (7.5.39) are thereby transformed into hot-plasma density inhomogeneities, and then into cold-matter density inhomogeneities. The corresponding density inhomogeneities result in so-called *adiabatic* perturbations of the metric, in contrast to the *isothermal* perturbations associated with inhomogeneities of the metric that arise at constant temperature.

The appearance of long-wave density (metric) perturbations is necessary for the subsequent formation of the large scale structure of the universe (galaxies, clusters of galaxies, voids, and so on). Another possible mechanism for generating density perturbations is related to the theory of cosmic strings produced during phase transitions in a hot universe. But it is very difficult to get along without an inflationary stage, and therefore the prospect of obtaining inhomogeneities of the type required simply by virtue of quantum effects at the time of inflation, without invoking any additional mechanisms, seems especially interesting. The fact that the contribution to $\langle\varphi^2\rangle$ (7.3.12) due to integration over the fixed interval $\Delta \ln \frac{k}{H}$ is independent of k leads to a flat spectrum $\delta\rho(k)$ (7.5.2) which does not depend on k (as the momentum varies on a logarithmic scale). This is exactly the sort of spectrum suggested earlier by Harrison and Zeldovich [76] (see also [214]) as the initial perturbation spectrum required for the subsequent formation of galaxies. If we normalize this spectrum so that $\delta\rho(k)$ denotes the contribution to $\delta\rho$ from all perturbations per unit interval in $\ln \frac{k}{H}$, then the desired spectral amplitude should be

$$\frac{\delta\rho(k)}{\rho} \sim 10^{-4}\text{–}10^{-5} \tag{7.5.3}$$

on a galactic scale ($l_g \sim 10^{22}$ cm at the present epoch; $l_g \sim 10^{-5}$ cm at the instant that inflation began). Notice, however, that rather than referring to perturbations $\delta\rho$ at the inflationary stage (7.5.2), condition (7.5.3) relates to their progeny at a later stage, after reheating of the universe, when its equation of state becomes $p = \frac{\rho}{3}$ (or $p = 0$, when cold nonrelativistic matter predominates). The question of how these perturbations actually relate to the initial perturbations (7.5.2) is a very complicated one. Some important steps in the development of a theory of adiabatic density perturbations formed during the exponential expansion stage of the universe are to be found in [101, 215–217]. The corresponding problem for the inflationary universe scenario was first solved by Mukhanov and Chibisov [107] in their investigation of the Starobinsky model [52]. The quantity $\frac{\delta\rho}{\rho}$ for the new inflationary universe scenario was calculated by four other groups practically simultaneously [114]. The results obtained by all of these authors, using different approaches, agreed to within a numerical factor $C \sim O(1)$:

$$\frac{\delta\rho(k)}{\rho} = C\, \frac{H(\varphi)\,\delta\varphi(k)}{\dot{\varphi}}\bigg|_{k \sim H} . \tag{7.5.4}$$

What this expression means is that in order to calculate $\frac{\delta\rho(k)}{\rho}$ on a logarithmic scale in k, one should calculate the value of the function $\frac{H[\varphi(t)]}{\dot{\varphi}(t)}$ at a time when the corresponding wavelength k^{-1} is of the order of the distance to the horizon H^{-1} (that is, when the field $\delta\varphi(k)$ becomes quasiclassical). For $\delta\varphi(k)$ here we can take the rms value defined by (see (7.3.12))

$$[\delta\varphi(k)]^2 = \frac{H^2(\varphi)}{4\pi^2} \int_{\ln \frac{k}{H}}^{\ln \frac{k}{H} + 1} d \ln \frac{k}{H} = \frac{H^2(\varphi)}{4\pi^2} , \tag{7.5.5}$$

or in other words

$$|\delta\varphi(k)| = \frac{H(\varphi)}{2\pi} . \tag{7.5.6}$$

In the final analysis, these same results are found to be correct for the chaotic inflation scenario as well [218].

It would be hard to overestimate the significance of the first papers on the density perturbations produced during inflation [114]. However, the validity of some of the assumptions made in these papers is not obvious. Furthermore, it was not entirely clear what the connection was between the density perturbations occurring during the inflationary stage (7.5.2) and Eq. (7.5.4), and in fact the value of the parameter C in the latter equation differed somewhat in the different papers. This situation engendered an extensive literature on the problem; for a review, see [219]. In our opinion, a final clarification of the situation was particularly facilitated by Mukhanov [218]. In what follows, we will describe the basic ingredients of his work, and use his results to obtain an equation for $\frac{\delta\rho}{\rho}$ that will be valid for a large class of inflationary models.

Consider a region of the inflationary universe of initial size $\Delta l \gtrsim H^{-1}$, containing a sufficiently homogeneous field φ ($\partial_i\varphi\, \partial^i\varphi \ll V(\varphi)$). All initial inhomogeneities of this field die out exponentially, and the total field in this region can therefore be put in the form

$$\varphi(\boldsymbol{x}, t) \to \varphi(t) + \delta\varphi(\boldsymbol{x}, t) \,, \tag{7.5.7}$$

where inhomogeneities $\delta\varphi(\boldsymbol{x}, t)$ appear because of long-wave fluctuations that are generated with $k \lesssim H$. The leading contribution to $\delta\varphi(\boldsymbol{x}, t)$ comes from fluctuations with exponentially long wavelengths. Therefore, the main contribution to inhomogeneities in the mean energy-momentum tensor T_μ^ν on the large scales that we are interested in come from terms like $\partial_0\varphi \cdot \partial_0[\delta\varphi(\boldsymbol{x}, t)]$ or $\delta\varphi(\boldsymbol{x}, t) \cdot \frac{dV}{d\varphi}$, rather than from spatial gradients $(\partial_i[\delta\varphi(\boldsymbol{x}, t)])^2$ (the second of these is in fact the leading contribution at the time of inflation; see (7.5.2)). This implies that δT_μ^ν is diagonal to first order in $\delta\varphi$. For perturbations of this type, the corresponding perturbations of the metric in a flat universe can be represented by [220]

$$ds^2 = (1 + 2\Phi)\, dt^2 - (1 - 2\Phi)\, a^2(t)\, d\boldsymbol{x}^2 \,. \tag{7.5.8}$$

The function $\Phi(\boldsymbol{x}, t)$ plays a role similar to that of the Newtonian potential used to describe weak gravitational fields (compare the metric (7.5.8) with the Schwarzschild metric (7.2.8)). The coordinate system (7.5.8) is more convenient for the investigation of density perturbations than the more frequently used synchronous system [65], since after a synchronous system is chosen through the condition $\delta g_{i0} = 0$, one still has the freedom to change the coordinate system; this leads to the existence of two nonphysical perturbation modes, which makes the calculations and their interpretation rather complicated and ambiguous. However, these modes do not contribute to $\Phi(\boldsymbol{x}, t)$. Density inhomogeneities of the type we are considering, with wavelengths $k^{-1} > \mathrm{H}^{-1}$, are related to the function $\Phi(\boldsymbol{x}, t)$ in a very simple way,

$$\frac{\delta\rho}{\rho} = -2\Phi . \tag{7.5.9}$$

A more detailed discussion of the use of the relativistic potential $\Phi(\boldsymbol{x}, t)$ may be found in [220–222, 133]. Linearizing the Einstein equations and the equation for the field $\varphi(\boldsymbol{x}, t)$ in terms of $\delta\varphi$ and Φ, one can obtain a system of differential equations for $\delta\varphi$ and Φ:

$$\ddot{\Phi} + \left(\frac{\dot{a}}{a} - 2\frac{\ddot{\varphi}}{\dot{\varphi}}\right)\dot{\Phi} - \frac{1}{a^2}\Delta\Phi + 2\left(\frac{\ddot{a}}{a} - \left(\frac{\dot{a}}{a}\right)^2 - \frac{\dot{a}}{a}\frac{\ddot{\varphi}}{\dot{\varphi}}\right)\Phi = 0 , \tag{7.5.10}$$

$$\frac{1}{a}(a\Phi)^{\cdot}_{,\beta} = \frac{4\pi}{\mathrm{M}_{\mathrm{P}}^2}(\dot{\varphi}\,\delta\varphi)_{,\beta} , \tag{7.5.11}$$

$$\delta\ddot{\varphi} + 3\frac{\dot{a}}{a}\delta\dot{\varphi} - \frac{1}{a^2}\Delta\delta\varphi + \frac{d^2\mathrm{V}}{d\varphi^2}\delta\varphi + 2\frac{d\,\mathrm{V}}{d\varphi}\Phi - 4\dot{\varphi}\dot{\Phi} = 0 . \tag{7.5.12}$$

Here $\varphi(t)$ and $a(t)$ are the solutions of the unperturbed equations (see Section 1.7), and a dot signifies differentiation with respect to time. Using one of the consequences of the Einstein equations for $a(t)$,

$$\left(\frac{\dot{a}}{a}\right)^{\cdot} = -\frac{4\pi}{\mathrm{M}_{\mathrm{P}}^2}\dot{\varphi}^2 , \tag{7.5.13}$$

Eq. (7.5.10) can be cast in the form

$$u'' - \Delta u - \frac{(a'/a^2\varphi'')''}{(a'/a^2\varphi')} u = 0 , \tag{7.5.14}$$

where $u = \frac{a}{\varphi'}\Phi$, and primes in this (and only in this) equation stand for differentiation with respect to the conformal time $\eta = \int \frac{dt}{a(t)}$. In the long-wave limit ($k \ll H, k^2 \ll \frac{d^2V}{d\varphi^2}$), the solution of Eq. (7.5.14) can be put in the form

$$\Phi = C\left(1 - \frac{\dot{a}}{a^2}\int_0^t a\,dt\right), \tag{7.5.15}$$

where C is some constant, and $Ht \gg 1$. Thereupon, making use of (7.5.11), we obtain

$$\delta\varphi = C\,\dot{\varphi}\cdot\frac{1}{a}\int_0^t a\,dt. \tag{7.5.16}$$

The desired result follows from (7.5.15) and (7.5.16), namely the relationship between long-wave fluctuations of the field φ, perturbations of the metric Φ, and density inhomogeneities [218]:

$$\frac{\delta\rho}{\rho} = -2\Phi = -2\left[\frac{a}{\int_0^t a\,dt}\,\frac{\delta\varphi}{\dot{\varphi}}\right]\left(1 - \frac{\dot{a}}{a^2}\int_0^t a\,dt\right). \tag{7.5.17}$$

The quantity in square brackets is the constant C of (7.5.15), the value of which can be computed at any stage of inflation. This is most conveniently done when the wavelength of a perturbation $\delta\varphi(k)$ is equal to the distance to the horizon, $k \sim H$. The amplitude at that time can be estimated with the help of Eq. (7.5.6).

We now make use of the preceding results to compare Eqs. (7.5.2) and (7.5.4), and we apply these results to the calculation of $\frac{\delta\rho}{\rho}$ in the simple models. One can easily verify that during the stage of inflation, when $\dot{H} \ll H^2$, $\ddot{H} \ll H^3$, and $Ht \gg 1$,

$$\frac{a}{\int_0^t a\,dt} = H(t)\left(1 + \frac{\dot{H}}{H^2}\left[1 + O\left(\frac{\dot{H}}{H^2}, \frac{\ddot{H}}{H^3}\right)\right]\right). \tag{7.5.18}$$

The expression in square brackets in (7.5.17) is thus equal to

$$C = H(\varphi(t))\frac{\delta\varphi}{\dot{\varphi}}. \tag{7.5.19}$$

On the other hand, Eq. (7.5.18) implies that during the inflationary stage, the parenthesized expression in (7.5.17) equals $\frac{\dot{H}}{H^2} \ll 1$. In that event, it is readily verified using (7.5.17) and (7.5.19) that density inhomogeneities during inflation are given by

$$\frac{\delta\rho}{\rho} \approx \frac{\delta V}{V} = \frac{V'}{V}\delta\varphi, \tag{7.5.20}$$

as one would find from (7.5.2), and not by (7.5.4). The difference between (7.5.20) and (7.5.4) lies in a small factor that is $O(\frac{\dot{H}}{H^2}) \ll 1$.

However, if the universe is hot ($a \sim t^{1/2}$) or cold ($a \sim t^{2/3}$), Eqs. (7.5.17) and (7.5.19) lead to Eq. (7.5.4), with $C = -4/3$ and $C = -6/5$ for these two respective cases [218, 220]. If for $\delta\varphi(k)$ we then take $\frac{H}{2\pi}$ as in (7.5.6) in order to find the rms value of $\frac{\delta\rho}{\rho}$ per unit interval in $\ln\frac{k}{H}$, we obtain

$$\frac{\delta\rho(k)}{\rho} = C\left.\frac{[H(\varphi)]^2}{2\pi\dot{\varphi}}\right|_{k\sim H}. \tag{7.5.21}$$

Expressing $\dot{\varphi}$ and $H(\varphi)$ in terms of $V(\varphi)$ during inflation, we find that at the stage when cold matter predominates (when galaxy formation presumably started),

$$\frac{\delta\rho(k)}{\rho} = \frac{48}{5}\sqrt{\frac{2\pi}{3}}\left.\frac{V^{3/2}}{M_P^3\frac{dV}{d\varphi}}\right|_{k\sim H(\varphi)} \tag{7.5.22}$$

(An irrelevant minus sign has been omitted from (7.5.22)). As an example of the use of this equation, one can obtain the amplitude of density inhomogeneities in the theory $\frac{\lambda}{4}\varphi^4$ for the chaotic inflation scenario,

$$\frac{\delta\rho(k)}{\rho} = \frac{6}{5}\sqrt{\frac{2\pi\lambda}{3}}\left(\frac{\varphi}{M_P}\right)^3\Bigg|_{k\sim H(\varphi)} \tag{7.5.23}$$

If we are to compare (7.5.23) with the value of $\frac{\delta\rho(k)}{\rho}$ on a galactic scale ($l_g \sim 10^{22}$ cm) or on the scale of the horizon ($l_H \sim 10^{28}$ cm), we must follow the behavior of a wave with momentum k during and after inflation. According to (1.7.25), a wave emitted at some value of φ will, over the course of inflation, increase in wavelength by a factor $\exp\left(\pi\frac{\varphi^2}{M_P^2}\right)$. After reheating to a temperature T_R and cooling to a temperature T_γ, where $T_\gamma \sim 3$ K is the present-day temperature of the microwave background radiation, the universe will again typically expand by another factor of $\frac{T_R}{T_\gamma}$. Presuming that reheating takes place immediately after the end of inflation (with $\varphi \sim \frac{M_P}{3}$), T_R will be of order $\left[V\left(\frac{M_P}{3}\right)\right]^{1/4} \sim \frac{\lambda^{1/4}}{10}M_P$. (The final results will be a very weak (logarithmic) function of the duration of reheating and the magnitude of T_R.) Thus, the present wavelength of a perturbation produced at the moment when the scalar field had some value φ is of order

$$\begin{aligned} l(\varphi) &\sim H^{-1}(\varphi)\frac{T_R}{T_\gamma}\exp\left(\frac{\pi\varphi^2}{M_P^2}\right) \\ &\sim M_P^{-1}\left(\frac{M_P}{\varphi}\right)\frac{M_P}{\lambda^{1/4}\,T_\gamma}\exp\left(\frac{\pi\varphi^2}{M_P^2}\right). \end{aligned} \tag{7.5.24}$$

Bearing in mind that 1 GeV corresponds approximately to 10^{13} K, $M_P \sim 10^{-33}$ cm, and $\varphi \sim 5\,M_P$ at the time of interest, we obtain (for $\lambda \sim 10^{-14}$; see below) whereupon

$$\varphi^2 \sim \frac{M_P^2}{\pi} \ln l, \tag{7.5.26}$$

where l here is measured in centimeters.

Equation (7.5.26) tells us that density perturbations on a scale $l_H \sim 10^{28}$ cm are produced at $\varphi = \varphi_H$, where

$$\varphi_H \sim 4.5\ M_P \sim 5.5 \cdot 10^{19}\ \text{GeV}, \tag{7.5.27}$$

and those on a galactic scale $l_g \sim 10^{22}$ cm come into existence at $\varphi = \varphi_g$, where

$$\varphi_g \sim 4\ M_P \sim 5 \cdot 10^{19}\ \text{GeV}. \tag{7.5.28}$$

Equations (7.5.23) and (7.5.26) yield a general equation for $\frac{\delta\rho}{\rho}$ in the theory $\frac{\lambda}{4}\varphi^4$:

$$\frac{\delta\rho}{\rho} \sim \frac{2\sqrt{6}}{5\pi}\sqrt{\lambda}\,\ln^{3/2} l\,(\text{cm}) \tag{7.5.29}$$

with the amplitude of inhomogeneities on the scale of the horizon being

$$\frac{\delta\rho}{\rho} \sim 150\sqrt{\lambda}\,, \tag{7.5.30}$$

while for those on a galactic scale,

$$\frac{\delta\rho}{\rho} \sim 110\sqrt{\lambda}\,. \tag{7.5.31}$$

The spectrum of $\frac{\delta\rho}{\rho}$ is evidently almost flat, increasing slightly (logarithmically) at long wavelengths.

We now discuss in somewhat more detail what the magnitude of the constant λ should be in order for the predicted inhomogeneities to be con-

mically) at long wavelengths.

We now discuss in somewhat more detail what the magnitude of the constant λ should be in order for the predicted inhomogeneities to be consistent with the observational data and the theory of galaxy formation.

Apparently, the most exacting constraints imposed by the cosmological data are not on $\frac{\delta\rho}{\rho}$ itself, but on the quantity A that determines the anisotropy of the microwave background $\frac{\Delta T}{T}$ produced by adiabatic perturbations of the metric [223–227],

$$\left(\frac{\Delta T}{T}\right)_l = \frac{A}{\sqrt{l(l+1)}} \frac{\mathcal{K}_l}{10\sqrt{\pi}}, \qquad (7.5.32)$$

where l is the order of the harmonic in the multipole expansion of $\frac{\Delta T}{T}$ ($l \geq 2$ in (7.5.32)). The relationship between A in (7.5.32) and metric perturbations is [220–222]

$$\frac{\delta\rho(k)}{\rho} = -2\Phi(k) = -\frac{\sqrt{2}}{\pi}\,\alpha\, A(k)\,. \qquad (7.5.33)$$

The numerical factors a and $\mathcal{K}_l$ in (7.5.32) and (7.5.33) depend on the specific assumptions made about the nature of the missing mass in the universe. The magnitude of $\mathcal{K}_l$ is usually of the order of one. As for α, that quantity is 2/3 for a hot universe, and 3/5 for a cold one. In either case,

$$A(k) \approx 16\,\pi \sqrt{\frac{\pi}{3}}\, \left.\frac{V^{3/2}}{M_P^3 \frac{dV}{d\varphi}}\right|_{k \sim H(\varphi)} \,. \qquad (7.5.34)$$

In particular, for the $\frac{\lambda}{4}\varphi^4$ theory,

$$A = \frac{2\sqrt{\lambda}}{3} \ln^{3/2} l(\mathrm{cm}) \sim 1.2\sqrt{\lambda} \ln^{3/2} l(\mathrm{cm}) \sim 6.10^2\sqrt{\lambda} \qquad (7.5.35)$$

on the scale of the horizon. From the observational constraints on $\frac{\Delta T}{T}$, it

$$5.10^{-5} \lesssim A \lesssim 5.10^{-4}, \tag{7.5.36}$$

depending on the physical nature of the dark matter in the universe [227]. Condition (7.5.2) is thus a consequence of (7.5.33) and (7.5.36) as well. To determine the constraints on λ, it is most convenient to use (7.5.35) and (7.5.36) directly:

$$0.5 \cdot 10^{-14} \lesssim \lambda \lesssim 0.5 \cdot 10^{-12}. \tag{7.5.37}$$

From here on, we assume for definiteness that

$$\lambda \sim 10^{-14}, \tag{7.5.38}$$

which is closer to estimates in the context of the theory of galaxy formation in a universe filled with cold dark matter. As the theory of large scale structure in the universe is developed and the observational limits on $\frac{\Delta T}{T}$ are refined [228], this estimate will improve.

Let us now consider another important example, the theory of a massive scalar field with $V(\varphi) = \frac{m^2}{2}\varphi^2$. For a cold Friedmann universe in this case,

$$\frac{\delta\rho(k)}{\rho} = \frac{24}{5}\sqrt{\frac{\pi}{3}}\frac{m}{M_P}\left(\frac{\varphi}{M_P}\right)^2\Bigg|_{k\sim H}. \tag{7.5.39}$$

Likewise, for either a hot or cold universe,

$$A(k) = 4\sqrt{2}\pi\sqrt{\frac{\pi}{3}}\frac{m}{M_P}\left(\frac{\varphi}{M_P}\right)^2\Bigg|_{k\sim H} \tag{7.5.40}$$

In this theory, both φ_H and φ_g are a factor of $\sqrt{2}$ less than in the $\frac{\lambda}{4}\varphi^4$ theory. The analog of Eq. (7.5.29) for the present theory is

$$\frac{\delta\rho}{\rho} \sim 0.8\frac{m}{M_P}\ln l\,[\mathrm{cm}], \tag{7.5.41}$$

and on the scale of the horizon, Eq. (7.5.35) for A becomes

$$A \sim 200 \frac{m}{M_P}, \tag{7.5.42}$$

whence

$$3.10^{12}\ \mathrm{GeV} \sim 2.5 \cdot 10^{-7} M_P \lesssim m \lesssim 2.5 \cdot 10^{-6} M_P \sim 3 \cdot 10^{13}\ \mathrm{GeV}. \tag{7.5.43}$$

Next, consider the more general theory with potential

$$V(\varphi) = \frac{\lambda \varphi^4}{n} \left(\frac{\varphi}{M_P} \right)^{n-4}. \tag{7.5.44}$$

For such a theory,

$$A = 16 \pi \sqrt{\frac{\pi}{3}} \left(\frac{V(\varphi)}{M_P^4} \right)^{\frac{1}{2}} \frac{\varphi}{n M_P} \tag{7.5.45}$$

and the field φ_H is

$$\varphi_H \sim 2 \sqrt{n}\, M_P. \tag{7.5.46}$$

Perturbations on the scale of the horizon are thus characterized by

$$A \sim 32 \pi \sqrt{\frac{\pi}{3n}} \left(\frac{V(\varphi_H)}{M_P^4} \right)^{\frac{1}{2}}. \tag{7.5.47}$$

Specifically, when $A \sim 10^{-4}$, we find from (7.5.46) that in the last stages of inflation, when the structure of the observable part of the universe had been formed, the value of the effective potential was of order

$$V(\varphi_H) \sim 10^{-12} n M_P^4 \sim n \cdot 10^{82}\ \mathrm{g \cdot cm^{-3}}. \tag{7.5.48}$$

The rate of expansion of the universe was then

$$H(\varphi_H) \sim 3 \cdot 10^{-6} \sqrt{n}\, M_P \sim 3.5 \sqrt{n} \cdot 10^{13}\ \mathrm{GeV}; \tag{7.5.49}$$

$$H(\varphi_H) \sim 3\cdot 10^{-6}\sqrt{n}\, M_P \sim 3.5\sqrt{n}\cdot 10^{13}\ \text{GeV}; \tag{7.5.49}$$

that is, the universe increased in size by a factor of e in a time

$$t \sim H^{-1} \sim n^{-1/2}\cdot 10^{-37}\ \text{sec}. \tag{7.5.50}$$

In such a theory, the constant λ should be (for $A \sim 5\cdot 10^{-5}$)

$$\lambda \sim 2.5\cdot 10^{-13}\, n^2 (4n)^{\frac{n}{2}}. \tag{7.5.51}$$

These results give a general impression of the orders of magnitude which might be encountered in realistic versions of the inflationary universe scenario. The estimate of $V(\varphi_H)$ deserves special attention: a similar estimate can also be obtained from an analysis of the theory of gravitational wave production at the time of inflation [117]. In the new inflationary universe scenario, an analogous result implies that *at all stages of inflation*, $V(\varphi)$ should be at least ten to twelve orders of magnitude less than M_P^4 [107, 229–231]. Within the framework of the chaotic inflation scenario, a similar statement is incorrect. The value of A, which is 10^{-4} when $\varphi \sim \varphi_H$, tends to increase in accordance with (7.5.45) at large φ, and the observational data impose no upper limits whatever on $V(\varphi)$. On the other hand, we can derive a rather general constraint on the magnitude of $V(\varphi)$ in the last stages of inflation from (7.5.34). In fact, at the end of inflation, the rate at which the potential energy $V(\varphi)$ decreases becomes large — the energy density $V(\varphi)$ is reduced by a quantity that is $O(V(\varphi))$ within a typical time $\Delta t = H^{-1}$. In other words, the criterion $\dot{H} \ll H^2$ is no longer satisfied. One can readily show that this means that at the end of inflation, $V' \sim \frac{V}{M_P}\sqrt{8\pi}$. In that case, (7.5.34) tells us that the quantity A, which is related to fluctuations of the field φ that are generated during the very last stage of inflation, is given to order of magnitude by

$$A \sim 10\sqrt{\frac{V(\varphi)}{M_P^4}} \tag{7.5.52}$$

With $A \lesssim 10^{-4}$, we then find that at the end of inflation,

$$V \lesssim 10^{-10} M_P^4. \tag{7.5.53}$$

the chaotic inflation scenario.

The formalism that we have employed in this chapter rests on an assumption of the relative smallness of $\frac{\delta\rho}{\rho}$. During the inflationary stage, as a rule, this condition is met. For example, in the $\frac{\lambda}{4}\varphi^4$ theory,

$$\frac{\delta\rho}{\rho} \sim \frac{V'\delta\varphi}{V} \sim \frac{4\delta\varphi}{\varphi} \sim \frac{2H(\varphi)}{\pi\varphi} \sim \frac{\sqrt{\lambda}\,\varphi}{M_P} \ll 1 \tag{7.5.54}$$

for $V(\varphi) \lesssim M_P^4$, $\lambda \ll 1$.

On relatively small scales ($l \sim H^{-1}$), gradient terms $\partial_i(\delta\varphi)\,\partial^i(\delta\varphi) \sim H^4$ make a sizable contribution to $\frac{\delta\rho}{\rho}$. We have not considered these terms, since in the last analysis we were interested in perturbations with exponentially long wavelengths. This contribution is also much less than $V(\varphi)$ when $V(\varphi) \ll M_P^4$.

However, density perturbations produced at large φ become large after inflation. In particular, according to (7.5.23), $\frac{\delta\rho}{\rho} \sim 1$ in the $\frac{\lambda}{4}\varphi^4$ theory for perturbations produced when $\varphi = \varphi^*$, where

$$\varphi^* \sim \lambda^{-1/6} M_P. \tag{7.5.55}$$

By (7.5.25), this means that after inflation ends, the universe only looks like a homogeneous Friedmann space on a scale

$$l_* \lesssim \exp\left(\pi\lambda^{-1/3}\right) \text{cm} \sim 10^{6\cdot 10^4}\ \text{cm}. \tag{7.5.56}$$

for $\lambda \sim 10^{-14}$. This is many orders of magnitude larger than the observable part of the universe, with $l_H \sim 10^{28}$ cm, so for a present-day observer such inhomogeneities would lie beyond his radius of visibility. From the standpoint of the global structure of the universe, however, nonuniformity on scales $l \gg l_*$ is exceedingly important, as we have discussed in Chapter 1. We shall return to this question in Chapter 10.

We make one more remark in closing. We have been accustomed to calling the quantity $l_H \sim 3\,t \sim 10^{28}$ cm the distance to the horizon, as in the usual Friedmann model (see (1.4.11)). But strictly speaking, the distance to the actual particle horizon in the inflationary universe is exponentially large. Denoting this distance by R_H (so as to distinguish it from $l_H \sim 3\,t$), we may use (1.4.10) and (1.7.28) for the $\frac{\lambda}{4}\varphi^4$ theory to obtain

$$R_H \sim M_P^{-1} \exp \frac{\pi}{\sqrt{\lambda}} \sim 10^{10^7}\ \text{cm} \tag{7.5.57}$$

(see also (1.7.39)). Nevertheless, this quantity can only tentatively be called the horizon. The photons which presently enable us to view the universe only permit us to see back to $t \gtrsim 10^5$ years after the end of inflation in our part of the universe, the reason being that the hot plasma that filled the universe at $t \lesssim 10^5$ years was opaque to photons. Thus, the size of that part of the universe accessible to electromagnetic observations is in fact l_H to high accuracy. A similar argument holds for neutrino astrophysics as well. We can proceed a bit further, studying metric perturbations [136]. According to the standard hot universe theory, gravitational waves provide the opportunity to obtain information about any process in the universe that takes place at less than the Planck density, since the universe is transparent to gravitational waves when $T \lesssim M_P$. This is not true in the inflationary universe scenario, however.

Let us consider a gravitational wave with wavelength $l \lesssim l_H$ (since these are the only waves we can study experimentally). At the stage of inflation, when the scalar field φ was equal to φ_H, the wavelength of this gravitational wave would have been of order $l \sim H^{-1} \sim 10^5 M_P^{-1}$ (7.5.48), whereas at $\varphi \gtrsim 1.05\ \varphi_H$, its wavelength would have been less than M_P^{-1}. Quantum fluctuations of the metric on this length scale are so large that no measurements of gravitational waves inside the present horizon (at $l \lesssim l_H$) could give us any information about the structure of the universe with $\varphi \gtrsim 1.05\ \varphi_H$. In that sense, the range $\varphi \gtrsim 1.05\ \varphi_H$, corresponding to scales $l \gtrsim l_H \cdot \frac{M_P}{H} \sim 10^5\, l_H$, is "opaque" even to gravitational waves. Thus, by analyzing perturbations of the metric, we can in principle study phenomena beyond the visibility horizon (at $l > l_H$), but here we cannot progress beyond a factor of $\frac{M_P}{H(\varphi_H)} \sim 10^5$. The energy density at the corresponding epoch (with $\varphi \sim \varphi_H$) was seven orders of magnitude less than the Planck density (7.5.48). What this means is that we cannot obtain information about the initial stages of inflation (with $V(\varphi) \sim M_P^4$) — that is, the present state of the observable part of the universe is essentially independent of the choice of initial conditions in the inflationary universe.

7.6 Are scale-free adiabatic perturbations sufficient to produce the observed large scale structure of the universe?

The creation of the theory of adiabatic perturbations in inflationary cosmology has been an unqualified success. Beginning in 1982, when the theory was constructed in broad outline, theoretical investigations of the formation of large scale structure in the inflationary universe have, as a rule, been based on two assumptions:

1) to high accuracy, the parameter $\Omega = \frac{\rho}{\rho_0}$ is presently equal to unity (the universe is almost flat);

2) initial density perturbations leading to galaxy formation were adiabatic perturbations with a flat (or almost flat) spectrum, $\frac{\delta\rho}{\rho} \sim 10^{-5}$.

The possibility of describing all of the existing data on large scale structure of the universe on the basis of these simple assumptions is quite attractive, but we should recall at this point the analogy between the universe and a giant accelerator. Experience has taught us that the correct description of a large body of diverse experimental data is seldom provided by the simplest possible theory. For example, the simplest description of the weak and electromagnetic interactions would be given by the Georgi–Glashow model [232], which is based on the symmetry group O(3). But the experimental discovery of neutral currents forced us to turn to the far more complicated Glashow–Weinberg–Salam model [1], based on the symmetry group $SU(2) \times U(1)$. The latter contains about 20 different parameters whose values are not grounded in any esthetic considerations at all. For instance, almost all coupling constants in this theory are $O(10^{-1})$, while the coupling constant for interaction between the electron and the scalar (Higgs) field is $2 \cdot 10^{-6}$. The reason for the appearance of such a small coupling constant (just like the reason for the appearance of the constant $\lambda \sim 10^{-14}$ in the simplest versions of the inflationary universe scenario) is as yet unclear.

It seems unlikely that cosmology will turn out to be a much simpler science than elementary particle theory. After all, the number of different types of large-scale objects in the universe (quasars, galaxies, clusters of galaxies, filaments and voids, etc.) is very large. The sizes of these objects form a hierarchy of scales that is absent from the flat spectrum of the initial perturbations. In principle, some of these scales may be related to the properties of the dark matter comprising most of the mass of the universe; see, for example, [224, 235, 236]. Nevertheless, it is not at all obvious how to consistently describe the formation of a large number of diverse large-scale objects, starting with the simple assumptions (1) and (2). The requisite theory encounters a number of difficulties [235] which, while not insurmountable, have nonetheless stimulated a search for alternative versions of the theory of formation of large scale structure; see, for example, [236].

Another potential problem that a theory based on assumptions (1) and (2) may encounter is related to measurements of the anisotropy $\frac{\Delta T(\theta)}{T}$ of the microwave background radiation, where θ is the angular scale of observation. So far, only a dipole anisotropy $\frac{\Delta T}{T}$ associated with the earth's motion through the microwave background has been detected, and neither a quadrupole anisotropy nor a small-angle anisotropy in $\frac{\Delta T}{T}$ has been found at a level $\frac{\Delta T}{T} \gtrsim 2 \cdot 10^{-5}$ [228]. Meanwhile, flat-spectrum adiabatic density perturbations should lead to $\frac{\Delta T}{T} = C \cdot 10^{-5}$ [223–227], where the function $C(\theta) = O(1)$ depends on the angle θ and on the properties of dark matter. The function $C(\theta)$ is especially large at large angles θ. The comparison between experimental constraints on $\frac{\Delta T}{T}$ and theoretical predictions of quadrupole anisotropy is therefore a particularly important question, with a bearing on perturbations $\frac{\delta\rho}{\rho}$ on a scale $l \sim l_H \sim 10^{28}$ cm. The complexity of the situation is exacerbated by the fact that the inflationary universe scenario gives an adiabatic perturbation spectrum that is not perfectly flat. In most models, $\frac{\delta\rho}{\rho}$ grows with increasing l. For example, in a theory with $V(\varphi) \sim \frac{\lambda}{4}\varphi^4$, as we progress from the galactic scale l_g to the size of the horizon l_H, the quantity $\frac{\delta\rho}{\rho}$ increases by a factor of about 1.4 (see (7.5.29) and (7.5.30)), resulting in a concomitant enhancement of the quadrupole anisotropy $\frac{\Delta T}{T}$. An assessment of the predictions of anisotropy $\frac{\Delta T}{T}$ in the simplest versions of the inflationary universe scenario has already made it possible to discard the simplest models of baryonic dark matter, and they cast some doubt upon the validity of models in which the missing mass is concentrated in massive neutrinos [225, 226]. However, in the cold dark matter models, in which the dark matter consists of axion fields [233, 234], Polonyi fields [46, 15], or any weakly interacting nonrelativistic particles, the theoretical estimates of $\frac{\Delta T}{T}$ are perfectly consistent with current observational limits [225, 226].

Thus, it remains possible that a theory of the formation of the large scale structure of the universe can be completely assembled within the framework of the very simple assumptions (1) and (2) (that is, a flat universe with a flat spectrum of adiabatic perturbations). However, as the inhabitants of new housing projects know only too well, the simplest project is almost never the most successful. It would therefore be well to understand whether we can somehow modify assumptions (1) and (2) while remaining within the framework of the inflationary universe scenario. Specifically, we would like to single out five basic questions.

1) Is it possible to get away from the condition $\Omega = 1$?

2) Is it possible to obtain nonadiabatic perturbations after inflation?

3) Is it possible to obtain perturbations with a spectrum that decreases at $l \sim l_H$, so as to reduce the quadrupole anisotropy of $\frac{\Delta T}{T}$?

4) Is it possible to obtain perturbations with a spectrum having one or perhaps several maxima, which would help to explain the origin of the hierarchy of scales (galaxies, clusters, ...)?

5) Is it possible to produce the large scale structure of the universe through nonperturbative effects associated with inflation?

For the time being, the answer to the first question is negative: we know of no way to obtain $\Omega \neq 1$ in a natural manner within the context of inflationary cosmology. Even if we could, it would most likely be only for some special choice of the potential $V(\varphi)$ and after painstaking adjustment of the parameters, for which there is as yet no particular justification.

Building a model in which the spectrum of adiabatic perturbations falls off monotonically at long wavelengths is possible in principle, but rather difficult. The only reasonable theory of this type that we are aware of is the Shafi–Wetterich model, based on a study of inflation in the Kaluza–Klein theory [237]. A peculiar feature of this model is that inflation and the evolution of a scalar field φ (the role of which is played by the logarithm of the compactification radius) are described by two different effective potentials, $V(\varphi)$ and $W(\varphi)$. Unfortunately, it is difficult to realize the initial conditions required for inflation in this model — see Chapter 9. Another suggestion that has been made is to study the spectra produced by double inflation, driven first by one scalar field φ, then by another Φ [238]. For the most natural initial conditions, however, the last stages of inflation are governed by the field with the flattest potential (the smallest parameters m^2 and λ). As a rule, therefore, rather than leading to cutoff, two-stage inflation will lead to a more abrupt rise in $\frac{\delta\rho}{\rho}$ at the long wavelengths generated during the stage when the "heavier" field φ is dominant.

Nevertheless, all of the questions posed above (except the first) can be answered in the affirmative. There is a rather broad class of models which, in addition to adiabatic perturbations, can also produce isothermal perturbations [239, 240], with spectra that fall off at long wavelengths [239, 241]. Particularly interesting effects are associated with phase transitions, which can occur at the later stages of inflation (when the universe still has another factor of e^{50}–e^{60} left to expand). In particular, such phase transitions can result in density perturbations having a spectrum with one or several maxima [242], and to the appearance of exponentially large strings, domain walls, bubbles, and other objects that can play a significant role in the formation of the large scale structure of the universe [125, 243]. We shall discuss some of the possibilities mentioned above in the next two sections.

7.7 Isothermal perturbations and adiabatic perturbations with a nonflat spectrum

The theory of the formation of density perturbations discussed in Section 7.5 was based upon a study of the simplest models, describing only a single scalar field φ responsible for the dynamics of inflation. In realistic elementary particle theories, there exist many scalar fields Φ_i of various kinds. To understand how inflation comes into play and what sorts of density inhomogeneities arise in such theories, let us consider first the simplest model, describing two noninteracting fields φ and Φ [239]:

$$L = \frac{1}{2}(\partial_\mu\varphi)^2 + \frac{1}{2}(\partial_\mu\Phi)^2 - \frac{m_\varphi^2}{2}\varphi^2 - \frac{m_\Phi^2}{2}\Phi^2 - \frac{\lambda_\varphi}{4}\varphi^4 - \frac{\lambda_\Phi}{4}\Phi^4. \quad (7.7.1)$$

We assume for simplicity that $\lambda_\varphi \ll \lambda_\Phi \ll 1$, and $m_\varphi^2, m_\Phi^2 \ll \lambda_\varphi M_P^2$. Then for large φ and Φ, terms quadratic in the fields can be neglected. The only constraint on the initial amplitudes of the fields φ and Φ is

$$V(\varphi) + V(\Phi) \sim \frac{\lambda_\varphi}{4}\varphi^4 + \frac{\lambda_\Phi}{4}\Phi^4 \lesssim M_P^4. \quad (7.7.2)$$

This means that the most natural initial values of the fields φ and Φ are $\varphi \sim \lambda_\varphi^{-1/4} M_P$, $\Phi \sim \lambda_\Phi^{-1/4} M_P$; that is, initially, $V(\varphi) \sim V(\Phi) \sim M_P^4$, $\varphi \gg \Phi \gg M_P$. Since the curvature of the potential $V(\Phi)$ is much greater than that of $V(\varphi)$, it is clear that under the most natural initial conditions, the field Φ and its energy $V(\Phi)$ fall off much more rapidly than the field φ and its energy $V(\varphi)$. The total energy density therefore quickly becomes equal to $V(\varphi)$; i.e., the Hubble parameter $H(\varphi, \Phi)$ becomes

$$H(\varphi, \Phi) \sim H(\varphi) = \sqrt{\frac{2\pi\lambda_\varphi}{3}}\,\frac{\varphi^2}{M_P}. \quad (7.7.3)$$

Thus, within a short time, inflation will be governed solely by the field φ having the potential $V(\varphi)$ with the least curvature (the smallest coupling constant λ_φ). For this reason, the field φ can be called the inflaton. It evolves just as if the field Φ did not exist (see (1.7.21)):

$$\varphi(t) = \varphi_0 \exp\left(-\sqrt{\frac{\lambda_\varphi}{6\pi}}\, M_P t\right). \quad (7.7.4)$$

In that case, the equation

$$\ddot{\Phi} + 3H\dot{\Phi} = -\lambda_\Phi \Phi^3 \quad (7.7.5)$$

implies that during the inflationary stage

$$\Phi(t) = \sqrt{\frac{\lambda_\varphi}{\lambda_\Phi}}\, \varphi(t), \quad (7.7.6)$$

and therefore

$$m^2(\varphi) = m^2(\Phi) = \frac{3M_P\sqrt{3\lambda_\varphi}}{\sqrt{2\pi}} H(\varphi), \qquad (7.7.7)$$

where (for $m_\Phi^2 \ll \lambda_\varphi M_P^2$)

$$m^2(\varphi) = \frac{d^2V}{d\varphi^2} = 3\lambda_\varphi\varphi^2,$$
$$m^2(\Phi) = \frac{d^2V}{d\Phi^2} = 3\lambda_\Phi\Phi^2. \qquad (7.7.8)$$

In the last stage of inflation, $\varphi \sim M_P$ and $\Phi \sim \sqrt{\frac{\lambda_\varphi}{\lambda_\Phi}}\, M_P$. The perturbations of the φ and Φ fields have equal amplitudes (see (7.5.6)):

$$\delta\varphi = \delta\Phi = \frac{H}{2\pi} = \sqrt{\frac{\lambda_\varphi}{6\pi}}\,\frac{\varphi^2}{M_P} \sim \sqrt{\lambda_\varphi}\, M_P. \qquad (7.7.9)$$

At that time, however, the contribution $\delta\rho_\Phi$ that the field Φ makes to density inhomogeneities $\delta\rho$ is much less than $\delta\rho_\varphi$, the corresponding contribution from φ:

$$\delta\rho_\Phi = \frac{dV}{d\Phi}\delta\Phi = \sqrt{\frac{\lambda_\varphi}{\lambda_\Phi}}\,\lambda_\varphi\varphi^3\delta\varphi = \sqrt{\frac{\lambda_\varphi}{\lambda_\Phi}}\,\delta\rho_\varphi \ll \delta\rho_\varphi. \qquad (7.7.10)$$

It is therefore precisely the fluctuations $\delta\rho_\varphi$ of the inflaton field that govern the amplitude of adiabatic density perturbations. At the stage of inflation with $\varphi \sim M_P$,

$$\frac{\delta\rho_\varphi}{\rho_\varphi} \approx \frac{\delta\rho}{\rho} = \frac{1}{V}\frac{dV}{d\varphi}\delta\varphi \sim 4\frac{\delta\varphi}{\varphi} \sim \sqrt{\lambda_\varphi}, \qquad (7.7.11)$$

where $\rho = \rho_\varphi + \rho_\Phi \approx \rho_\varphi$. Meanwhile,

$$\frac{\delta\rho_\Phi}{\rho_\Phi} \sim \frac{4\,\delta\Phi}{\Phi} \sim \sqrt{\lambda_\Phi}\,. \tag{7.7.12}$$

After inflation, the inhomogeneities (7.7.11) produce the adiabatic perturbations (7.5.29), which will in fact remain the dominant density perturbations *if* upon further expansion of the universe, the quantity ρ_Φ falls off in the same way as ρ_φ. However, in some cases this condition is not satisfied, since the evolution of ρ_Φ and ρ_φ depends on the interaction of these fields with other fields, and on the shape of $V(\varphi)$ and $V(\Phi)$. Let us assume, for example, that the field Φ interacts very weakly with other fields. Such weakly interacting scalar fields do exist in many realistic theories (axions, Polonyi fields, etc.). If the field φ interacts strongly with other fields, its energy is quickly transformed into heat, $\rho_\varphi \to T_R^4$, and falls off with the expansion of the universe as $T^4 \sim a^{-4}$. At the same time, the field Φ, without decaying, oscillates in the neighborhood of the point $\Phi = 0$ with frequency $\kappa_0 = m_\Phi$. As it does so, its energy falls off in the same way as the energy of nonrelativistic particles, $\rho_\Phi \sim a^{-3}$ (see Section 7.9), i.e., much more slowly than the energy of the decay products from the field φ. In the later stages of evolution of the universe, the energy of the field Φ can therefore become greater than the energy of decay products of the inflaton field, $\rho = \rho_\varphi + \rho_\Phi \approx \rho_\Phi$.

This is just the effect that underlies the possibility, discussed in [49], that the axion field Θ may be responsible for the missing mass of the universe at the present epoch.

Prior to the stage at which the field Φ is dominant, both the mean density ρ_Φ and the quantity $\rho_\Phi + \delta\rho_\Phi$ fall off in the same way, $\rho_\Phi \sim \rho_\Phi + \delta\rho_\Phi \sim a^{-3}$; the quantity $\frac{\delta\rho_\Phi}{\rho_\Phi} \sim \sqrt{\lambda_\Phi}$ therefore remains constant. From the start, inhomogeneities $\delta\rho_\Phi$ are in no way associated with the temperature inhomogeneities δT of decay products of the field φ, and in that sense they are isothermal. They might also be called *isoinflaton* inhomogeneities, as they are independent of fluctuations of the inflaton field φ. Consequently, due to the increasing fraction of the overall matter density ρ accounted for by ρ_Φ, isothermal perturbations $\delta\rho_\Phi$, with $\sqrt{\lambda_\Phi} \gtrsim 10^2\sqrt{\lambda_\varphi}$, begin to dominate, generating adiabatic perturbations

$$\frac{\delta\rho}{\rho} \approx \frac{\delta\rho_\Phi}{\rho_\Phi} \sim \sqrt{\lambda_\Phi} \tag{7.7.13}$$

in the process. Note that in (7.7.13), there is no enhancement factor $O(10^2)$ associated with the transition from the inflationary stage to the expansion with $a \sim t^{1/2}$ or $a \sim t^{2/3}$.

Thus, even in the simplest theory of two noninteracting fields, the process by which density perturbations are generated can unfold in a fairly complicated manner: in addition to adiabatic density perturbations, isothermal perturbations can also come about, and for $\lambda_\varphi \ll 10^{-14}$, $\lambda_\Phi \gtrsim 10^{-10}$, the latter can dominate.

Even more interesting possibilities are revealed when we allow for interactions between the fields φ and Φ. Consider, for example, a theory with the effective potential

$$V(\varphi, \Phi) = \frac{m_\varphi^2}{2}\varphi^2 + \frac{\lambda_\varphi}{4}\varphi^4 - \frac{m_\Phi^2}{2}\Phi^2 + \frac{\lambda_\Phi}{4}\Phi^4 + \frac{\nu}{2}\varphi^2\Phi^2 + V(0). \quad (7.7.14)$$

Let us suppose that $0 < \lambda_\varphi \ll \nu \ll \lambda_\Phi$ and $\lambda_\varphi \lambda_\Phi > \nu^2$, and assume also that $m_\varphi^2 \ll \lambda_\varphi M_P^2$ and $m_\Phi^2 \ll C\nu M_P^2$, where $C = O(1)$. Just as in the theory (7.7.1), the most natural initial values for φ and Φ satisfy the conditions $\varphi \gg M_P, \varphi \gg \Phi$. With $\varphi \gg M_P$, the minimum of $V(\varphi, \Phi)$ is located at $\Phi = 0$, and the effective mass of the field Φ at $\Phi = 0$ is

$$m_\Phi^2(\varphi, 0) = \left.\frac{\partial^2 V}{\partial \Phi^2}\right|_{\Phi=0} = \nu\,(\varphi^2 - C M_P^2) \sim \nu\,\varphi^2. \quad (7.7.15)$$

This is much greater than the mass of the field φ,

$$m_\varphi^2(\varphi, 0) = m_\varphi^2 + 3\lambda_\varphi\varphi^2 \sim 3\lambda\varphi^2 \ll \nu\varphi^2. \quad (7.7.16)$$

The field Φ is therefore rapidly dumped into the minimum of $V(\varphi, \Phi)$, and as in the theory (7.7.1), inflation becomes driven by the field φ.

In the last stage of inflation, when the field φ becomes less than

$$\varphi_c = \sqrt{C}\, M_P, \quad (7.7.17)$$

the minimum of $V(\varphi, \Phi)$ is located at

$$\Phi^2 = \frac{m_\Phi^2 - \nu\varphi^2}{\lambda_\Phi} = \nu\,\frac{C M_P^2 - \varphi^2}{\lambda_\Phi}, \quad (7.7.18)$$

and the effective mass of the field Φ is then

$$m_\Phi^2(\varphi, \Phi) = 2\,\nu\,(CM_P^2 - \varphi^2)\,. \tag{7.7.19}$$

Notice that both when $\varphi \gg \varphi_c$ and $\varphi \ll \varphi_c$, the effective mass of the field Φ is much greater than the Hubble constant $H \sim \frac{\sqrt{\lambda_\varphi}\,\varphi^2}{M_P}$. Long-wave fluctuations $\delta\Phi$ of the field Φ are therefore generated only in some neighborhood of the phase transition point at $\varphi \sim \varphi_c$. In studying the density perturbations produced in this model, it turns out to be important that the amplitude of fluctuations $\delta\Phi$ displays a variety of temporal behaviors, depending on what precisely the value of the field φ was when these fluctuations arose. Numerical calculations [242] taking this fact into account have shown that for certain relationships among the parameters of the theory (7.7.14), the spectrum of adiabatic perturbations produced during inflation may have a reasonably narrow maximum that is slightly shifted with respect to $l \sim \exp\left(\frac{\pi\varphi_c^2}{M_P^2}\right)$.

It should be pointed out here that many different types of scalar fields figure into realistic elementary particle theories. It would therefore be hard to doubt that phase transitions should actually take place at the time of inflation, and in fact most likely not one, but many. The only question is whether these phase transitions take place fairly late, when the field φ is changing between φ_H (7.5.27) and φ_g (7.5.28). This is a condition that is satisfied, given an appropriate choice of parameters in the theory. The actual parameter values chosen (like the parameters used in building the theories of the weak and electromagnetic interactions) should be based on experimental data, rather than on some *a priori* judgment about their naturalness (since according to that criterion one could reject the Glashow–Weinberg–Salam model — see the preceding section). In the case at hand, these data consist of observations of the large scale structure of the universe and the anisotropy of the cosmic microwave radiation background. In our opinion, the possibility of studying the phase structure of unified theories of elementary particles and determining the parameters of these theories through astrophysical observations is extremely interesting.

To conclude this section, let us briefly deal with the production of isothermal perturbations in axion models. To this end, we examine the theory of a complex scalar field Φ which interacts with an inflaton field φ:

$$V(\varphi, \Phi) = \frac{m_\varphi^2}{2}\varphi^2 + \frac{\lambda_\varphi}{4}\varphi^4 - m_\Phi^2 \Phi^*\Phi + \lambda_\Phi(\Phi^*\Phi)^2 + \frac{\nu}{2}\varphi^2\Phi^*\Phi + V(0). \tag{7.7.20}$$

After the spontaneous symmetry breaking which occurs at $\varphi < \varphi_c = \frac{m_\Phi}{\sqrt{\nu}}$, the field Φ can be represented in the form

$$\Phi(x) = \Phi_0 \exp\left(\frac{i\Theta(x)}{\sqrt{2}\,\Phi_0}\right), \tag{7.7.21}$$

where $\Phi_0 = \frac{m_\Phi}{\sqrt{\lambda_\Phi}}$ for $\varphi \ll \varphi_c$. The field $\Theta(x)$ is a massless Goldstone scalar field [244] with vanishing effective potential, $V(\Theta) = 0$.

In contrast to the familiar Goldstone field described above, the axion field is not massless. Because of nonperturbative corrections to $V(\varphi, \Phi)$ associated with strong interactions, the effective potential $V(\Theta)$ becomes [233, 234]

$$V(\Theta) = C\, m_\pi^4 \left(1 - \cos\frac{N\Theta}{\sqrt{2}\,\Phi_0}\right). \tag{7.7.22}$$

Here $C \sim O(1)$, and N is an integer that depends on the detailed structure of the theory; for simplicity, we henceforth consider only the case $N = 1$. We thus see from (7.7.22) that axions can now have a small mass $m_\Theta \sim \frac{m_\pi^2}{\Phi_0} \sim 10^{-2}\,\mathrm{GeV}^2/\Phi_0$

From the standpoint of elementary particle theory, the main reason for considering the axion field Θ is that the field value that minimizes $V(\Theta)$ automatically leads to cancellation of the effects of the strong CP violation that are associated with the nontrivial vacuum structure in the theory of strong interactions [233, 234]. Cosmologists became interested in this field for another reason. It turns out that at temperatures $T \gg 10^2$ MeV, the nonperturbative effects resulting in nonvanishing $V(\Theta)$ are strongly inhibited. Therefore, the field Θ is equally likely to take any initial value in the range $-\sqrt{2}\,\pi\Phi_0 \leq \Theta \leq \sqrt{2}\,\pi\Phi_0$. As the temperature of the universe drops to $T \lesssim 10^2$ MeV, the effective potential $V(\Theta)$ takes the form (7.7.22), so that in the mean, the field Θ acquires an energy density of order $m_\pi^4 \sim 10^{-4}\ \mathrm{GeV}^4$. This field interacts with other fields extremely weakly, and its mass is extraordinarily small ($m_\Theta \sim 10^{-5}$ eV for the realistic value $\Phi_0 \sim 10^{12}$ GeV; see below). It therefore mainly loses its energy not through radiation, but through damping of its oscillations near $\Theta = 0$ as the universe expands (by virtue of the term $3H\dot{\Theta}$ in the equation for the field Θ). As we have already said, the energy density of any noninteracting massive field, which oscillates near the minimum of its effective potential, falls off in the same way as the energy density of a gas of nonrelativistic particles, $\rho_\Theta \sim a^{-3}$, i.e., more slowly than that of a relativistic gas. As a result, the relative contribution of the axion field to the total energy density increases.

The present-day value of the ratio $\frac{\rho_\Theta}{\rho}$ depends on the value of Φ_0. For $\Phi_0 \sim 10^{12}$ GeV, most of the total energy density of the universe should presently be concentrated in an almost homogeneous, oscillating axion field, which would then account for the missing mass. As has been claimed in [49], a value of $\Phi_0 \gg 10^{12}$ GeV would be difficult to reconcile with the available cosmological data (see Section 10.5, however). When $\Phi_0 \ll 10^{12}$ GeV, the relative contribution of axions to the energy density of the universe falls off as $(\Phi_0/10^{12}\ \mathrm{GeV})^2$.

If the phase transition with symmetry breaking and the creation of the pseudo-Goldstone field Θ takes place during the inflationary stage, then inflation leads to fluctuations of the field Θ; as before, $\delta\Theta = \frac{H}{2\pi}$ per unit interval $\Delta \ln k$. When $T < 10^2$ MeV, density inhomogeneities $\frac{\delta\rho_\Theta}{\rho_\Theta} \sim \frac{\delta V(\Theta)}{V(\Theta)}$ appear, which are associated with these fluctuations. But because of the periodicity of the potential $V(\Theta)$, these inhomogeneities are related in a much more complicated way to the magnitude of the fluctuations in the field Θ. Let us assume, for example, that after a phase transition, inflation continues long enough that the rms value $\sqrt{\langle\Theta^2\rangle} = \frac{H}{2\pi}\sqrt{Ht}$ becomes much greater than Φ_0. The classical field Θ will then take on any value in the range $-\sqrt{2}\,\pi\Phi_0 \leq \Theta \leq \sqrt{2}\,\pi\Phi_0$ with uniform probability. The addition of a

greater than Φ_0. The classical field Θ will then take on any value in the range $-\sqrt{2}\,\pi\Phi_0 \leq \Theta \leq \sqrt{2}\,\pi\Phi_0$ with uniform probability. The addition of a constant field $\delta\Theta$ to the field Θ will rotate the distribution of Θ by an angle $\delta\Theta$, but it will not change the mean value of $V(\Theta)$ anywhere within the volume considered. This phenomenon lies at the root of the cutoff effect for long-wave isothermal perturbations in axion field theory [239]. Detailed study of this effect [241, 125] leads to the following expression for $\frac{\delta\rho_\Theta}{\rho}$

$$\frac{\delta\rho_\Theta(l)}{\rho} \sim \frac{\rho_\Theta}{\rho}\sqrt{2\beta}\,\sin\frac{\Theta(l)}{\sqrt{2}\,\phi_0}\cdot l^{-\beta}, \tag{7.7.23}$$

where l is the present scale size of inhomogeneities, measured in centimeters, $\beta = \left(\frac{H}{4\pi\Phi_0}\right)^2$, $\frac{\rho_\Theta}{\rho} \sim (\Phi_0/\,10^{12}\,\text{GeV})^2$, and H is the Hubble parameter in the last stages of inflation. For $H \sim 10^{12}$–10^{13} GeV, adiabatic density perturbations will be of order 10^{-5}–10^{-6} (see (7.5.33), (7.5.46)), and isothermal perturbations with the spectrum (7.7.23), gradually falling off toward longer wavelengths, will be the main contributors. This falloff becomes even steeper than the $l^{-\beta}$ trend of (7.7.23) at wavelengths $l \gtrsim \exp\left(\frac{\pi\varphi_c^2}{M_P^2}\right)$, where φ_c is the critical field (7.7.17); this is where the Goldstone axion field Θ first makes its appearance. The reason for this is that when $\varphi > \varphi_c$, no semiclassical long-wave fluctuations of the field Φ are produced, due to the large mass of this field.

We should also draw attention to the factor $\sin\frac{\Theta(l)}{\sqrt{2}\,\phi_0}$ in (7.7.23). Here $\Theta(l)$ is the initial value of the field Θ averaged over the length scale l. This field takes on different values at different points, but values of $\Theta(l)$ are correlated at a scale separation $\exp\left(\frac{2\pi\Phi_0}{H}\right)$ cm, resulting in an additional degree of ordering of the large scale structures in the universe [125].

Thus, due to the nontrivial nature of the relationship between $\delta\Theta(x)$ and $\delta\rho(x)$ in axion models, the distribution of inhomogeneities $\delta\rho$ in these models is significantly different from the usual Gaussian distribution studied in Section 7.5. An additional modification of the distribution of inhomogeneities $\delta\rho(x)$ is discussed in [245, 361].

7.8 Nonperturbative effects: strings, hedgehogs, walls, bubbles, ...

In the preceding sections, we have studied mechanisms for generating small density perturbations in the inflationary universe. But phase transitions at the time of inflation can lead not just to small density perturbations; they can also produce nontrivial structures of exponentially large size. Herewith, we present some examples.

1. *Strings.* The theory of the formation of density inhomogeneities during the evolution of cosmic strings [81] was long considered to be the only real alternative to inflationary theory for the formation of flat-spectrum adiabatic perturbations. It is now quite clear that there exists a wide range of other possibilities — see Section 7.7 and the discussion below. Furthermore, without inflation, string theory provides no help in solving the problems of standard Friedmann cosmology, and the formation of superheavy strings through high-temperature phase transitions following inflation is complicated by the fact that in most models, the universe after inflation is not hot enough. But it is perfectly possible to produce strings during phase transitions in the inflationary stage [125, 246, 247]. About the simplest model in which one could treat such a process would be a theory describing the interaction of an inflaton φ with a complex scalar field Φ having effective potential (7.7.20). In the early stages of inflation, when $\varphi^2 > \frac{m_\Phi^2}{\nu}$, symmetry is restored in the theory (7.7.20). As the field φ falls off toward $\varphi = \varphi_c = \frac{m_\Phi}{\sqrt{\nu}}$, the symmetry-breaking phase transition leading to the production of strings takes place, as happens in the case of a phase transition with a decrease in temperature; see Section 6.2. The difference here is that during inflation, the typical size of the strings produced increases by a factor of $\exp\left(\frac{\pi\varphi_c^2}{M_P^2}\right) = \exp\left(\frac{\pi m_\Phi^2}{\nu M_P^2}\right)$. If this factor is not too large, then most of the results obtained in the theory of formation of density inhomogeneities due to strings [81] remain valid.

2. *Hedgehogs.* Phase transitions at the time of inflation also lead to the production of hedgehog–antihedgehog pairs (see Section 6.2). A typical separation r_0 between a hedgehog and antihedgehog is of order $H^{-1}(\varphi_c)$, but as a result of inflation, this separation increases exponentially.

the size of the horizon $\sim t$ grows to a size comparable with the distance between the hedgehog and antihedgehog. This gives rise to density inhomogeneities $\frac{\delta\rho}{\rho}$ of order $\frac{\Phi_0^2}{M_P^2}$, for the same reason as in the theory of strings (6.2.3). In the present case, however, the spectrum of density inhomogeneities will have a pronounced maximum at a wavelength of the order of the typical distance between hedgehogs, $\sim \exp\left(\frac{\pi\varphi_c^2}{M_P^2}\right)$.

3. *Monopoles.* Monopoles can be produced as a byproduct of phase transitions at the time of inflation. The density of such monopoles will be reduced by factors like $\exp\left(-\frac{3\pi\varphi_c^2}{M_P^2}\right)$, but for sufficiently small φ_c, attempts to detect them experimentally may have some chance of succeeding.

4. *Monopoles connected by strings.* Such objects also crop up in certain theories. Just as for hedgehogs, such monopoles appear in a confinement phase, and in the hot universe theory, where the typical distance between monopoles is of order T_c^{-1}, they are rapidly annihilated [81]. In the inflationary universe scenario, they can lead to approximately the same consequences as hedgehogs.

5. *Domain walls bounded by strings.* The axion theory discussed in the preceding section qualifies as one of a number of theories in which strings are produced following symmetry breaking. With currently accepted model parameter values, axion strings, in and of themselves, are too light to induce large enough density inhomogeneities. But a more careful analysis shows that every axion string is boundary of a domain wall [43, 81]. This is related to the fact that in proceeding around a string, with the quantity $\frac{\Theta(x)}{\sqrt{2}\,\Phi_0}$ changing by 2π, we necessarily pass through a maximum of $V(\Theta)$ (7.7.22). Energetically, the most favorable configuration of the field $\Theta(x)$ is that in which the field Θ does not change as one proceeds around the string; this corresponds to having a minimum of $V(\Theta)$ everywhere except at a wall whose thickness is of order m_Θ^{-1}, and having the quantity $\frac{\Theta(x)}{\sqrt{2}\,\Phi_0}$ change by 2π when the latter is traversed. The surface energy of the wall is of order $m_\pi^2\Phi_0$.

Analysis of the evolution of a system of strings that acts like a wire frame supporting a soap film composed of domain walls shows that the initial field configuration resembles a single infinitely curved surface containing a large number of holes. Moreover, there also exist isolated surfaces of finite size, but they contribute negligibly to the total energy of the universe [81]. Portions of these surfaces eventually begin to intersect and tear one another over a surface region of small extent, and resemble frothy "pancakes," which subsequently oscillate and radiate their energy away in the form of gravitational waves. If these surfaces form as a result of phase transitions in a hot universe, the pancakes turn out to be extremely small, and they quickly disappear. But surfaces created at the time of inflation produce pancakes that are exponentially large [81, 125]. The possible role of such objects in the formation of large scale structure in the universe requires further investigation.

6. *Bubbles*. In studying the cosmological consequences of the phase transitions occurring at the time of inflation, we have implicitly assumed that they were soft transitions, with no tunneling through barriers, as in the second-order phase transition considered in Section 7.7. Meanwhile, the phase transitions can also be first-order — see Section 7.4. Bubbles of the field Φ could then be produced. During inflation, there is much less energy in the fields Φ than in the inflaton field φ. For that reason, the appearance of such bubbles has practically no effect on the rate of expansion of the universe, and after inflation, the sizes of bubbles of the field Φ wind up being exponentially large; more specifically, all bubbles turn out to have a typical size of order $\exp\left(\frac{\pi\varphi_c^2}{M_P^2}\right)$ cm. If the rate of bubble production is high, then the resulting distribution of the field Φ will resemble a foam (cells), with maximum energy density on the walls of adjoining bubbles and with voids within them. If the bubble creation rate is low, then mutually separated regions will arise within which the matter density is lower than that outside. In the later stages of the evolution of the universe, when the energy of the field Φ may become dominant, the corresponding density contrast may turn out to be quite high [125, 240, 243].

7. *Domains*. Especially interesting effects can transpire when the universe acquires a domain structure at the time of inflation. As the simplest example of this we consider the possible kinetics of the SU(5)-breaking phase transition at that epoch. As we already noted in the preceding chapter, as the temperature T drops, phase transitions in the SU(5) theory entail the formation of bubbles which can contain a field Φ that corresponds to any one of four different types of symmetry breaking: SU(3) × SU(2) × U(1), SU(4) × U(1), SU(3) × U(1) × U(1), or SU(2) × SU(2) × U(1) ×

U(1). An analogous phase transition can also take place during inflation, but in the latter event, inflation ensures that bubbles of the different phases becomes exponentially large. This results in the formation of large domains filled with matter in different phases — that is, with slightly different density. In the standard SU(5) model, only the phase SU(3) × SU(2) × U(1) is stable after inflation, so ultimately the entire universe is transformed to this phase, and the domain walls that separate the different phases disappear. However, the corresponding density inhomogeneities that appeared during the epoch when domains were still present somehow remain imprinted on the subsequent density distribution of matter in the universe.

If the probability of bubble formation is significantly different for bubbles containing matter in different phases, the universe will eventually consist of islands of reduced or enhanced density superimposed on a relatively uniform background. In principle, we could associate such islands with galaxies, clusters of galaxies, or even the insular structure of the universe proposed in [248].

If on the other hand the universe simultaneously spawns comparable numbers of bubbles in different phases, the resulting density distribution takes on a sponge-like structure. Specifically, there will be cells containing phases of different density, but a significant fraction of the cells of a given phase will be connected to one another, so that one could pass from one part of the universe to another through cells that are all of the same type (percolation). Concepts involving a sponge-like universe have lately become rather popular.

Recent results [249] indicating that the universe effectively consists of contiguous bubbles 50–100 Mpc ($1.5 \cdot 10^{26}$–$3 \cdot 10^{26}$ cm) in size containing few luminous entities, so that galaxies are basically concentrated at bubble walls, have been especially noteworthy. Particularly interesting in that regard is the fact that such structures may appear as a natural consequence of phase transitions at the time of inflation [125, 244].

In the context of the present model, the advent of regions of the universe containing most of the luminous (baryon) matter is not at all necessarily connected with density enhancements above the mean. Firstly, post-inflation baryon production (see the next section) proceeds entirely differently in the different phases (SU(3) × SU(2) × U(1) or SU(4) × U(1)). It could turn out, in principle, that baryons are only produced in those regions filled with a phase whose density lies below the mean, and these would then be just the regions in which we would see galaxies. Secondly, if galaxy formation is associated with isothermal perturbations of the field Φ, then one should take into account that the amplitude of such perturbations will also depend on the phase inside each of the domains. Isothermal perturbations can therefore only be large enough for subsequent galaxy formation in those domains filled with some particular phase, and these are precisely the regions in which galaxies, clusters of galaxies, and so forth should form. Thus, depending on the specific elementary particle theory chosen, galaxies will preferentially form in regions of either enhanced or reduced density, and either in the space outside bubbles (for example, at bubble walls) or within them.

If certain phases remain metastable after inflation, then as a rule their characteristic decay time will turn out to be much greater than the age of the observable part of the universe, $t \sim 10^{10}$ yr. In that event, the universe should be partitioned right now into domains that contain matter in a variety of phase states. This is just the situation in supersymmetric SU(5) models, where the minima corresponding to SU(5), SU(3) × SU(2) × U(1), and SU(4) × U(1) symmetries are of almost identical depth and are separated from one another by high potential barriers [91-93]. During inflation, the universe is partitioned into exponentially large domains, each of which contains one of the foregoing phases, and we happen to live in one such domain corresponding to the SU(3) × SU(2) × U(1) phase [211]. If inflation were to go on long enough after the phase transition (that is, if the phase transition had taken place with $\varphi_c \gtrsim 5\ M_P$ in the $\frac{\lambda}{4}\varphi^4$ theory), then there would be not a single domain wall in the observable part of the universe. In the opposite case, domains would be less than 10^{28} cm in size. If regions containing different phases have the same probability of forming (as in a theory that is symmetric with respect to the interchange $\varphi \to -\varphi$), then for $\varphi_c \lesssim 5\ M_P$, we encounter the domain wall problem discussed in Section 6.2. However, the probability of producing bubbles containing matter in different phases depends on the height of the walls separating different local minima of $V(\Phi)$, and in general differs significantly among the phases. The universe is therefore mostly filled with just one of its possible phases, and the other phases are present in the form of widely-spaced, exponentially large, isolated domains. Those domains containing energetically unfavorable phases later collapse. As we have already remarked in Section 7.4, those regions whose probability of formation is fairly low should be close to spherical, and the collapse of such regions proceeds in an almost perfectly spherically symmetric manner. The entire gain in potential energy due to compression of a bubble of a metastable phase would then be converted into kinetic energy of its compressed wall. If the wall is comprised of a scalar field Φ that interacts strongly enough both with itself and with other fields, the bubble wall energy will be transformed after compression into the energy of those elementary particles created at the instant of wall collapse. The particles thus produced fly off in all directions, forming a spherical shell. We thereby have yet another mechanism capable of producing a universe with bubble-like structure. The process will be more complicated if the original bubble is significantly nonspherical, and the cloud of newly-created particles will also no longer be spherically symmetric.

This model resembles the model of Ostriker and Cowie [250] for the explosive formation of the large scale structure of the universe. However, the physical mechanism discussed above differs considerably from that suggested in Ref. 250.

The investigation of nonperturbative mechanisms for the formation of the large scale structure of the universe has just begun, but it is already apparent from the foregoing discussion how many new possibilities the study of the cosmological consequences of phase transitions during the inflationary stage carries with it. The overall conclusion is that inflation can result in the appearance of various exponentially large objects. The latter may be of interest not just as the structural material out of which galaxies could subsequently be built, but, for example, as a possible source of intense radio emission [251]. They could also turn into supermassive black holes, and finally they might turn out to be responsible for anomalous exoergic processes in the universe. This abundance of new possibilities does not mean that anything goes, but it does substantially expand the horizons of those seeking the correct theory of formation of the large scale structure of the universe.

7.9 Reheating of the universe after inflation

The production of inhomogeneities in the inflationary universe has elicited a great deal of interest of late, as this process is directly reflected in the structure of the observable part of the universe. Of no less value is the study of the process whereby the universe is reheated and its baryon asymmetry generated, since this process is a mandatory connecting link between the inflationary universe in its vacuum-like state and the hot

Friedmann universe. In the present section, we study the reheating process through the example of the simplest theory of a massive scalar field φ that interacts with a scalar field χ and a spinor field ψ, with the Lagrangian

$$L = \frac{1}{2}(\partial_\mu\varphi)^2 - \frac{m_\varphi^2}{2}\varphi^2 + \frac{1}{2}(\partial_\mu\chi)^2 - \frac{m_\chi^2}{2}\chi^2 + \bar{\psi}(i\gamma_\mu\partial_\mu - m_\psi)\psi + \nu\,\sigma\varphi\chi^2 - h\bar{\psi}\psi\varphi - \Delta V(\varphi,\chi). \tag{7.9.1}$$

Here ν and h are small coupling constants, and σ is a parameter with the dimensionality of mass. In realistic theories, the constant part of the field φ, for example, can play the role of σ. What we mean by $\Delta V(\varphi,\chi)$ is that part of $V(\varphi,\chi)$ that is of higher order in φ^2 and χ^2. We shall assume (allowing for $\Delta V(\varphi,\chi)$) that in the last stages of inflation, the role of the inflaton field is taken on by the field φ, and then go on to investigate the process by which the energy of this field is converted into particles χ and ψ. We suppose for simplicity that $m_\varphi \gg m_\chi$, m_ψ, and that at the epoch of interest, $\nu\sigma\varphi \ll m_\chi^2, h\,\varphi \ll m_\psi$.

If we ignore effects associated with particle creation, the field φ after inflation will oscillate near the point $\varphi = 0$ at a frequency $k_0 = m_\varphi$. The oscillation amplitude will fall off as $[a(t)]^{-3/2}$, and the energy of the field φ will decrease in the same way as the density of nonrelativistic φ particles of mass m_φ : $\rho_\varphi = V(\varphi) = \frac{m_\varphi^2}{2}\varphi^2 \sim a^{-3}$, where φ is the amplitude of oscillations of the field [252]. The physical meaning of this is that a homogeneous scalar field φ, oscillating at frequency m_φ, can be represented as a coherent wave of φ-particles with vanishing momenta and particle density $n_\varphi = \frac{\rho_\varphi}{m_\varphi} = \frac{m_\varphi}{2}\varphi^2$. If the total number of particles $\sim n_\varphi a^3$ is conserved (no pair production), the amplitude of the field φ will fall off as $a^{-3/2}$. The equation of state of matter at that time is $p = 0$; i.e., $a(t) \sim t^{2/3}$, $H = \frac{2}{3t}$, $\varphi \sim a^{-3/2} \sim t^{-1}$.

In order to describe the particle production process with its concomitant decrease in the amplitude of the field φ, let us consider the quantum corrections to the equation of motion for the homogeneous field φ, oscillating at a frequency $k_0 = m_\varphi \gg H(t)$:

$$\ddot{\varphi} + 3\,\mathrm{H}(t)\,\dot{\varphi} + \left[m_{\varphi}^{2} + \Pi\,(k_0)\right]\varphi = 0\,. \tag{7.9.2}$$

Here $\Pi\,(k_0)$ is the polarization operator for the field φ at a four-momentum $k = (\,k_0, 0, 0, 0)$, $k_0 = m_{\varphi}$.

The real part of $\Pi\,(k_0)$ gives only a small correction to m_{φ}^{2}, but when $k_0 > 2\,m_{\chi}$ (or $k_0 > 2\,m_{\psi}$), $\Pi\,(k_0)$ acquires an imaginary part $\mathrm{Im}\,\Pi\,(k_0)$. For $m_{\varphi}^{2} \gg \mathrm{H}^2$ and $m_{\varphi}^{2} \gg \mathrm{Im}\,\Pi$, and neglecting the time-dependence of H, we obtain a solution of Eq. (7.9.2) that describes the damped oscillations of the field φ near the point $\varphi = 0$:

$$\varphi = \varphi_0 \exp\,(\,i m_{\varphi} t)\cdot \exp\left[-\frac{1}{2}\left(3\,\mathrm{H} + \frac{\mathrm{Im}\,\Pi\,(m_{\varphi})}{m_{\varphi}}\right)t\right]. \tag{7.9.3}$$

From the unitarity relations [10, 124], it follows that

$$\mathrm{Im}\,\Pi\,(m_{\varphi}) = m_{\varphi}\Gamma_{tot}\,, \tag{7.9.4}$$

where Γ_{tot} is the total decay probability for a φ-particle. Hence, when $\Gamma_{tot} \gg 3\,\mathrm{H}$, the energy density of the field φ decreases exponentially in a time less than the typical expansion time of the universe $\Delta t \sim \mathrm{H}^{-1}$:

$$\rho_{\varphi} = \frac{m^2\varphi^2}{2} \sim \rho_0\, e^{-\Gamma_{tot} t}. \tag{7.9.5}$$

This is exactly the result one would expect on the basis of the interpretation of the oscillating field φ as a coherent wave consisting of (decaying) φ-particles.

The probability of decay of a φ-particle into a pair of χ-particles or ψ-particles is known — see, for example, [10, 122, 123]. For $m_{\varphi} \gg m_{\chi}, m_{\psi}$,

$$\Gamma(\varphi \to \chi\chi) = \frac{\nu^2\sigma^2}{8\pi m_{\varphi}}, \tag{7.9.6}$$

$$\Gamma(\varphi \to \bar{\psi}\psi) = \frac{h^2 m_{\varphi}}{8\,\pi}. \tag{7.9.7}$$

If the constants $\nu\sigma$ and h^2 are small, then initially

$\Gamma_{tot} = \Gamma(\varphi \to \chi\chi) + \Gamma(\varphi \to \bar{\psi}\psi) < 3\,\mathrm{H}(t) = \frac{2}{t}$. Basically, in that event, the energy density of the field φ decreases from the outset simply due to the expansion of the universe, $\frac{m^2\varphi^2}{2} \sim t^{-2}$. The fraction of the total energy converted into energy of the particles that are produced remains small right up to the time t^* at which $3\mathrm{H}(t^*)$ becomes less than Γ_{tot}. Particles produced prior to this time can also be thermalized, in principle, and their temperature in certain situations can be even higher than the final temperature $\mathrm{T_R}$ [253]. But the contribution of newly-created particles to the overall matter density becomes significant only starting at the time t^*, after which practically all the energy of the field φ is transformed into the energy of newly-created χ- and ψ-particles within a time $\Delta t \sim t^* \lesssim \mathrm{H}^{-1}$. The condition $3\,\mathrm{H}(t^*) \sim \Gamma_{tot}$ tells us that the energy density of these particles at the time t^* is of order

$$\rho^* \sim \frac{\Gamma_{tot}^2 \mathrm{M_P^2}}{24}. \tag{7.9.8}$$

If the χ- and ψ-particles interact strongly enough with each other, or if they can rapidly decay into other species, then thermodynamic equilibrium quickly sets in, and matter acquires a temperature $\mathrm{T_R}$, where according to (1.3.17) and (7.9.8)

$$\rho^* \sim \frac{\pi \mathrm{N(T_R)}}{30}\,\mathrm{T_R^4} \sim \frac{\Gamma_{tot}^2 \mathrm{M_P^2}}{24}. \tag{7.9.9}$$

Here $\mathrm{N(T_R)}$ is the effective number of degrees of freedom at $\mathrm{T = T_R}$, with $\mathrm{N(T_R)} \sim 10^2$–$10^3$, so that

$$\mathrm{T_R} \sim 10^{-1}\sqrt{\Gamma_{tot}\mathrm{M_P}}\,. \tag{7.9.10}$$

Note that as we said earlier, $\mathrm{T_R}$ does not depend on the initial value of the field φ, and is determined solely by the parameters of the elementary particle theory.

Let us now estimate $\mathrm{T_R}$ numerically. In order for adiabatic inhomogeneities $\frac{\delta\rho}{\rho} \sim 10^{-5}$ to appear in the present theory, it is necessary that m_φ be of order $10^{-6}\mathrm{M_P} \sim 10^{13}$ GeV. One can readily verify that quantum corrections investigated in Chapter 2 do not significantly alter the form of $V(\varphi)$ at $\varphi \lesssim M_P$ only if $h^2 \lesssim \frac{8\, m_\varphi}{M_P} \sim 10^{-5}$ and $\nu\sigma \lesssim 5\, m_\varphi \sim 10^{14}$ GeV. Under these conditions,

$$\Gamma(\varphi \to \chi\chi) \lesssim m_\varphi \sim 10^{-6} M_P, \tag{7.9.11}$$

$$\Gamma(\varphi \to \bar{\psi}\psi) \lesssim \frac{m_\varphi^2}{M_P} \sim 10^{-12} M_P. \tag{7.9.12}$$

For completeness, we note that in theories like the Starobinsky model or supergravity, the magnitude of Γ for φ-field decays induced by gravitational effects is usually [135, 286]

$$\Gamma_g \sim \frac{m_\varphi^3}{M_P^2} \sim 10^{-18} M_P. \tag{7.9.13}$$

Thus, if direct decay of the field φ into scalar χ-particles is possible, then one might expect that in general this will be the leading process [123]. We see from (7.9.11) that the rate at which the φ-field decays into χ-particles can be of the same order of magnitude as the rate at which the φ-field oscillates; it can therefore divest itself of most of its energy in several cycles of oscillation (or even simply in the time it takes to roll down from $\varphi \sim M_P$ to $\varphi = 0$ [254]). Since $H(\varphi) \sim m_\varphi$ at the end of inflation, the universe has almost no time to expand during reheating, and almost all the energy stored in the field φ can be converted into energy for the production of χ-particles. This same result follows from (7.9.8):

$$\rho^* = \frac{m_\varphi^2}{2}\varphi^2 \lesssim \frac{m_\varphi^2 M_P^2}{24}, \tag{7.9.14}$$

whence $\varphi(t^*) \lesssim M_P$ and

$$T_R \lesssim 10^{-1}\sqrt{m_\varphi M_P} \sim 10^{15}\ \text{GeV}. \tag{7.9.15}$$

Reheating to $T_R \sim 10^{15}$ GeV occurs only for a special choice of parameters. Moreover, in some models the universe cannot be heated up to a temperature much higher than m_φ; see Section 7.10. Nevertheless, one must keep in mind the possibility of such an efficient reheating, which can take place immediately after inflation ends, despite the weakness of the interaction between the φ- and χ-fields. A similar possibility can come into play if the potential $V(\varphi)$ takes on a more complicated form — for example, if the curvature of $V(\varphi)$ near its minimum is much greater than at $\varphi \sim M_P$ [255].

If the field φ can only decay into fermions, then we find from (7.9.12) and (7.9.10) that in the simplest models, the temperature of the universe after reheating will be at least three orders of magnitude lower,

$$T_R \lesssim 10^{-1} m_\varphi \sim 10^{12}\ \text{GeV}, \tag{7.9.16}$$

and if gravitational effects are dominant, then

$$T_R \lesssim 10^{-1} m_\varphi \sqrt{\frac{m_\varphi}{M_P}} \sim 10^9\ \text{GeV}. \tag{7.9.17}$$

The foregoing estimates have been made using the simplest model and assuming that the oscillating field is small. But if the field φ is large ($\nu\sigma\varphi > m_\chi^2$ or $h\varphi > m_\psi$), it is not enough to calculate just the polarization operator; one must then either calculate the imaginary part of the effective action $S(\varphi)$ in the external field $\varphi(t)$ [122, 256], or employ methods based on the Bogolyubov transformation [74].

We shall not discuss this point in detail here, since for the theory (7.9.1), an investigation of the case in which $\nu\sigma\varphi > m_\chi^2$ and $h\varphi > m_\psi$ results only in a change in the numerical coefficients in (7.9.6) and (7.9.7). More important changes arise in the theories with Lagrangians that lack ternary interactions like $\varphi\chi^2$ and $\varphi\bar{\psi}\psi$, having only vertices like φ^4, $\varphi^2\chi^2$, or $\varphi^2 A_\mu^2$, with a field φ that has no classical part φ_0.

Thus, for example, in the $\frac{\lambda}{4}\varphi^4$ theory of the massless field, evaluation of the imaginary part of the effective Lagrangian $L(\varphi)$ leads to an expression for the probability of pair production [122]:

$$P \sim 2\ \text{Im}\, L(\varphi) \sim \lambda^2 \varphi^4 \cdot O(10^{-3}). \tag{7.9.18}$$

An analogous expression holds for a $\lambda\varphi^2\chi^2$ theory. The energy density of particles created in a time $\Delta t \sim H^{-1}$ is

$$\Delta\rho \sim 10^{-3}\lambda^2\varphi^4 \cdot \sqrt{\lambda}\,\varphi \cdot H^{-1} \sim 10^{-3}\lambda^2\varphi^3 M_P, \tag{7.9.19}$$

where the effective mass of the φ- and χ-fields is $O(\sqrt{\lambda}\,\varphi)$. This quantity becomes comparable to the total energy density $\rho(\varphi) \sim \frac{\lambda}{4}\varphi^4$ when

$$\varphi \lesssim 10^{-2}\lambda M_P, \tag{7.9.20}$$

that is, when

$$\rho(\varphi) \sim 10^{-8}\lambda^5 M_P^4, \tag{7.9.21}$$

whereupon

$$T_R \lesssim 10^{-3}\lambda^{5/4} M_P. \tag{7.9.22}$$

For $\lambda \sim 10^{-14}$,

$$T_R \lesssim 3 \cdot 10^{-21} M_P \sim 3 \cdot 10^{-2}\,\mathrm{GeV}. \tag{7.9.23}$$

If the field φ in the theory $\frac{\lambda}{4}\varphi^4$ has a nonvanishing mass m_φ, the reheating of the universe becomes ineffectual at $\varphi \lesssim \frac{m_\varphi}{\sqrt{\lambda}}$, since at small φ, the value of $\Delta\rho$ from (7.9.19) always turns out to be less than $\rho(\varphi) \sim \frac{m_\varphi^2}{2}\varphi^2$. In such a situation, the energy of the field φ basically falls due to the expansion of the universe, $\rho(\varphi) \sim a^{-3}$, rather than due to the field decay. This implies that after expansion of the universe to its present state, even a strongly interacting oscillating classical field φ ($10^{-14} \ll \lambda \lesssim 1$) can turn out to be largely undecayed into elementary particles, and can make a sizable contribution to the density of dark matter in the universe.

7.10 The origin of the baryon asymmetry of the universe

As we have already noted, the elaboration of feasible mechanisms for generating an excess of baryons over antibaryons in the inflationary universe [36–38] was one of the most important stages in the development of modern cosmology. The baryon asymmetry problem served to demonstrate quite clearly that questions which to many had seemed meaningless, or at best metaphysical ("Why is the universe structured as it is, and not otherwise?"), could actually have a physical answer. Without a solution of the baryogenesis problem, the inflationary universe scenario would be impossible, since the density of baryons that exist at the earliest stages of evolution of the universe becomes exponentially small after inflation. The generation of a baryon asymmetry in the universe is therefore just as indispensable an element of the inflationary universe scenario as the reheating of the universe discussed in the previous section.

As the first treatment of the origin of baryons in the universe made clear [36], an asymmetry between the number of baryons and antibaryons arises when three conditions are satisfied:

1. The processes involved violate baryon charge conservation.
2. These processes also violate C and CP invariance.
3. Baryon production processes take place in a nonequilibrium thermodynamic state. One example would be the decay of particles with mass $M \gg T$.

The need for the first condition is obvious. The second is needed in order for the decay of particles and antiparticles to produce different numbers of baryons and antibaryons. The third condition is primarily needed to prevent inverse processes which might destroy baryon asymmetry.

Genuine interest in the possibility of generating the baryon asymmetry of the universe was kindled by the advent of grand unified theories, in which baryons could freely transform into leptons prior to symmetry breaking between the strong and electroweak interactions. After the symmetry breaking, superheavy scalar and vector particles (Φ, H, X, and Y) decay into baryons and leptons. If the decay of these particles takes place in a state far removed from thermodynamic equilibrium, so that the inverse processes of baryon and lepton reversion to superheavy particles are inhibited, and if C and CP invariance is violated, then the decays will produce slightly different numbers of baryons and antibaryons. This difference, after annihilation of baryons and antibaryons, is exactly what produces the baryonic matter that we see in our universe. The small number $\frac{n_B}{n_\gamma} \sim 10^{-9}$ comes about as a product of the gauge coupling constant in grand unified theories, the constant related to the strength of CP violation, and the relative abundance of the particles that produce baryon asymmetry after their decay [38].

We shall not dwell here on a detailed description of this mechanism for baryogenesis, referring the reader instead to the excellent reviews in [105, 257, 258]. What is important for us is that theories leading to the desired result $\frac{n_B}{n_\gamma} \sim 10^{-9}$ actually do exist. A similar mechanism can also operate as a part of the inflationary universe scenario, where it is even more effective, since the universe is reheated after inflation in what is essentially a nonequilibrium process, and during this process, superheavy particles can be produced with masses much greater than the temperature of the universe after reheating T_R [122]. However, in the minimal SU(5) theory with a single family of Higgs bosons H and the most natural relationship between coupling constants, $\frac{n_B}{n_\gamma}$ turns out to be many orders of magnitude less than 10^{-9}. In order to get $\frac{n_B}{n_\gamma} \sim 10^{-9}$, it is necessary either to introduce two additional families of Higgs bosons, or to consider the possibility of a complex sequence of phase transitions as the universe cools in the SU(5) theory [259]. Furthermore, it is far from easy to obtain the fairly large number of superheavy bosons needed to implement this mechanism at the time of post-inflation reheating. This is especially difficult in supergravity-type theories, where the temperature to which the universe is reheated is usually 10^{12} GeV at most. Finally, one more potential problem became apparent fairly recently. It was found that nonperturbative effects lead to efficient annihilation of baryons and leptons at a temperature that is higher than or of the same order as the phase transition temperature $T_c \sim 200$ GeV in the Glashow–Weinberg–Salam model [129]. This means that if equal numbers of baryons and leptons are generated in the early stages of evolution of the universe, so that $B - L = 0$, where B and L are the baryon and lepton charges respectively (and this is exactly the situation in the simplest models of baryogenesis [38]), then the entire baryon asymmetry of the universe arising at $T > 10^2$ GeV subsequently vanishes. If this is so, then either it is necessary to have theories that begin with an asymmetry $B - L \neq 0$, which makes these theories even more complicated, or mechanisms for baryogenesis which could operate efficiently even at a temperature $T \lesssim 10^2$ GeV must be worked out. Several possible mechanisms of this kind have been proposed in recent years. Below we describe one of them, the details of which are probably the closest of any to the inflationary universe scenario.

The basic idea behind that scheme was proposed in a paper by Affleck and Dine [97]; their mechanism was implemented in the context of the inflationary universe scenario [98]. Later, it was demonstrated that this mechanism could work in models based on superstring theory [260]. Referring the reader to the original literature for details, we discuss here the basic outlines of this new mechanism for baryogenesis.

As an example, consider a supersymmetric SU(5) grand unification theory. In this theory there exist squarks and sleptons, which are scalar fields, the superpartners of quarks and leptons. An analysis of the shape of the effective squark and slepton potential shows that it has valleys — flat directions — in which the effective potential approaches zero [97]. We will refer to the corresponding linear combinations of squark and slepton fields in the flat directions as the scalar field φ. After supersymmetry breaking in the model in question, the value of the effective potential $V(\varphi)$ in the valleys rises slightly, and the field φ acquires an effective mass $m \sim 10^2$ GeV. Excitations of this field consist of electrically neutral unstable particles having baryon and lepton charge $B = L = \pm 1$. The baryon charge of each such particle is not conserved by their interactions, but the difference $B - L$ is. These particles interact among themselves via the same gauge coupling constant g as do the quarks. The coupling constant of the baryon-nonconserving interactions of the φ-particles is $\lambda = O\left(\frac{m^2}{M_X^2}\right)$, where M_X is the X-boson mass in the SU(5) theory. For large values of the classical field φ, many of the particles that interact with it acquire a very high mass that is $O(g\varphi)$. However, there are also light particles such as quarks, leptons, W mesons, and so forth, that interact only indirectly with the field φ (via radiative corrections), with an effective coupling constant $\tilde{\lambda} \sim \left(\frac{\alpha_s}{\pi}\right)^2 \frac{m^2}{\varphi^2}$, where $\alpha_s = \frac{g^2}{4\pi}$. In a rigorous treatment, it would be necessary to consider the dynamics of the two fields v and a, corresponding to different combinations of squark–slepton fields in the valley of the effective potential [97]. A thorough study of a system of such fields in the SU(5) theory would be fairly complicated, but fortunately, in the most important instances, it can be reduced to the study of one simple model that describes the complex scalar field $\varphi = \frac{1}{\sqrt{2}}(\varphi_1 + i\varphi_2)$, with the somewhat unusual potential [97, 98]

$$V(\varphi) = m^2\varphi^*\varphi + \frac{i}{2}\lambda\left[\varphi^4 - (\varphi^*)^4\right]. \tag{7.10.1}$$

The quantity $j_\mu = -i\,\varphi^*\overleftrightarrow{\partial}_\mu\varphi = \frac{1}{2}(\varphi_1\partial_\mu\varphi_2 - \varphi_2\partial_\mu\varphi_1)$ corresponds to the baryon current of scalar particles in the SU(5) model, while j_0 is the baryon charge density n_B of the field φ. The equations of motion of the fields φ_1 and φ_2 are

$$\ddot{\varphi}_1 + 3H\dot{\varphi}_1 = -\frac{\partial V}{\partial\varphi_1} = -m^2\varphi_1 + 3\lambda\varphi_1^2\varphi_2 - \lambda\varphi_2^3, \tag{7.10.2}$$

$$\ddot{\varphi}_2 + 3H\dot{\varphi}_2 = -\frac{\partial V}{\partial\varphi_2} = -m^2\varphi_2 - 3\lambda\varphi_2^2\varphi_1 + \lambda\varphi_1^3. \tag{7.10.3}$$

During inflation, when H is a very large quantity, the fields φ_i evolve slowly, so that as usual the terms $\ddot{\varphi}_i$ in (7.10.2) and (7.10.3) can be neglected. This then leads to an expression for the density n_B at the time of inflation:

$$n_B \equiv j_0 = \frac{1}{3H}\left(\varphi_1\frac{\partial V}{\partial\varphi_2} - \varphi_2\frac{\partial V}{\partial\varphi_1}\right) = \frac{\lambda}{3H}\left(\varphi_1^4 - 6\varphi_1^2\varphi_2^2 + \varphi_2^4\right). \tag{7.10.4}$$

If, for example, we take the initial conditions to be $\varphi_2 \gtrsim \frac{1}{4}\varphi_1 > 0$, $\lambda\varphi_i^2 \ll m^2$, then we find from (7.10.2)–(7.10.4) that during inflation the field φ_1 evolves very slowly, and it remains much smaller than φ_2, so that during the inflationary stage n_B is approximately constant in magnitude and equal to its initial value, $n_B \sim \frac{\lambda}{3H}\varphi_2^4$.

To clarify the physical meaning of this result, let us write down the equation for the partially conserved current in our model in the following form:

$$a^{-3}\,\frac{d\left(n_B a^3\right)}{dt} \equiv \dot{n}_B + 3n_B H = i\left(\varphi^*\frac{\partial V}{\partial\varphi} - \varphi\frac{\partial V}{\partial\varphi^*}\right), \tag{7.10.5}$$

where $a(t)$ is the scale factor. If there were no term $\sim i\lambda\left(\varphi^4 - (\varphi^*)^4\right)$ in (7.10.1) leading to nonconservation of baryon charge, the total baryon charge of the universe $B \sim n_B a^3$ would be constant, and the baryon charge density n_B would become exponentially small at the time of inflation. In our example, however, the right-hand side of (7.10.5) does not vanish, and serves as a source of baryon charge. Bearing in mind then that all fields vary very slowly during inflation, so that $\dot{n}_B \ll 3 n_B H$, (7.10.5) again implies (7.10.4), as obtained previously.

In other words, due to the presence of the last term in (7.10.1), the baryon charge density varies very slowly during inflation, as do the fields φ_i (see (7.10.4)), while the total baryon charge of the part of the universe under consideration grows exponentially. The baryon charge density and its sign depend on the initial values of the fields φ_i, and will be different in different parts of the universe.

When the rate of expansion of the universe becomes low, the field begins to oscillate in the vicinity of the minimum of $V(\varphi)$ at $\varphi = 0$. While this is going on, the gradual decrease in the amplitude of oscillation reduces terms $\sim \lambda \varphi^4$ in (7.10.5) which are responsible for the nonconservation of baryon charge; the total baryon charge of the field φ will then be conserved, and its density will fall off as $a^{-3}(t)$. Notice that at that point the energy density of the scalar field $\rho \sim \frac{m^2\varphi^2}{2}$ will also fall off as $a^{-3}(t)$. This coincidence has a very simple meaning. As we have already discussed, a homogeneous field φ oscillating with frequency m can be represented as a coherent wave consisting of particles of the field φ with particle density $n_\varphi = \frac{\rho}{m} = \frac{m}{2}\varphi^2$, where φ is the amplitude of the oscillating field. Some of these particles have baryon charge $B = +1$, and some have $B = -1$. The baryon charge density n_B is thus proportional to n_φ, so that the ratio $\frac{n_B}{n_\varphi}$ is time-independent and cannot be bigger than unity in absolute value:

$$\frac{|n_B|}{n_\varphi} = \text{const} \leq 1 . \qquad (7.10.6)$$

This ratio $\frac{n_B}{n_\varphi}$ is determined by the initial conditions. The oscillatory regime sets in at $H \sim m$, so (7.10.4) implies that at that stage

$$\frac{n_B}{n_\varphi} \sim \frac{\lambda \tilde{\varphi}_2^4}{3m} , \qquad (7.10.7)$$

where $\tilde{\varphi}_2$ is the value of the field φ_2 at the onset of the oscillation stage. In the realistic SU(5) model, Eq. (7.10.7) also contains a factor cos 2θ, where θ is the angle between the v and a fields in the complex plane. The φ-particles are unstable and decay into leptons and quarks. The temperature of the universe rises at that point, but it cannot rise much higher than m, since at high temperature the quarks have an effective mass $m_q \sim g\mathrm{T} \sim \mathrm{T}$, so that the field φ cannot decay when $\mathrm{T} \gg m$. The field φ therefore oscillates and decays gradually, rather than suddenly, and in the process it warms the universe up to a constant temperature $\mathrm{T} \sim m \sim 10^2$ GeV. By the end of this stage, the entire baryon charge of the scalar field has been transformed into the baryon charge of the quarks, and for every quark or antiquark produced through the decay of a φ-particle, there is approximately one photon of energy $\mathrm{E} \sim \mathrm{T} \sim m$. This means that the density of photons n_γ produced by the decaying field φ is of the same order of magnitude as n_φ. The baryon asymmetry of the universe thus engendered is

$$\frac{n_\mathrm{B}}{n_\varphi} \sim \frac{n_\mathrm{B}}{n_\gamma} \sim \cos 2\theta \cdot \frac{\lambda\, \tilde{\varphi}_2^2}{m^2} \sim \cos 2\theta \cdot \frac{\tilde{\varphi}_2^2}{\mathrm{M}_\mathrm{X}^2}. \qquad (7.10.8)$$

Note that this equation is only valid when $\lambda\, \tilde{\varphi}_2^2 \ll m$, that is, when $\tilde{\varphi}_2 \ll \mathrm{M_X}$, so only from that point onward can the violation of baryon charge conservation be neglected, with the quantity $\frac{n_\mathrm{B}}{n_\varphi}$ becoming constant. As expected from (7.10.6), the baryon asymmetry of the universe as given by (7.10.8) then turns out to be less than unity. However, from (7.10.8), it follows that the mechanism of baryogenesis discussed above may even be too efficient. For example, for $\tilde{\varphi}_2 \sim \mathrm{M_X}$, Eq. (7.10.8) yields $\frac{n_\mathrm{B}}{n_\gamma} = \mathrm{O}(1)$. We must therefore try to understand what $\tilde{\varphi}_2$ should be equal to, and how to reduce $\frac{n_\mathrm{B}}{n_\varphi}$ to the desirable value $\frac{n_\mathrm{B}}{n_\varphi} \sim 10^{-9}$.

Research into this question has shown that just like money, baryon asymmetry is hard to come by but easy to get rid of [98]. One mechanism for reducing the baryon asymmetry is the previously cited nonperturbative scheme [129]. If, for example, the temperature of the universe following decay of the field φ exceeds approximately 200 GeV, then virtually the entire baryon asymmetry that has been produced will disappear, with the exception of a small part resulting from processes that violate B – L invariance. This residual can in fact account for the observed asymmetry

$\frac{n_B}{n_\varphi} \sim 10^{-9}$. Another possibility is that the temperature is less than 200 GeV when decay of the φ-field ends; that is, the baryons do not burn up, but the initial value of the field φ is fairly small. This could happen, for instance, if the fields φ_i were to vanish due to high-temperature effects or interaction with the fields responsible for inflation. The role played by these fields would then be taken up by their long-wave quantum fluctuations with an amplitude proportional to $\frac{H}{2\pi}\sqrt{H\,t}$ (see (7.3.12)), which could be several orders of magnitude less than M_X.

Finally, the Anthropic Principle provides one more plausible explanation of why $\frac{n_B}{n_\gamma}$ is so small in the observable part of the universe. The fields φ_i and the quantity cos 2θ take on all possible values in different regions of the universe. In most such regions, φ can be extremely large, and $|\cos 2\theta| \sim 1$. But these are regions with $\frac{n_B}{n_\gamma} \gg 10^{-9}$, and life of our type it is impossible. The reason is that for a given amplitude of perturbations $\frac{\delta\rho}{\rho}$, elevating the baryon density by just two to three orders of magnitude results in the formation of galaxies having extremely high matter density and a completely different complement of stars. It is therefore not inconsistent to think that there are relatively few regions of the universe with small initial values of φ and cos 2θ — but these are just the regions with the highest likelihood of supporting life of our type. We will discuss this problem in a more detailed way in Chapter 10.

In addition to the mechanism discussed above, several more that could operate at temperatures $T \lesssim 10^2$ GeV have recently been proposed [130, 131, 178, 261–263]. It is still difficult to say which of these are realistic. One important point is that many ways have been found to explain the baryon asymmetry of the universe, and superhigh temperatures $T \sim M_X \sim 10^{14}$–10^{15} GeV, which occur only after extremely efficient reheating, are not at all mandatory. In principle, baryon asymmetry could even occur if the temperature of the universe never exceeded 100 GeV! This then substantially facilitates the construction of realistic models of the inflationary universe. On the other hand, the realization that it is possible to construct a consistent theory of the evolution of the universe in which the temperature may never exceed $T \sim 10^2$ GeV $\sim 10^{-17} M_P$ leads us yet again to ponder the extent to which our notions have changed over the past few years, and to wonder what surprises might await us in the future.

8

The New Inflationary Universe Scenario

8.1 Introduction. The old inflationary universe scenario

In the previous chapter, we described the building blocks needed for a complete theory of the inflationary universe. It is now time to demonstrate how all parts of the theory that we have described thus far may be combined into a single scenario, implemented in the context of some of recently developed theories of elementary particles. As we have already pointed out, however, there are presently two significantly different fundamental versions of inflation theory, namely the new inflationary universe scenario [54, 55], and the chaotic inflation scenario [56, 57]. Although we lean toward the latter, in view of its greater naturalness and simplicity, it is still too soon to render a final decision. Moreover, many of the results obtained in the course of constructing the new inflationary universe scenario will prove useful, even if the scenario itself is to be abandoned. We therefore begin our exposition with a description of the various versions of the new inflationary universe scenario, and in the next chapter we turn to a description of the chaotic inflation scenario. Our description of the former would be incomplete, however, if we did not say a few words about the old inflationary universe scenario proposed in the important paper by Guth [53].

As stated in Chapter 1, the old scenario was based on the study of phase transitions from a strongly supercooled unstable phase $\varphi = 0$ in grand unified theories. The theory of such phase transitions had been worked out long before Guth's effort (see Chapter 5), but nobody had attempted to use that theory to resolve such cosmological problems as the flatness of the universe or the horizon problem.

Guth drew attention to the fact that upon strong supercooling, the energy density of relativistic particles, being proportional to T^4, becomes negligible in comparison with the vacuum energy $V(\varphi)$ in the vacuum state $\varphi = 0$. This then means that in the limit of extreme supercooling, the energy density ρ of an expanding (and cooling) universe tends to V(0) and ceases to depend on time. At large t, then, according to (1.3.7), the universe expands exponentially,

$$a(t) \sim e^{Ht}, \tag{8.1.1}$$

where the Hubble constant at that time is

$$H = \sqrt{\frac{8\pi V(0)}{3M_P^2}} . \tag{8.1.2}$$

If all of the energy is rapidly transformed into heat at the time of a phase transition to the absolute minimum of $V(\varphi)$, the universe will be reheated to a temperature $T_R \sim [V(0)]^{1/4}$ after the transition, regardless of how long the previous expansion went on (this circumstance was exploited earlier by Chibisov and the present author to construct a model of the universe which could initially be cold, but would ultimately be reheated by a strongly exoergic phase transition; this model has been reviewed in Refs. [24, 105]).

Since the temperature T_R to which the universe is reheated after the phase transition does not depend on the duration of the exponential expansion stage in the supercooled state, the only quantity that depends on the length of that stage is the scale factor $a(t)$, which grows exponentially at that time. But as we have already remarked, the universe becomes flatter and flatter during exponential expansion (inflation). This is an especially clear-cut effect when one considers why the total entropy of the universe is so high, $S \gtrsim 10^{87}$ (as noted in Chapter 1, this problem is closely related to the flatness problem).

Prior to the phase transition, the total entropy of the universe could be fairly low. But afterwards, it increases markedly, with

$$S \gtrsim a^3 T_R^3 \sim a^3 [V(0)]^{3/4},$$

where a^3 can be exponentially large. For example, let the exponential expansion begin in a closed universe at a time when its radius is $a_0 = c_1 M_P^{-1}$, and the vacuum energy is $V(0) = c_2 M_P^4$, where c_1 and c_2 are certain constants. In realistic theories, c_1 lies between 1 and 10^{10}, and c_2 is of order 10^{-10}; we will soon see that the quantity of interest depends very weakly on c_1 and c_2. Following exponential expansion lasting for a period Δt, the total entropy of the universe becomes

$$S \sim a_0^3 e^{3H\Delta t} T_R^3 \sim c_1^3 c_2^{3/4} e^{3H\Delta t}, \tag{8.1.3}$$

whereupon S exceeds 10^{87} if

$$\Delta t \gtrsim H^{-1}(67 - \ln c_1 c_2^{1/4}). \tag{8.1.4}$$

Under typical conditions, the absolute value of $\ln c_1 c_2^{1/4}$ will be 10 at most. The implication is that in order to solve the flatness problem, it is necessary that the universe be in a supercooled state $\varphi = 0$ for a period

$$\Delta t \gtrsim 70\, H^{-1} = 70\, M_P \sqrt{\frac{3}{8\pi V(0)}}\,. \tag{8.1.5}$$

It must be noted here that if Δt is much greater than $70\, H^{-1}$ (as will happen in any realistic version of the inflationary universe scenario), then after inflation and reheating, the universe will be almost perfectly flat, with $\Omega = \frac{\rho}{\rho_c} = 1$. Allowing for moderate local variations of ρ on the scale of the observable part of the universe, this is one of the most important observational predictions of the inflationary universe scenario.

It can readily be shown that the condition $(aT)^3 \gtrsim 10^{87}$ means that the "radius" of the universe $a \sim c_1 M_P^{-1}$ after expansion up through the present epoch will exceed the size of the observable part of the universe, $l \sim 10^{28}$ cm (see the preceding chapter). But what this means is that in a time only slightly greater (by $H^{-1} \ln c_1$) than $70\, H^{-1}$, any region of space of size $\Delta l \sim M_P^{-1}$ would have inflated so much that by the present epoch it would be larger than the observable part of the universe.

If we then bear in mind that we are considering processes taking place in the post-Planckian epoch ($\rho < M_P^4$, $T < M_P$, $t > M_P^{-1}$), it becomes clear that a region $\Delta l \sim M_P^{-1}$ in size at the onset of exponential expansion must necessarily be causally connected. Thus, in this scenario, the entire observable part of the universe results from the inflation of a single causally connected region, and the horizon problem is thereby solved.

The primordial monopole problem could in principle also be solved within the framework of the proposed scenario. Primordial monopoles are produced only at the points of collisions of several different bubbles of the field φ that are formed during the phase transition. If the phase transition is significantly delayed by supercooling, then the bubbles will become quite large by the time they begin to fill the entire universe, and the density of the monopoles produced in the process will be extremely low.

Unfortunately, however, as noted by Guth himself, the scenario that he had proposed led to a number of undesirable consequences with regard to the properties of the universe after the phase transition. Specifically, within the bubbles of the new phase, the field φ rapidly approached the equilibrium field φ_0 corresponding to the absolute minimum of $V(\varphi)$, and all the energy of the unstable vacuum with $\varphi = 0$ within the bubble was transformed into kinetic energy of the walls, which moved away from the center of the bubble at close to the speed of light. Reheating of the universe after the phase transition would have to result from collisions of the bubble walls, but due to the large size of the bubbles in this scenario, the universe after collisions between bubble walls would become highly inhomogeneous and anisotropic, a result flatly inconsistent with the observational data.

Despite all the problems encountered by the first version of the inflationary universe scenario, it engendered a great deal of interest, and in the year following the publication of Guth's work this scenario was diligently studied and discussed by many workers in the field. These investigations culminated in the papers by Hawking, Moss, and Stewart [112] and Guth and Weinberg [113], where it was stated that the defects inherent in this scenario could not be eliminated. Fortunately, the new inflationary universe scenario had been suggested by then [54, 55]; it was not only free of some of the shortcomings of the Guth scenario, but also held out the possibility of solving a number of other cosmological problems enumerated in Section 1.5.

8.2 The Coleman–Weinberg SU(5) theory and the new inflationary universe scenario (initial simplified version)

The first version of the new inflationary universe scenario was based on the study of the phase transition with the symmetry breaking SU(5) → SU(3) × SU(2) × U(1) in the SU(5)-symmetric Coleman–Weinberg theory (2.2.16). The theory of this phase transition is very complicated. We therefore start by giving somewhat of a simplified description of this phase transition, so as to elucidate the general idea behind the new scenario.

First of all, we examine how the effective potential in this theory behaves with respect to the symmetry breaking SU(5) → SU(3) × SU(2) × U(1) (2.2.16) at a finite temperature.

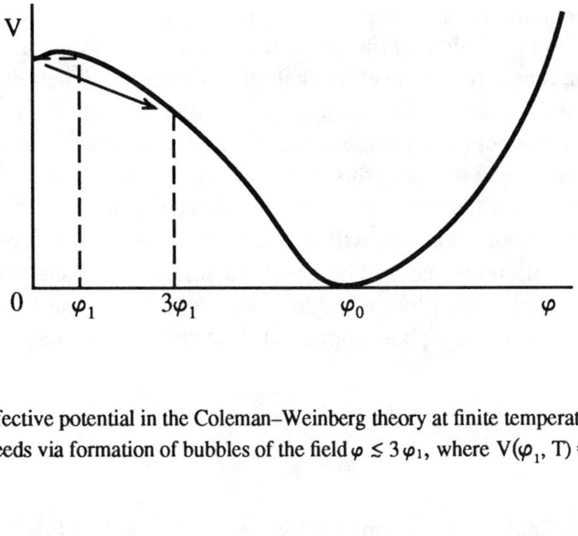

Fig. 34. Effective potential in the Coleman–Weinberg theory at finite temperature. Tunneling proceeds via formation of bubbles of the field $\varphi \lesssim 3\varphi_1$, where $V(\varphi_1, T) = V(0, T)$.

As we said in Chapter 3, symmetry is restored in gauge theories, as a rule, at high enough temperatures. It can be shown in the present case that when $T \gg M_X$, the function $V(\varphi, T)$ in the Coleman–Weinberg theory becomes

$$V(\varphi, T) = \frac{5}{8} g^2 T^2 \varphi^2 + \frac{25 g^4 \varphi^4}{128\pi^2}\left(\ln\frac{\varphi}{\varphi_0} - \frac{1}{4}\right) + \frac{9M_X^4}{32\pi^2} + cT^4 . \quad (8.2.1)$$

where c is some constant of order 10. An analysis of this expression shows that at high enough temperature T, the only minimum of $V(\varphi, T)$ is the one at $\varphi = 0$; that is, symmetry is restored. When $T \ll M_X \sim 10^{14}$ GeV, all high-temperature corrections to $V(\varphi)$ at $\varphi \sim \varphi_0$ vanish. However, the masses of all particles in the Coleman–Weinberg theory tend to zero as $\varphi \to 0$, so in the neighborhood of the point $\varphi = 0$, Eq. (8.2.1) for $V(\varphi, T)$ holds for $T \ll 10^{14}$ GeV as well. This means that the point $\varphi = 0$ remains a local minimum of the potential $V(\varphi, T)$ at any temperature T, regardless of the fact that the minimum at $\varphi \sim \varphi_0$ is much deeper when $T \ll M_X$ (Fig. 34).

In an expanding universe, a phase transition from the local minimum at $\varphi = 0$ to a global minimum at $\varphi = \varphi_0$ takes place when the typical time required for the multiple production of bubbles with $\varphi \neq 0$ becomes less than the age of the universe t. Study of this question has led many researchers to conclude that the phase transition in the Coleman–Weinberg theory is a long, drawn-out affair that takes place only when the temperature T of the universe has fallen to approximately $T_c \sim 10^6$ GeV (this is not an entirely correct statement, but for simplicity we shall temporarily assume that it is, and return to this point in Section 8.3). It is clear, however, that at such a low temperature, the barrier separating the minimum at $\varphi = 0$ from the minimum at $\varphi = \varphi_0$ will be located at $\varphi \ll \varphi_0$ (see Fig. 34), and the bubble formation process will be governed solely by the shape of $V(\varphi, T)$ near $\varphi = 0$, rather than by the value of φ_0. As a result, the field φ within the bubbles of the new phase formed in this way is at first very small,

$$\varphi \lesssim 3\varphi_1 \approx \frac{12\pi T_c}{g\sqrt{5 \ln \dfrac{M_X}{T_c}}} \ll \varphi_0 , \tag{8.2.2}$$

where the field φ_1 is determined by the condition $V(0,T) = V(\varphi_1,T)$ (see figure 26). With this value of the field, the curvature of the effective potential is relatively small:

$$|m^2| = \left|\frac{d^2V}{d\varphi^2}\right| \lesssim 75\, g^2 T_c^2 \sim 25\, T_c^2 . \tag{8.2.3}$$

The field φ within the bubble will clearly grow to its equilibrium value $\varphi \sim \varphi_0$ in a time $\Delta t \gtrsim |m^{-1}| \sim 0.2\, T_c^{-1}$. For most of this time, the field φ

will remain much smaller than φ_0. This means that over a period of order $0.2\, T_c^{-1}$, the vacuum energy of $V(\varphi, T)$ will remain almost exactly equal to $V(0)$, and consequently the part of the universe inside the bubble will continue to expand exponentially, just as at the beginning of the phase transition. Here we have the fundamental difference between the new inflationary universe scenario and the scenario of Guth, in which it is assumed that exponential expansion ceases at the moment that bubbles are formed.

When $\varphi \ll \varphi_0$ and $M_X \sim 5 \cdot 10^{14}$ GeV, the Hubble constant H is given by

$$H = \sqrt{\frac{8\pi}{3M_P^2} V(0)} = \frac{M_X^2}{2M_P}\sqrt{\frac{3}{\pi}} \approx 10^{10}\ \text{GeV}. \qquad (8.2.4)$$

In a time $\Delta t \sim 0.2\, T_c^{-1}$, the universe expands by a factor $e^{H\Delta t}$, where

$$e^{H\Delta t} \sim e^{0.2HT_c^{-1}} \sim e^{2000} \sim 10^{800}. \qquad (8.2.5)$$

To order of magnitude, the typical size of a bubble at the instant it is formed is $T_c^{-1} \sim 10^{-20}$ cm. After expansion, this size becomes $\sim 10^{800}$ cm, which is enormously greater than the size of the observable part of the universe, $l \sim 10^{28}$ cm. Thus, within the scope of this scenario, the entire observable part of the universe should lie *within a single bubble*. We therefore see no inhomogeneities that might arise from bubble wall collisions.

As in the Guth scenario, exponential expansion by a factor of more than e^{70} (8.2.5) enables one to resolve the horizon and flatness problems. But more than that, it makes it possible to explain the large-scale homogeneity and isotropy of the universe (see Chapter 7).

Since bubble sizes exceed the dimensions of the observable part of the universe, and since monopoles and domain walls are only produced near bubble walls, there should be not a single monopole or domain wall in the observable part of the universe, which removes the corresponding problems discussed in Section 1.5.

Note that the curvature of the effective potential (8.2.1) grows rapidly with increasing field φ. The slow-growth stage of the field φ, which is accompanied by exponential expansion of the universe, is therefore replaced by a stage with extremely rapid attenuation of the field φ to its equilibrium value $\varphi = \varphi_0$, where it oscillates about the minimum of the effective potential. In the model in question, the oscillation frequency is equal to the mass of the Higgs field φ when $\varphi = \varphi_0$, $m = \sqrt{V''(\varphi_0)} \sim 10^{14}$ GeV. The typical period of oscillation $\sim m^{-1}$ is evidently many orders of magnitude less than the characteristic expansion time of the universe H^{-1}. In studying oscillations of the field φ near the point φ_0, one can therefore neglect the expansion of the universe. This means that at the stage we are considering, all of the potential energy V(0) is transformed into the energy of the oscillating scalar field. The oscillating classical field φ produces Higgs bosons and vector bosons, which quickly decay. In the end, all of the energy of the oscillating field φ is transformed into the energy of relativistic particles, and the universe is reheated to a temperature [123, 124]

$$T_R \sim [\,V(0)]^{1/4} \sim 10^{14}\ \text{GeV}.$$

The mechanism for reheating the universe in the new scenario is thus quite different from the corresponding mechanism in the Guth scenario.

The baryon asymmetry of the universe is produced when scalar and vector mesons decay during the reheating of the universe [36–38]. Because of the fact that processes taking place at that time are far from equilibrium, however, the baryon asymmetry is produced much more efficiently in this model than in the standard hot universe theory [123].

We see, then, that the fundamental idea behind the new inflationary universe scenario is quite simple: it requires that symmetry breaking due to growth of the field φ proceed fairly slowly at first, giving the universe a chance to inflate by a large factor, and that in the later stages of the process, the rate of growth and oscillation frequency of the field φ near the minimum of $V(\varphi)$ be large enough to ensure that the universe is reheated efficiently after the phase transition. This idea has been used both in a refined version of the new scenario, which we discuss next, and in all subsequent variants of the inflationary universe scenario.

8.3 Refinement of the new inflationary universe scenario

The description of the new inflationary universe scenario in the previous section was oversimplified, its main drawback being our neglect of the effects of exponential expansion of the universe on the kinetics of a phase transition. When $T \gg H \sim 10^{10}$ GeV, such a simplification is completely admissible, but according to the discussion in Section 8.2, the phase transition can only begin when $T_c \ll H$. In that event, high-temperature effects exert practically no influence on the kinetics of the phase transition. Indeed, the typical time over which bubbles might be formed at a temperature T_c must certainly be greater than

$$m^{-1}(\varphi = 0, T = T_c) \sim (gT_c)^{-1} \gg H^{-1}.$$

But in that much time, the universe expands by a factor of approximately e^{H/gT_c}, and the temperature falls from $T = T_c$ practically to zero. Thus, the role of high-temperature effects is just to place the field φ at the point $\varphi = 0$, and one can then neglect all high-temperature effects in describing the formation of bubbles of the field φ and the process by which φ rolls down to φ_0. It is necessary, however, to take account of effects related to the rapid expansion of the universe, since at that time $H \gg T$.

The resulting refinement of the scenario takes place in several steps.

1) In studying the evolution of the field φ in an inflationary universe, one must make allowance for the fact that the equation of motion of the field is modified, and takes the form

$$\ddot{\varphi} + 3H\dot{\varphi} - \frac{1}{a^2}\nabla^2\varphi = -\frac{dV}{d\varphi}. \tag{8.3.1}$$

If the effective potential is not too steep, the $\ddot{\varphi}$ term in (8.3.1) can be discarded, so that the homogeneous field φ satisfies the equation

$$\dot{\varphi} = -\frac{1}{3H}\frac{dV}{d\varphi}. \tag{8.3.2}$$

In particular, (8.3.2) implies that with $H = \text{const} \gg m$ in a theory with $V = V(0) + \frac{m^2}{2}\varphi^2$,

$$\varphi \sim \varphi_0 \exp\left(-\frac{m^2}{3H}t\right), \tag{8.3.3}$$

and in a theory with $V = V(0) - \frac{m^2}{2}\varphi^2$,

$$\varphi \sim \varphi_0 \exp\left(+\frac{m^2}{3\mathrm{H}}t\right). \tag{8.3.4}$$

This means, in particular, that the curvature of the effective potential at $\varphi = 0$ need not necessarily be zero. To solve the flatness and horizon problems, it is sufficient that the field φ (as well as the quantity $V(\varphi)$) vary slowly over a time span $\Delta t \gtrsim 70\,\mathrm{H}^{-1}$. In conjunction with (8.2.4), this condition leads to the constraint

$$|m^2| \lesssim \frac{\mathrm{H}^2}{20}. \tag{8.3.5}$$

It will also be useful to investigate the evolution of a classical field in the theory described by

$$V(\varphi) = V(0) - \frac{\lambda}{4}\varphi^4. \tag{8.3.6}$$

In that case, it follows from (8.3.2) that

$$\frac{1}{\varphi_0^2} - \frac{1}{\varphi^2} = \frac{2\lambda}{3\mathrm{H}}(t - t_0), \tag{8.3.7}$$

where φ_0 is the initial value of the field φ. This means that the field becomes infinitely large in a finite time

$$t - t_0 = \frac{3\mathrm{H}}{2\lambda\varphi_0^2}. \tag{8.3.8}$$

If $\lambda\varphi_0^2 \ll \mathrm{H}^2$, then $t - t_0 \gg \mathrm{H}^{-1}$, and φ will spend most of this time span in its slow downhill roll. It is only at the end of the interval (8.3.8) that the field quickly rolls downward, with $\varphi \to \infty$, in a time $\Delta t \sim \mathrm{H}^{-1}$. For $\lambda\varphi_0^2 \ll \mathrm{H}^2$, therefore, the duration of the inflationary stage in the theory (8.3.6) as the field φ rolls down from $\varphi = \varphi_0$ is $\frac{3\mathrm{H}}{2\lambda\varphi_0^2}$ (8.3.8) (to within $\Delta t \sim \mathrm{H}^{-1}$). This result will shortly prove useful.

2) Corrections to the expression (8.2.1) for $V(\varphi)$ arise in de Sitter space. If we limit attention, as before, to the contribution to $V(\varphi)$ from heavy vector particles (see Chapter 2), then for small φ ($e\varphi \ll \mathrm{H}$), $V(\varphi)$ takes the form [264, 265]

$$V(\varphi, R) = \frac{\mu_1^2}{2} R + \frac{e^2 R}{64\pi^2} \varphi^2 \ln \frac{R}{\mu_2^2} + \frac{3e^4 \varphi^4}{64\pi^2} \ln \frac{R}{\mu_3^2} + V(0, R) , \quad (8.3.9)$$

where R is the curvature scalar ($R = 12\, H^2$), and the μ_i are some normalization factors with dimensions of mass, whose magnitude is determined by the normalization conditions imposed on $V(\varphi, R)$. When $V(\varphi) \ll M_P^4$, the corresponding corrections to the effective potential $V(\varphi)$ itself are extremely small, although they can induce significant corrections to the quantity $m^2 = \frac{d^2V}{d\varphi^2}\big|_{\varphi=0}$ that are of order e^2H^2, and these can prevent (8.3.5) from being satisfied. Fortunately, there does exist a choice of normalization conditions (i.e., a redefinition of the Coleman–Weinberg theory in curved space) for which this does not happen, and for which m^2 remains equal to zero. We shall not pursue this problem any further here, referring the reader to Ref. [265] for a discussion of the renormalization of $V(\varphi, R)$ for the Coleman–Weinberg theory in de Sitter space.

3) The most important refinement of the scenario has to do with the first stage of growth of the field φ. As stated earlier, some time $\tau \sim O\,(H^{-1})$ after the temperature of the universe has dropped to $T \sim H$, the temperature and effective mass of the field φ at the point $\varphi = 0$ become exponentially small. At that time, the effective potential $V(\varphi)$ (8.2.1) in the neighborhood of interest around $\varphi = 0$ (with $H \lesssim \varphi \lesssim \frac{H}{\sqrt{\lambda}}$) can be approximated by (8.3.6), where

$$\lambda \sim \frac{25 g^4}{32\pi^2}\left(\ln\frac{H}{\varphi_0} - \frac{1}{4}\right),\ V(0) = \frac{9M_X^4}{32\pi^2}. \quad (8.3.10)$$

According to Eq. (8.3.8), the classical motion of the field φ, starting out from the point $\varphi_0 = 0$, would go on for an infinitely long time. As we noted in Section 7.3, however, quantum fluctuations of the field φ in the inflationary universe engender long-wave fluctuations in the field, and on a scale $l \sim H^{-1}$, these look like a homogeneous classical field. Making use of (7.3.12), the rms value of this field (averaged over many independent regions of size $l \gtrsim H^{-1}$), is

$$\varphi \sim \frac{H}{2\pi}\sqrt{H(t-t_0)}\,. \tag{8.3.11}$$

In the case at hand, t_0 is the time at which the effective mass squared of the field φ at $\varphi = 0$ becomes much less than H^2.

Long-wave fluctuations of the field φ can play the role of the initial nonzero field φ in Eq. (8.3.7). Here, however, we must voice an important reservation. In different regions of the universe, the fluctuating field φ will take on different values; in particular, there will always be regions in which φ doesn't decrease at all, giving rise to a self-regenerating inflationary universe [266, 267, 204] analogous to that of the chaotic inflation scenario [57, 132, 133] (see Section 1.8). Further on, we shall discuss the average behavior of the fluctuating field φ (8.3.11).

During the first stage of the process, fluctuating (diffusive) growth of the field φ takes place more rapidly than the classical rolling:

$$\dot{\varphi} \sim \frac{H^2}{4\pi\sqrt{H(t-t_0)}} \gg \frac{\lambda\varphi^3}{3H} \sim \frac{\lambda H^2[H(t-t_0)]^{3/2}}{6\pi\sqrt{2\pi}}\,. \tag{8.3.12}$$

This stage lasts for a time

$$\Delta t = t - t_0 \sim \frac{\sqrt{2}}{H\sqrt{\lambda}}\,, \tag{8.3.13}$$

during which the mean field φ (8.3.11) rises to

$$\varphi_0 \sim \frac{H}{2\pi}\left(\frac{2}{\lambda}\right)^{\frac{1}{4}}. \tag{8.3.14}$$

To a good approximation, subsequent evolution of the field φ may be described by Eq. (8.3.7), where we must substitute $t_0 + \Delta t$ for t_0. The overall duration of the rolling of the field φ from $\varphi = \varphi_0$ to $\varphi = \infty$ is

$$t - (t_0 + \Delta t) = \frac{3H}{2\lambda\varphi_0^2} = \frac{3\sqrt{2}\,\pi}{\sqrt{\lambda}\,H}\,, \tag{8.3.15}$$

and the total duration of inflation is given by

$$t - t_0 \sim \frac{4\sqrt{2}\,\pi}{\sqrt{\lambda}\,H}. \tag{8.3.16}$$

During that time, the size of the universe grows by approximately a factor of

$$\exp\left(H(t - t_0)\right) \sim \exp\left(\frac{4\sqrt{2}\,\pi}{\sqrt{\lambda}}\right). \tag{8.3.17}$$

The condition $H(t - t_0) \gtrsim 70$ leads to the constraint [265, 128, 134, 135]

$$\lambda \lesssim \frac{1}{20}, \tag{8.3.18}$$

which can also be satisfied, in principle, in the SU(5) Coleman–Weinberg theory.

With minor modifications, the foregoing discussion also applies to the case in which $m^2 \equiv V''(0) < 0$, $|m^2| \ll H^2$, as well as to the case in which the effective potential has a shallow local minimum at $\varphi = 0$ — that is, when $0 < m^2 \ll H^2$.

In the first of these two instances, the process whereby the field rolls down from the point $\varphi = 0$ is analogous to the previous situation. In the second, diffusion of the field φ looks like tunneling, the theory of which was discussed in Section 7.4.

Clearly, the details of the behavior of the scalar field φ as it undergoes a phase transition from the point $\varphi = 0$ to a minimum of $V(\varphi)$ at $\varphi = \varphi_0$ differ from the description given in the preceding section. Nevertheless, most of the qualitative conclusions having to do with the existence of an inflationary regime in the Coleman–Weinberg theory remain valid.

Unfortunately, however, the original version of the new inflationary universe scenario, based on the theory (8.2.1), is not entirely realistic, the point being that fluctuations of the scalar field φ that are generated during the inflationary stage give rise to large density inhomogeneities by the time inflation has ended. Specifically, according to (7.5.22), after the inflation, reheating, and subsequent cooling of the universe, density inhomogeneities

$$\frac{\delta\rho(\varphi)}{\rho} = \frac{48}{5}\sqrt{\frac{2\pi}{3}}\frac{[V(\varphi)]^{3/2}}{M_P^3 V'(\varphi)} \tag{8.3.19}$$

will be produced. In this expression, φ is the value of the field at the time when the corresponding fluctuations $\delta\varphi$ had a wavelength $l \sim k^{-1} \sim H^{-1}$. In the new inflationary universe scenario, $V(\varphi) \sim V(0)$ at the time of inflation. Let us estimate the present-day wavelength of a perturbation whose wavelength was previously $l \sim [H(\varphi)]^{-1}$. Equation (8.3.8) tells us that after the field becomes equal to φ, the universe still has an inflation factor of $\exp\left(\frac{3H^2}{2\lambda\varphi^2}\right)$ to go. Estimates analogous to those made in the previous chapter then show that after inflation and the subsequent stage of hot universe expansion, the wavelength $l \sim [H(\varphi)]^{-1}$ typically increases to

$$l \sim \exp\left(\frac{3H^2}{2\lambda\varphi^2}\right) \text{cm.} \tag{8.3.20}$$

From (8.3.19) and (8.3.20), we obtain

$$\frac{\delta\rho}{\rho} \sim \frac{9}{5\pi}\frac{H^3}{\lambda\varphi^3} \sim \frac{2\sqrt{6}}{5\pi}\sqrt{\lambda}\,\ln^{3/2} l\,[\text{cm}], \tag{8.3.21}$$

just as in the chaotic inflation scenario (7.5.29). At the galactic scale $l_g \sim 10^{22}$ cm,

$$\frac{\delta\rho}{\rho} \sim 110\sqrt{\lambda}\,. \tag{8.3.22}$$

This means that $\frac{\delta\rho}{\rho} \sim 10^{-5}$ when

$$\lambda \sim 10^{-14}, \tag{8.3.23}$$

again as in the chaotic inflation scenario; see (7.5.38). In the original version of the new inflationary universe scenario, the condition (8.3.23) was not satisfied. This made it necessary to seek other more realistic models in which the new inflationary universe scenario might be realizable, and we now turn to a discussion of the models proposed.

8.4 Primordial inflation in N = 1 supergravity

The main reason why the new inflationary universe scenario has not been fully implemented in the Coleman–Weinberg SU(5) theory is that the scalar field interacts with vector particles, and as a result it acquires an effective coupling constant $\lambda \sim g^4 \gg 10^{-14}$. The conclusion to be drawn, then, is that the (inflaton) field φ responsible for the inflation of the universe must interact both with itself and with other fields extremely weakly. In particular, it must not interact with vector fields, or in other words it ought to be a singlet under gauge transformations in grand unified theories.

A long list of requirements has been formulated which must be satisfied in order for a theory to provide a feasible setting for the new inflationary universe scenario [268]. Specifically, the effective potential at small φ must be extremely flat (as can be seen from (8.3.5) and (8.3.23)), and near its minimum at $\varphi = \varphi_0$, it must be steep enough to ensure efficient reheating of the universe. Next, after the main requirements for a theory have been formulated, the search for a realistic elementary particle theory of the desired type begins. Since the step following the construction of grand unified theories was the development of phenomenological theories based on N = 1 supergravity, there has been a great deal of work attempting to describe inflation within the scope of these theories (for example, see [269]–[271]).

In N = 1 supergravity, the inflaton field φ responsible for inflation of the universe is represented by the scalar component z of an additional singlet of the chiral superfield Σ. In the theories considered, the Lagrangian for this field can be put into the form [272]

$$L = G_{zz^*} \partial_\mu z \partial^\mu z^* - V(z, z^*) , \qquad (8.4.1)$$

$$V(z, z^*) = e^G (G_z G_{zz^*}^{-1} G_{z^*} - 3) , \qquad (8.4.2)$$

where G is an arbitrary real-valued function of z and z^*, G_z is its derivative with respect to z, and G_{zz^*} is its derivative with respect to z and z^*. In the minimal versions of the theory, one imposes the constraint $G_{zz^*} = 1/2$ on G so that the kinetic term in (8.4.1) takes on the standard (minimal) form

$\partial_\mu z \partial^\mu z^*$ (up to a factor 1/2), while the form chosen for G itself is

$$G(z, z^*) = \frac{z z^*}{2} + \ln|g(z)|^2, \tag{8.4.3}$$

where $g(z)$ is an arbitrary function of the field z, called the superpotential; all dimensional terms in (8.4.3) are expressed in units of $\frac{M_P}{\sqrt{8\pi}}$. The effective potential is then given by

$$V(z, z^*) = e^{zz^*/2}\left(2\left|\frac{dg}{dz} + \frac{z^2}{2}g\right|^2 - 3|g^2|\right). \tag{8.4.4}$$

The function g is subject to two constraints, namely $V(z_0) = 0$ and $g(z_0) \ll 1$, where z_0 is the point at which $V(z, z^*)$ has its minimum. The first condition means that the vacuum energy vanishes at the minimum of $V(z, z^*)$, and the second is required in order that the mass of the gravitino $m_{3/2}$, which is proportional to $g(z_0)$, be much lower than the other masses that occur in the theory. This requirement is necessary for the solution of the mass hierarchy problem in the context of N = 1 supergravity [15].

The superpotential $g(z)$ can be expressed as a product $\mu^3 f(z)$, where μ is some parameter with dimensions of mass. The potential $V(z, z^*)$, and thus the effective coupling constants for the z and z^* fields, are consequently proportional to μ^6. The choice $\mu \sim 10^{-2}$–10^{-3}, which seems to be a fairly natural one, therefore results in the appearance of extremely small effective coupling constants $\lambda \sim 10^{-12}$–10^{-18}, which is just what is needed to obtain the desired amplitude $\frac{\delta\rho}{\rho} \sim 10^{-4}$–$10^{-5}$ if inflation takes place in the theory (8.4.4). This variant of the new inflationary universe scenario was called the primordial inflation scenario by its authors [270], since it was expected to be played out on an energy scale far exceeding that of the grand unified theories. In fact, it has turned out that the corresponding energy scales are virtually identical.

The development of the primordial inflation scenario was party to many interesting ideas and considerable ingenuity. Unfortunately, however, no realistic versions of this scenario (or indeed of any other versions of the new inflationary universe scenario) have yet been suggested. The principal reason for this is that particles of the field z, interacting very weakly (either gravitationally or through a coupling constant $\mu^6 \sim 10^{-14}$) with one another and with other fields, were not in a state of thermodynamic equilibrium in the early universe. Furthermore, even if they had been, the corresponding corrections of the type $\lambda z z^* T^2$ to $V(z, z^*)$ are so small that they are incapable of changing the initial value of the field z; that is, in most models of this kind, they cannot raise the field z to a maximum of the potential $V(z, z^*)$, as required for the onset of inflation in this scenario [115, 116] (this point will be treated in more detail in Section 8.5). Meanwhile, as we shall show in Chapter 9, the chaotic inflation scenario can be implemented in N = 1 supergravity [273, 274].

8.5 The Shafi–Vilenkin model

It was Shafi and Vilenkin who came closest to a consistent implementation of the new inflationary universe scenario [275] (see also [276]). They returned to a consideration of the SU(5)-symmetric theory of Coleman and Weinberg, with symmetry breaking due to the Coleman–Weinberg mechanism occurring not in the field Φ, which interacts with vector bosons through a gauge coupling constant g, but in a new, specially introduced field χ, an SU(5) singlet, which interacts very weakly with the superheavy Φ and H_5 Higgs fields. The effective potential in this model is

$$V = \frac{1}{4} a \,\mathrm{Tr}(\Phi^2)^2 + \frac{1}{2} b \,\mathrm{Tr}\,\Phi^4 - \alpha (H_5^+ H_5)\,\mathrm{Tr}\,\Phi^2 + \frac{\gamma}{4} (H_5^+ H_5)^2$$
$$-\beta H_5^+ \Phi^2 H_5 + \frac{\lambda_1}{4} \chi^4 - \frac{\lambda_2}{2} \chi^2 \,\mathrm{Tr}\,\Phi^2 + \frac{\lambda_3}{2} \chi^2 H_5^+ H_5 \qquad (8.5.1)$$
$$+ A \chi^4 \left(\ln \frac{\chi^2}{\chi_0^2} + C \right) + V(0)\,.$$

where a, b, α, and γ are all proportional to g^2; C is some normalization constant; $0 < \lambda_i \ll g^2$, $\lambda_1 \ll \lambda_2^2, \lambda_3^2$, and the magnitude of A is determined by radiative corrections associated with the interaction of the field χ with the fields Φ, H_5, and (indirectly) with the X and Y vector mesons.

In the present case, it is not an entirely trivial matter to calculate A, and the procedure requires some explanation. Spontaneous SU(5) symmetry breaking takes place when the nonzero classical field χ emerges, thanks to the term $-\frac{1}{2}\lambda_2\chi^2\,\mathrm{Tr}\,\Phi^2$ in (8.5.1). The symmetry breaks down to SU(3) × SU(2) × U(1) by virtue of the emergence of the field

$$\Phi = \sqrt{\frac{2}{15}}\,\varphi\cdot\mathrm{diag}\left(1,\,1,\,1,\,-\frac{3}{2},\,-\frac{3}{2}\right)$$

(see (1.1.19)), where

$$\varphi^2 = \frac{2\lambda_2}{\lambda_c}\chi^2 \tag{8.5.2}$$

and $\lambda_c = a + \frac{7}{15}b$. The time needed for the field φ to grow to the value (8.5.2) is then $\tau \sim \left(\sqrt{\lambda_2}\,\chi\right)^{-1}$, which is much less than the typical time scale of variations in χ at the time of inflation (see below). The field φ thus continuously follows the behavior of the field χ. Consequently, not only does a change in χ alter the masses of those particles with which this field interacts directly (such as Φ and H_5), but also those that interact with the field φ — in particular, the X and Y vector mesons. Here the behavior of the H_5 boson masses is especially interesting. The first two components of H_5 play the role of the Higgs field doublet in SU(2) × U(1) symmetry breaking. These should be quite light, with $m_2 \sim 10^2$ GeV $\ll m_3$, M_X, M_Y, ... To lowest order, one may therefore put $m_2 = 0$ at the minimum of $V(\varphi, \chi)$.

The general expression for the doublet and triplet masses of the H fields follows from (8.5.1):

$$m_2^2 = \lambda_3\chi^2 - (\alpha + 0.3\,\beta)\,\varphi^2, \tag{8.5.3}$$

$$m_3^2 = m_2^2 + \frac{\beta}{6}\varphi^2. \tag{8.5.4}$$

Making use of (8.5.2), one obtains

$$\lambda_3 = \frac{2\lambda_2}{\lambda_c}(\alpha + 0.3\,\beta). \tag{8.5.5}$$

This implies that not just at the minimum of $V(\varphi, \chi)$, but anywhere along a trajectory along which the field χ varies,

$$m_2^2 = 0,\ m_3^2 = \frac{\beta}{6}\varphi^2. \tag{8.5.6}$$

The constant λ_3 is thus not independent, and the value of m_3^2 is proportional to $\frac{\beta}{6}\varphi^2$, rather than $\lambda_3\chi^2$. A calculation of the radiative corrections to $V(\varphi, \chi)$ in the vicinity of a trajectory down which the field χ is rolling, with $\lambda_i, \beta \ll g^2$, finally gives [277]

$$A = \frac{\lambda_2^2}{16\pi^2}\left(1 + \frac{25\,g^4}{16\lambda_c^2} + \frac{14\,b^2}{9\lambda_c^2}\right). \tag{8.5.7}$$

(This expression differs slightly from the one given in Ref. [275].) If one takes for simplicity $a \sim b \sim g^2$, then Eq. (8.5.7) yields

$$A \sim 1.5 \cdot 10^{-2}\lambda_2^2. \tag{8.5.8}$$

The effective potential $V(\varphi, \chi)$ in the theory (8.5.1) looks like

$$V = \frac{\lambda_c}{16}\varphi^4 - \frac{\lambda_2}{4}\varphi^2\chi^2 + \frac{\lambda_1}{4}\chi^4 + A\chi^4\left(\ln\frac{\chi}{M} + C\right) + V(0), \tag{8.5.9}$$

where M and C are some normalization parameters. To determine M, C, and V(0), one should use Eq. (8.5.2):

$$V = -\frac{\lambda_2^2}{4\lambda_c}\chi^4 + A\chi^4\left(\ln\frac{\chi}{M} + C\right) + V(0). \tag{8.5.10}$$

With an appropriate choice of the normalization constant C, the effective potential (8.5.10) can be put in the standard form

$$V(\chi) = A\,\chi^4\left(\ln\frac{\chi}{\chi_0} - \frac{1}{4}\right) + \frac{A\chi_0^4}{4}, \tag{8.5.11}$$

where χ_0 gives the position of the minimum of $V(\chi)$. The corresponding minimum in φ is located at $\varphi_0 = \sqrt{\frac{2\lambda_2}{\lambda_c}}\,\chi_0$ (see (8.5.2)), and the mass of the X boson is equal to

$$M_X = \sqrt{\frac{5}{3}}\,\frac{g\varphi_0}{2} \sim 10^{14}\ \text{GeV}.$$

Hence,

$$\chi_0 \sim \frac{M_X}{g}\sqrt{\frac{6\lambda_c}{5\lambda_2}},$$

and

$$V(0) = \frac{A}{4}\chi_0^4 \sim M_X^4.$$

The high-temperature correction to the effective potential (8.5.11) is given by

$$\Delta V(\chi, T) = \left(\frac{5}{12}\lambda_3 - \lambda_2\right)T^2\chi^2, \tag{8.5.12}$$

which for $\lambda_3 > \frac{12}{5}\lambda_2$ could lead to the restoration of symmetry, $\chi \to 0$ (see the next section, however). Upon cooling, the process of inflation would begin, which would be very similar to the process described in Section 8.3.

To determine the numerical value of the parameter A, one should first determine the value of $\ln\left(\frac{\chi_0}{\chi}\right)$ at which the observable structure of the universe is actually formed — this occurs when a time $t \sim 60\ H^{-1}$ remains prior to the end of inflation. According to (8.2.15), the magnitude of the field χ at that point is given by

$$\chi^2 \sim \frac{H^2}{40\,\lambda(\chi)} \tag{8.5.13}$$

where for $\ln\left(\frac{\chi_0}{\chi}\right) \gg 1$, the effective coupling constant $\lambda(\chi)$ and the Hubble constant H are

$$\lambda(\chi) \approx 4A \ln\left(\frac{\chi_0}{\chi}\right) \sim 10^{-14},$$
$$H = \sqrt{\frac{8\pi V(0)}{3M_P^2}} \sim 3\frac{M_X^2}{M_P} \sim 3\cdot 10^9 \text{ GeV}, \tag{8.5.14}$$

(see (8.3.23)), whereupon

$$\chi \sim 5\cdot 10^{15} \text{ GeV}. \tag{8.5.15}$$

Inserting these values, $\ln\left(\frac{\chi_0}{\chi}\right)$ is found to be of order 3 (see below). From (8.5.8) and (8.5.14), it follows that

$$\lambda_2 \sim 3.10^{-6}, \tag{8.5.16}$$

$$\chi_0 \sim \frac{M_X}{g}\sqrt{\frac{6\lambda_c}{5\lambda_2}} \sim 10^{17} \text{ GeV}. \tag{8.5.17}$$

According to (8.5.11), the value of λ_3 should be greater than $\frac{12\lambda_2}{5}$. But λ_3 cannot be *much* greater than λ_2, since it can be shown that if it were, φ would not vanish at high temperature [275]. In accordance with [275], we shall therefore assume that $\lambda_3 \sim 3\cdot 10^{-6}$, like λ_2.

We have from (8.3.17) that in the present model, a typical inflation factor is of order

$$\exp\left(\frac{4\sqrt{2}\,\pi}{\sqrt{\lambda(\chi)}}\right) \sim 10^{10^8}, \tag{8.5.18}$$

which is more than adequate.

Unfortunately, both reheating and the baryon asymmetry production in the post-inflation universe are rather inefficient in this model. After inflation, the field χ oscillates about the minimum of $V(\chi)$ at $\chi = \chi_0$ at a very low frequency,

$$m_\chi = 2\sqrt{A}\,\chi_0 \sim 10^{11} \text{ GeV}. \tag{8.5.19}$$

The principal decay mode of the field χ is $\chi\chi \to H_3^+ H_3$, where H_3 is the triplet of heavy Higgs bosons. Subsequent decay of the H_3 bosons gives rise to the baryon asymmetry of the universe. The corresponding part of the effective Lagrangian responsible for decay of the field χ is of the form $\frac{\beta\lambda_2}{6\lambda_c}\chi^2 H_3^+ H_3$ But such a process is only possible if $m_3 < m_\chi \sim 10^{11}$ GeV, and if H_3 had such a mass, the proton lifetime would be unacceptably short, making the entire scheme unrealistic.

Let us digress from this problem for a moment, since in any case the SU(5) model in question is in need of modification — it gives a high probability of proton decay even when $m_3 \gg m_\chi$. In order for the decay $\chi\chi \to H_3^+ H_3$ to occur, let us take $m_\chi \sim m_{H_3^+}$, that is, $\beta \sim 10^{-6}$. In that event,

$$\Gamma(\chi\chi \to H_3^+ H_3) \sim \frac{(10^{-11}\chi)^2}{m_\chi} \cdot O(10^{-2}) \sim 10^{-2}\ \text{GeV}, \qquad (8.5.20)$$

and so, according to (7.9.10),

$$T_R \sim 10^{-1}\sqrt{\Gamma M_P} \sim 3 \cdot 10^7\ \text{GeV}. \qquad (8.5.21)$$

Baryon asymmetry formation is a possible process in this model, since H_3 bosons are created and destroyed, but each decay generates $O\left(\frac{m_3}{T_R}\right) \sim 3 \cdot 10^3$ photons of energy $E \sim T$. This tends to make the occurrence of a baryon asymmetry less likely by a factor of $3 \cdot 10^3$. To circumvent this problem, one should either invoke alternative baryon production mechanisms (see Chapter 7), or modify the Shafi–Vilenkin model. We shall return to this question in the next chapter; for the time being, let us try to analyze the main results obtained above, and evaluate the prospects for further development of the new inflationary universe scenario.

8.6 The new inflationary universe scenario: problems and prospects

As we have already seen, the basic problems with the new inflationary universe scenario have to do with the need to obtain small density inhomogeneities in the observable part of the universe after inflation. This question deserves more detailed discussion.

1. As we mentioned in Section 7.5, the condition $A \lesssim 10^{-4}$ imposes

an overall constraint on $V(\varphi)$ in the new inflationary universe scenario,

$$V(\varphi) \lesssim 10^{-10} M_P^4. \tag{8.6.1}$$

This means that in any version of this scenario, including the primordial inflation scenario, the process of inflation can only begin at a time

$$t \gtrsim H^{-1} \sim \sqrt{\frac{3M_P^2}{8\pi V}} \sim 10^{-36} \text{ sec},$$

or in other words, six orders of magnitude later than the Planck time $t_P \sim M_P^{-1} \sim 10^{-43}$ sec. Bearing in mind, then, that a typical total lifetime for a hot, closed universe is of order $t \sim t_P$ (see Chapter 1), it is clearly almost always the case that a closed universe simply fails to survive until the beginning of inflation; i.e., the flatness problem cannot be solved for a closed universe. The new inflationary universe scenario can therefore only be realized in a topologically nontrivial or a noncompact (infinite) universe, and only in those parts of the latter in which both don't collapse and are sufficiently large ($l \gtrsim 10^5 M_P^{-1}$) the moment when the matter density therein becomes less than $V(\varphi) \sim 10^{-10} M_P^4$.

2. Let us now direct our attention to the fact that the new inflationary universe scenario can only be realized in theories in which the potential energy $V(\varphi)$ takes on a highly specific form where, as we have seen, the coupling constants are strongly interrelated. Considerable ingenuity is required to construct such theories, with the result that the original simplicity underlying the idea of inflation of the universe is gradually lost in the profusion of conditions and reservations required for its implementation.

3. The basic difficulty of the new inflationary universe scenario relates to the question of how the field φ reaches the maximum of the effective potential $V(\varphi)$ at $\varphi = 0$. This problem turned out to be an especially serious one as soon as it was realized that the field φ ought to interact extremely weakly with other fields.

To get to the heart of the problem, let us examine some region of a hot universe in which the field φ has the initial value $\varphi \sim \varphi_0$. Suppose that high-temperature corrections lead to a correction to $V(\varphi)$ of the form

$$\Delta V \sim \frac{\alpha^2}{2} \varphi^2 T^2. \tag{8.6.2}$$

The age of the hot universe equals $\frac{H^{-1}}{2}$ (1.4.6):

$$t = \frac{H^{-1}}{2} = \frac{M_P}{2}\sqrt{\frac{3}{8\pi\rho}} < \frac{M_P}{2}\sqrt{\frac{3}{8\pi\Delta V}} \sim \frac{M_P}{4\alpha\varphi T}. \tag{8.6.3}$$

Over this time span, the corrections (8.6.2) can only change the initial value $\varphi = \varphi_0$ if the typical time $\tau = (\Delta m)^{-1}(T) \sim (\alpha T)^{-1}$ is less than the age of the universe t, whereupon we obtain the condition

$$\varphi_0 \lesssim \frac{M_P}{3}. \tag{8.6.4}$$

High-temperature corrections can thus affect the initial value of the field φ only if the latter is less than $\frac{M_P}{3}$. Meanwhile, theories with $V(\varphi) \sim \varphi^n$ place no constraints on the initial value of the field φ except to require that $V(\varphi) \lesssim M_P^4$. For example, in the Shafi–Vilenkin theory (as in a $\frac{\lambda\varphi^4}{4}$ theory with $\lambda \sim 10^{-14}$), the constraint $V(\chi) \lesssim M_P^4$ implies that the field χ can initially take on any value in the range

$$-10^4 M_P \lesssim \chi \lesssim 10^4 M_P, \tag{8.6.5}$$

and only less than 10^{-4} of this interval is pertinent to values of χ for which high-temperature corrections can play any role.

For the case $\varphi \lesssim \frac{M_P}{3}$ one can make another estimate. In a hot universe with N different particle species,

$$t \lesssim \frac{1}{4\pi}\sqrt{\frac{45}{\pi N}}\,\frac{M_P}{T^2}, \tag{8.6.6}$$

(see (1.3.21)). Comparing t from (8.6.6) with $\tau \sim (\alpha T)^{-1}$, we see that high-temperature effects only begin to change the field φ at a temperature

$$T \lesssim T_1 \sim \frac{\alpha M_P\sqrt{45}}{4\pi\sqrt{\pi N}} \sim \frac{\alpha M_P}{50}, \tag{8.6.7}$$

when the overall energy density of hot matter is

$$\rho(T_1) \sim \frac{\pi^2}{30} N T_1^4 \lesssim \alpha^4 M_P^4 \sim \frac{3 \cdot 10^{-3} \alpha^4 M_P^4}{N^2} \sim 10^{-7} \alpha^4 M_P^4 \qquad (8.6.8)$$

for N ~ 200 (as would be the case in grand unified theories). Notice, however, that the process whereby the field φ decreases can continue only so long as the total density $\rho(T)$ is not comparable to V(0), since soon afterward, high-temperature effects become exponentially small due to inflation. This leads to the constraint

$$10^{-7} \alpha^4 M_P^4 > V(0) . \qquad (8.6.9)$$

In the theory of a field φ that interacts only with itself (with a coupling constant $\lambda \sim 10^{-14}$), and in primordial inflation models, the parameter α^2 is of order 10^{-14}, so that (8.6.9) then becomes

$$V(0) \lesssim 10^{-35} M_P^4 . \qquad (8.6.10)$$

This value is much smaller than the actual value of V(0) in all realistic models of new inflation.

The situation is somewhat better in the Shafi–Vilenkin model. There, α^2 is of order 10^{-7}, and (8.6.9) remains valid. We must ascertain, however, whether high-temperature effects can make χ smaller than $5 \cdot 10^{15}$ GeV (8.5.15), which would be necessary for the universe to inflate by a factor of e^{60}–e^{70}.

For this to happen, the field χ must be reduced to $5 \cdot 10^{15}$ GeV at the time when the quantity

$$\frac{d\Delta V(\chi, T)}{d\chi} \sim \alpha^2 T^2 \chi$$

becomes less than

$$\frac{dV(\chi)}{d\chi} \sim 4 A \chi^3 \ln \frac{\chi_0}{\chi} .$$

This occurs at a temperature

$$T_2 \sim 10^{12}\,\text{GeV}. \tag{8.6.11}$$

While the temperature T drops from T_1 to T_2, the field χ oscillates in the potential $\Delta V(\chi, T) \sim \frac{\alpha^2}{2}\chi^2 T^2$ at a frequency $m_\chi \sim \alpha T$. The rate of production of pairs by this field is very low (8.5.20), so that its oscillation amplitude in the early universe decreases mainly by virtue of the expansion of the universe. It is readily shown that in the present case (with $\Delta V(\chi, T) \sim \frac{\alpha^2}{2}\chi^2 T^2$), the falloff in the field χ is proportional to the temperature T. As the temperature drops from $T_1 \sim 10^{14}$ GeV to $T_2 \sim 10^{12}$ GeV, the original amplitude of the field χ is reduced by a factor of 10^2, and it becomes less than $\sim 5 \cdot 10^{15}$ GeV only if the initial field χ was less than $5 \cdot 10^{17}$ GeV.

Thus, in order to implement the new inflationary universe scenario in the Shafi–Vilenkin model, the field χ must originally be a factor of 20 less than M_P, a requirement that is quite unnatural.

It must be understood that the foregoing estimates are model-dependent. There do exist theories in which the effective potential $V(\varphi)$ rises so rapidly with increasing field φ that it becomes greater than M_P^4 for $\varphi \gtrsim M_P$. In that event, the condition $\varphi_0 \lesssim M_P$ may be warranted. In general, it is possible to suggest mechanisms whereby a field $\varphi \lesssim M_P$ in the early universe rapidly drops to $\varphi \ll M_P$. But the examples considered above show that it is indeed difficult to obtain consistency among all the requirements needed for a successful implementation of the new inflationary universe scenario. As a result, a consistent implementation of this scenario within the framework of a realistic elementary particle theory is still lacking.

Of course, one cannot rule out the possibility that some future elementary particle theory will automatically satisfy all the necessary conditions. But for now, there is no need to insist that all of these conditions be satisfied, as there is another scenario amenable to realization over a much wider class of theories, namely the chaotic inflation scenario.

9

The Chaotic Inflation Scenario

9.1 Introduction. Basic features of the scenario. The question of initial conditions

The general underpinnings of the chaotic inflation scenario were described in some detail in Chapter 1. Rather than reiterating what has already been said, we shall attempt here to review the basic features of this scenario, which may perhaps stand out in better relief against the backdrop of the preceding discussion of the new inflationary universe scenario.

The basic idea behind this scenario is simply that one need no longer assume that the field φ lies at a minimum of its effective potential $V(\varphi)$ or $V(\varphi, T)$ from the very outset in the early universe. Instead, it is only necessary to study the evolution of φ for a variety of fairly natural initial conditions, and check to see whether or not inflation sets in.

If one requires in addition that a solution to the flatness problem be accessible within the context of this scenario even when the universe is closed, then it becomes necessary that inflation be able to start with $V(\varphi) \sim M_P^4$. As demonstrated in Chapter 1, this requirement is satisfied by a broad class of theories in which the effective potential $V(\varphi)$ increases no faster than a power of the field φ in the limit $\varphi \gg M_P$. In principle, inflation can also take place in theories for which $V(\varphi) \sim \exp\left(\alpha \frac{\varphi}{M_P}\right)$ when $\varphi \gg M_P$ if α is sufficiently small $\alpha \lesssim 5$. The general criterion for the onset of inflation follows from the condition $\dot{H} \ll H^2 = \frac{8\pi V}{3M_P^2}$ and (1.7.16):

$$\frac{d\ln V}{d\varphi} \ll \frac{4\sqrt{\pi}}{M_P}. \tag{9.1.1}$$

As we stated in Chapter 1, the most natural initial conditions for the field φ on a scale $l \sim H^{-1} \sim M_P^{-1}$ are that $\partial_0\varphi\,\partial^0\varphi \sim \partial_i\varphi\,\partial^i\varphi \sim V(\varphi) \sim M_P^4$. The probability of formation of an inflationary region of the universe then becomes significant — we might estimate it to be perhaps 1/2 or 1/10. For our purposes, the only important consideration is that the probability is not reduced by a factor like $\exp(-1/\lambda)$ [118]. Meanwhile, it has been argued by some authors (see [258, 278], for example) that the probability of inflation in the $\frac{\lambda\varphi^4}{4}$ theory might actually be suppressed by a factor of this kind. A thorough investigation of this question is necessary for a proper understanding of those changes that the inflationary universe scenario has introduced into our conception of the world about us. We will discuss this question here, following the arguments put forth in [118].

1. First, let us try to understand whether it is actually necessary to assume that $\frac{\dot\varphi^2}{2} \ll V(\varphi)$ from the very outset. For simplicity, we consider Eqs. (1.7.12) and (1.7.13) in a flat universe ($k = 0$) with a uniform field φ, and with $\dot\varphi^2 \gg V(\varphi)$. Then (1.7.12) and (1.7.13) imply that $\ddot\varphi \gg V'(\varphi)$ and

$$\ddot\varphi = \frac{2\sqrt{3\pi}}{M_P}\dot\varphi\,, \tag{9.1.2}$$

so that

$$\dot\varphi = -|\dot\varphi_0|\left(1 + \frac{2\sqrt{3\pi}}{M_P}|\dot\varphi_0|\,t\right)^{-1} \tag{9.1.3}$$

$$\varphi = \varphi_0 - \frac{M_P}{2\sqrt{3\pi}}\ln\left(1 + \frac{2\sqrt{3\pi}}{M_P}|\dot\varphi_0|\,t\right). \tag{9.1.4}$$

This means that when $t \gtrsim H^{-1} \sim \frac{M_P}{|\dot\varphi_0|}$, the kinetic energy of the field φ falls off according to a power law, as $\dot\varphi^2 \sim t^{-2}$, while the magnitude of the field itself (and thus of $V(\varphi) \sim \varphi^n$) falls off only logarithmically. The kinetic energy of the field φ therefore drops rapidly, and after a short time (just several times the value of H^{-1}), the field enters the asymptotic regime $\dot{\varphi}^2 \ll V(\varphi)$ [118, 110].

The thrust of this result, a more general form of which was derived in [279, 280] for both an open and closed universe, is quite simple. When $\dot{\varphi}^2 > V(\varphi)$, the energy-momentum tensor has the same form as the energy-momentum tensor of matter whose equation of state is $p = \rho$. The energy density of such matter rapidly decreases as the universe expands, while the value of a sufficiently flat potential $V(\varphi)$ changes very slowly.

Let us estimate the fraction of initial values of $\dot{\varphi}$ for which the universe fails to enter the inflationary regime in the $\frac{\lambda\varphi^4}{4}$ theory. This requires that $\dot{\varphi}^2$ remains larger than $V(\varphi)$ until φ becomes smaller than $\frac{M_P}{3}$. The initial value of $\frac{\dot{\varphi}^2}{2}$ is of order M_P^4 (prior to that point, it is impossible to describe the universe classically), and the initial value φ_0 of the field φ can take on any value in the range $-\lambda^{1/4} M_P \lesssim \varphi \lesssim \lambda^{1/4} M_P$. In that event, we see from (9.1.4) that the total time needed for the field φ to decrease from φ_0 to $\varphi \sim M_P$ is of order $\frac{1}{2\sqrt{6\pi}\, M_P} \exp\left(\frac{2\sqrt{3\pi}\,\varphi_0}{M_P}\right)$. In this time span, $\dot{\varphi}$ is reduced in magnitude by approximately a factor $\exp\left(\frac{2\sqrt{3\pi}\,\varphi_0}{M_P}\right)$. We then find that when $\lambda \ll 1$, $\dot{\varphi}^2$ can remain larger than $V(\varphi)$ during the whole process only if $\varphi_0 \sim M_P$. The probability that the field φ, which initially can take any value in the range from $-\lambda^{1/4} M_P$ to $\lambda^{1/4} M_P$, winds up being of order M_P can be estimated to be $\lambda^{1/4} \sim 3 \cdot 10^{-4}$ for $\lambda \sim 10^{-14}$. It is therefore practically inevitable that a homogeneous, flat universe passes through the inflationary regime in the $\frac{\lambda\varphi^4}{4}$ theory [280, 110, 118]. One comes to the same conclusion for an open universe as well. For a closed universe, the corresponding probability is of order 1/4 [280]. The reason that the probability is smaller for a closed universe is that it may collapse before $\dot{\varphi}^2$ becomes smaller than $V(\varphi)$. In any event, the probability of occurrence of an inflationary regime turns out to be quite significant, as we expected.

2. Next, let us discuss a more general situation in which the field φ is inhomogeneous. If the universe is closed, then its overall initial size l is $O(M_P^{-1})$ (when $l \ll M_P^{-1}$, the universe cannot be described in terms of classical space-time, and in particular one cannot say that its size l is much less than M_P^{-1}). If $\partial_0\varphi\,\partial^0\varphi$ and $\partial_i\varphi\,\partial^i\varphi$ are both severalfold smaller than $V(\varphi)$, which is not improbable, then the universe begins to inflate, and the gradients $\partial_i\varphi$ soon become exponentially small. Thus, the probability of formation of a closed inflationary universe remains almost as large as in the case considered above, even when possible inhomogeneity of the field φ is taken into account.

If the universe is infinite, then the probability that the conditions necessary for inflation actually come to pass would seem at first glance to be extremely low [258]. In fact, if a typical initial value of the field φ in a $\frac{\lambda\varphi^4}{4}$ theory is, as we have said, of order $\varphi_0 \sim \lambda^{-1/4}M_P \sim 3000\ M_P$, then the condition $\partial_i\varphi\,\partial^i\varphi \lesssim M_P^4$ might lead one to conclude that φ should remain larger than $\sim\lambda^{-1/4}M_P$ on a scale $l \gtrsim \lambda^{-1/4}M_P \sim 3000\ M_P^{-1}$. But this is highly improbable, since initially (i.e., at the Planck time $t_P \sim M_P^{-1}$) there can be no correlation whatever between values of the field φ in different regions of the universe separated from one another by distances greater than M_P^{-1}. The existence of such correlation would violate causality (see the discussion of the horizon problem in Chapter 1).

The response to this objection is very simple [118, 78, 79]. We have absolutely no reason to expect that the overall energy density ρ will *simultaneously* become less than the Planck energy M_P^4 in all causally disconnected regions of an infinite universe, since *that* would imply the existence of an acausal correlation between values of ρ in different domains of size $O(M_P^{-1})$. Each such domain looks like an isolated island of classical space-time, which emerges from the space-time foam independently of other such islands. During inflation, each of these islands acquires dimensions many orders of magnitude larger than the size of the observable part of the universe. If some gradually join up with others through connecting necks of classical space-time, then in the final analysis, the universe as a whole will begin to look like a cluster (or several independent clusters) of topologically connected mini-universes. However, such a structure may only come into being later (see Chapter 10 in this regard), and a typical *initial* size of a domain of classical space-time with $\rho \lesssim M_P^4$ is extremely small — of the order of the Planck length $l_P \sim M_P^{-1}$. Outside each of these domains the condition $\partial_i\varphi\,\partial^i\varphi \lesssim M_P^4$ no longer holds, and there is no correlation at all between values of the field φ in different disconnected regions of classical space-time of size $O(M_P^{-1})$. But such correlation is not really necessary for the realization of the inflationary universe scenario — according to the "no hair" theorem for de Sitter space, a sufficient condition for the existence of an inflationary region of the universe is that inflation take place *inside* a region whose size is of order $H^{-1} \sim M_P^{-1}$, which in our example is actually the case.

We wish to emphasize once again (and this will subsequently be of some importance) that the confusion that we have analyzed above, involving the correlation between values of the field φ in different causally disconnected regions of the universe, is rooted in the familiar notion of a universe that is *instantaneously* created from a singular state with $\rho \to \infty$, and *instantaneously* passes through a state with the Planck density $\rho \sim M_P^4$. The lack of justification for such a notion is the very essence of the horizon problem; see Section 1.5. Now, having disposed of the horizon problem with the aid of the inflationary universe scenario, we may possibly manage to familiarize ourselves with a different picture of the universe, a picture whose specific features are gradually becoming clear. We shall return to this question in the next chapter.

Evidently, the condition $\partial_i\varphi\partial^i\varphi < V(\varphi)$ invoked above can also be relaxed, in the same way that we relaxed the requirement $\partial_0\varphi\partial^0\varphi < V(\varphi)$. The basic idea here is that if the effective potential $V(\varphi)$ is flat enough, then during the expansion of the universe (in all regions which neither drop out of the general expansion process nor collapse), the gradients of the field φ fall off rapidly, whereas the mean value of φ decreases relatively slowly. The net result is that just as in the case of kinetic energy $\frac{\dot{\varphi}^2}{2}$, we ought to arrive at a situation in which the energy density associated with gradients of the field φ in a significant part of the universe will have fallen to much less than $V(\varphi)$; in other words, the conditions necessary for inflation appear. We shall not discuss this possibility any further, as the results obtained above suffice for our purposes.

To conclude this section, let us note that the foregoing question involving acausal correlation does not arise in realistic theories, in general, where apart from a "light" field φ with $\lambda \sim 10^{-14}$, there is at least one scalar field Φ with a bigger coupling constant $\lambda_\Phi \gtrsim 10^{-2}$. In such theories, the "acausal correlation length" between values of Φ in different regions is only marginally bigger than the distance to the horizon, so that even if the aforementioned arguments concerning the acausal correlation between values of Φ were true, the probability that inflation would be driven by the field Φ would not be noticeably suppressed. As demonstrated in [281], long-wave fluctuations of the light field φ that are generated at the time of inflation bring about self-regenerating inflationary regions (see Section 1.8) filled with a quasihomogeneous field φ for which $V(\varphi) \lesssim M_P^4$. The heavy field Φ rapidly decreases in these regions, so that the last stages of inflation are governed by the field φ with $\lambda \sim 10^{-14}$, as before.

Our principal conclusion, then, is that there exists a broad class of elementary particle theories within the scope of which inflation sets in under natural initial conditions.

9.2 The simplest model based on the SU(5) theory

The chaotic inflation scenario can be implemented within the framework of many models (and in particular the Shafi–Vilenkin model, which was originally designed to implement the new inflationary universe scenario). But one can achieve the same end using simpler models, since we no longer need to satisfy the numerous constraints imposed on the theory by the new inflationary universe scenario. To be specific, there is no need to invoke the Coleman–Weinberg mechanism; the superheavy Φ and H_5 field sector in SU(5) theory can be cast in a standard form; we can omit interactions between the χ-field and Φ-fields, and so on.

Consider, for example, a theory with the effective potential

$$V = \frac{1}{4}a\,\mathrm{Tr}(\Phi^2)^2 + \frac{1}{2}b\,\mathrm{Tr}\,\Phi^4 - \frac{M_\Phi^2}{2}\,\mathrm{Tr}\,\Phi^2 - \alpha\,(H_5^+ H_5)\,\mathrm{Tr}\,\Phi^2$$
$$+ \frac{\lambda}{4}(H_5^+ H_5)^2 - \beta\, H_5^+ \Phi^2 H_5 + m_5^2 H_5^+ H_5 \qquad (9.2.1)$$
$$- \frac{m^2}{2}\chi^2 + \frac{\lambda_1}{4}\chi^4 + \frac{\lambda_2}{2}\chi^2 H_5^+ H_5\,,$$

and assume that $a \sim b \sim \alpha \sim g^2$, $\lambda_1 \gg \lambda_2^2$, so that quantum corrections to λ_1 may be neglected. In this theory, in contrast to Eq. (8.5.3) and (8.5.4), the masses m_2 and m_3 are given by

$$m_2^2 = m_5^2 + \lambda_2\chi^2 - (\alpha + 0.3\,\beta)\,\varphi^2 \qquad (9.2.2)$$

$$m_3^2 = m_2^2 + \frac{\beta}{6}\varphi^2. \tag{9.2.3}$$

Inflation takes place during the time that the χ-field rolls down from $\chi \sim \lambda_1^{-1/4} M_P$ to the minimum of $V(\chi)$ at $\chi_0 = \frac{m}{\sqrt{\lambda_1}}$. We will assume for simplicity that $\chi_0 \lesssim M_P$; then $\frac{\delta\rho}{\rho} \sim 10^{-5}$ for $\lambda_1 \sim 10^{-14}$. The universe is reheated much more efficiently than in the Shafi–Vilenkin model, as the terms in the Lagrangian responsible for decay of the field χ are now of the form $\sim\lambda_2 \chi^2 H_5^+ H_5$ (there is no additional coefficient $\beta \sim 10^{-6}$ resulting from the simultaneity of oscillations of the φ and χ fields). This effect and additional energy transfer during oscillations of the H_1 and H_2 fields (which result from sign changes in m_2^2 as the field χ oscillates in the neighborhood of χ_0) lead to rapid reheating of the universe. This is also facilitated by an increase in oscillation frequency of the field χ. If, for example, one takes $m \sim 10^{12}$ GeV, then $\chi_0 \sim M_P$. The oscillation frequency of the field χ then becomes $\sqrt{2}\, m = 1.5 \cdot 10^{12}$ GeV. The reheating temperature T_R in this model can reach 10^{12}–10^{13} GeV. The decay $\chi\chi \to H_3^+ H_3$ takes place at $m_3 \lesssim 10^{12}$ GeV; the particular difficulties with proton decay which are related to the low mass of the m_3 do not appear in this model, and the temperature T_R is large enough to facilitate the standard baryogenesis mechanism based on the decay of H_3 particles.

The model presented here admits of a great many generalizations. For example, one can delete the terms $-\frac{m^2}{2}\chi^2$ and $\frac{\lambda_1}{4}\chi^4$ from (9.2.1), leaving only the last term $\frac{\lambda_2}{2}\chi^2 H_5^+ H_5$. Then due to radiative corrections, an induced term like $C\frac{\lambda_2^2 \chi^4}{64\pi^2}\left(\ln \frac{\chi}{\chi_0} - \frac{1}{4}\right)$ will be responsible for inflation. When $\lambda_2 \sim 10^{-6}$, this term gives rise to density inhomogeneities $\frac{\delta\rho}{\rho} \sim 10^{-5}$. This model is analogous to the Shafi–Vilenkin model, but it is much simpler, and it is also shares none of the latter's problems with baryogenesis. Likewise, the extremely small coupling constant $\lambda_1 \sim 10^{-14}$ once again need not be introduced beforehand — the constant $\lambda_2 \sim 10^{-6}$ is sufficient, which seems more natural, inasmuch as similar constants do appear in such popular models as the Glashow–Weinberg–Salam theory.

9.3 Chaotic inflation in supergravity

There are currently several different models that describe chaotic inflation in the context of supergravity [273, 274, 282]. Here we examine one of these that seems to us to be particularly simple, a model related to SU(n, 1) supergravity [283], several versions of which arise in the low-energy limit of superstring theory [17].

One of the major problems that comes up in constructing realistic models based on supergravity theory is how to make the effective potential $V(z)$ vanish at its minimum z_0. As a first step toward such a theory, one can attempt to find a general form of the function $G(z, z^*)$ for which the potential $V(z, z^*)$ in (8.4.2) is identically zero. This occurs when [284]

$$G(z, z^*) = -\frac{3}{2}\ln\left(g(z) + g^*(z)\right)^2, \tag{9.3.1}$$

where $g(z)$ is some arbitrary function. In that case, the Lagrangian is

$$L = G_{zz^*}\,\partial_\mu z\,\partial^\mu z = 3\,\frac{\partial_\mu g\,\partial^\mu g}{(g+g^*)^2}. \tag{9.3.2}$$

All such theories with different $g(z)$ are equivalent to one another after the transformation $g(z) \to z$. The Lagrangian

$$L = 3\,\frac{\partial_\mu z\,\partial^\mu z}{(z+z^*)^2} \tag{9.3.3}$$

is invariant under the group of SU(1, 1) transformations

$$z \to \frac{\alpha z + i\beta}{i\gamma z + \delta} \tag{9.3.4}$$

with real parameters α, β, γ, δ such that $\alpha\delta + \beta\gamma = 1$ [284]. Such theories have been called SU(1, 1) supergravity for that reason.

One possible generalization of the function $G(z, z^*)$ of (9.3.1) that leads to a potential $V(z, z^*, \varphi, \varphi^*) \geq 0$, where φ is the scalar (inflation) field responsible for inflation, is

$$G = -\frac{3}{2}\ln\left(z + z^* + h(\varphi, \varphi^*)\right)^2 + g(\varphi, \varphi^*)\,, \qquad (9.3.5)$$

where h and g are arbitrary real-valued functions of φ and φ^*. In the theory (9.3.5),

$$V = \frac{1}{|z + z^*|^2}\, e^{g}\, \frac{|g_\varphi|^2}{G_{\varphi\varphi^*}}\,, \qquad (9.3.6)$$

where $G_{\varphi\varphi^*} = g_{\varphi\varphi^*} + G_z h_{\varphi\varphi^*} \geq 0$ if the kinetic term for the field φ has the correct (positive) sign.

Cast in terms of the variables z and φ, the theory (9.3.5) looks rather complicated, but it can be simplified considerably by diagonalizing the kinetic part of the Lagrangian, reducing it to the form [285]

$$L_{kin} = \frac{1}{12}\partial_\mu\zeta\, \partial^\mu\zeta + \frac{3}{4} e^{\frac{2}{3}\zeta}\, I_\mu^2 + G_{\varphi\varphi^*}\, \partial_\mu\varphi^*\, \partial^\mu\varphi\,, \qquad (9.3.7)$$

where

$$\zeta = -\frac{3}{2}\ln\left(z + z^* + h(\varphi, \varphi^*)\right)^2,$$

$$I_\mu = i\,[\partial_\mu(z - z^*) + h_\varphi\, \partial_\mu\varphi - h_{\varphi^*}\, \partial_\mu\varphi^*]\,, \qquad (9.3.8)$$

$$G_{\varphi\varphi^*} = g_{\varphi\varphi^*} + G_z h_{\varphi\varphi^*} = g_{\varphi\varphi^*} - 3\, e^{\zeta/3}\, h_{\varphi\varphi^*}\,.$$

In terms of ζ, the potential becomes

$$V = e^{\zeta + g}\, \frac{|g_\varphi|^2}{G_{\varphi\varphi^*}}\,. \qquad (9.3.9)$$

As the simplest realization of the chaotic inflation scenario in this model [274], one can consider the theory (9.3.5) with

$$g(\varphi, \varphi^*) = (\varphi - \varphi^*)^2 + \ln|f(\varphi)|^2, \qquad (9.3.10)$$

while $h(\varphi, \varphi^*)$ satisfies

$$h_{\varphi\varphi^*} = (2a)^{-1}\, g_{\varphi\varphi^*} = -a^{-1}, \tag{9.3.11}$$

where a is a positive constant. Then

$$V_\zeta \equiv \frac{\partial V}{\partial \zeta} = V\, \frac{a - e^{\zeta/3}}{a - \frac{3}{2} e^{\zeta/3}}, \tag{9.3.12}$$

that is, $V_\zeta = 0$ when $e^{\zeta/3} = a$. Notice that at the extremum of V (i.e., at $V_\zeta = 0$), the field φ has the canonical kinetic term

$$G_{\varphi\varphi^*} = -\frac{1}{2}\, g_{\varphi\varphi^*} = 1, \tag{9.3.13}$$

and

$$V_{\zeta\zeta^*} = \frac{2}{3} V > 0. \tag{9.3.14}$$

This means that the potential $V(\varphi, \zeta)$ has a hollow located at $\zeta = 3 \ln a$, $-\infty < \varphi < \infty$. At the bottom of the hollow, the potential $V(\varphi, \zeta)$ is

$$V(\varphi) = a^3 e^g |g_\varphi|^2 = a^3 e^{-4\eta^2} |f_\varphi + 4 i\eta|^2, \tag{9.3.15}$$

where $\varphi = \xi + i\,\eta$. Equation (9.3.15) implies that for all real φ,

$$V = a^3 |f_\varphi|^2. \tag{9.3.16}$$

This resembles the expression for the effective potential in a globally supersymmetric theory with the superpotential $f(\varphi)$. Inflation takes place in this theory for a broad class of superpotentials, such as those with $f(\varphi) \sim \varphi^n$, $n > 1$. A complete description of inflation in this theory is quite complicated, particularly on account of the presence of non-minimal kinetic energy terms in (9.3.7). The third term in (9.3.7), for example, leads to an extra term $\sim a^{-1} e^{\zeta/3} |\partial_\mu \varphi|^2$ in V_ζ (9.3.12). Fortunately, $|\partial_\mu \varphi|^2 \ll V$ during inflation, and the corresponding correction turns out to be negligible.

In order to study the evolution of the universe in this model, we assume that φ is originally a fairly large field, $|\varphi| \gg 1$ (or $|\varphi| \gg \frac{M_P}{\sqrt{8\pi}}$ in conventional units). Then both the curvature $V_{\eta\eta} \sim a^3 |f|^2$ and the curvature $V_{\zeta\zeta} \sim a^3 |f_\varphi|^2$ are much greater than the curvature $V_{\xi\xi}$, which in the theory under consideration is of order $a^3 |f_\varphi|^2 \varphi^{-2}$. If $\zeta \neq 3 \ln a$ ($\zeta > 3 \ln \frac{2a}{3}$) from the outset and $\eta \neq 0$ ($|\eta| \lesssim 1$), then the fields ζ and φ quickly roll down to the bottom of the hollow, where $\zeta = 3 \ln a$ and $\eta = 0$, and the effective potential is given by (9.3.16). The field φ then has the usual kinetic energy term (9.3.13), and for $f = \mu^3 \varphi^n$,

$$V(\varphi) = n^2 a^3 \mu^6 \varphi^{2n-2}. \tag{9.3.17}$$

In particular, if $f = \mu^3 \varphi^3$,

$$V(\varphi) = 9\, a^3 \mu^6 \varphi^4. \tag{9.3.18}$$

The universe undergoes inflation as the field φ rolls down from $\varphi \gg 1$ to $\varphi \lesssim 1$. The density inhomogeneities that are produced in the theory (9.3.18) are of order $\frac{\delta\rho}{\rho} \sim 10^{-5}$ when $\sqrt{a}\,\mu \sim 10^{-2} - 10^{-3}$. There is thus no need to introduce any anomalously small coupling constants like $\lambda \sim 10^{-14}$; in this scenario, the combination $a^3 \mu^6$ takes on that role. A typical inflation factor for the universe in this model is of order 10^{10^7}. The process whereby the universe is reheated depends on the manner in which the field φ interacts with the matter fields. As a rule, reheating to a temperature $T_R \gtrsim 10^8$ GeV is readily accomplished [286], enabling baryon asymmetry production by the mechanisms described in Chapter 7.

9.4 The modified Starobinsky model and the combined scenario

In all of the models discussed thus far, inflation is driven by an elementary scalar field φ. However, the role of this field can also be played by a condensate of fermions $\langle \bar{\psi}\psi \rangle$ or vector particles $\langle G^a_{\mu\nu} G^a_{\mu\nu} \rangle$, or simply by the curvature scalar R itself. The latter possibility provided the basis for the Starobinsky model [52], which could be considered the first version of the inflationary universe scenario, and which predated even the model of Guth. In its original form, this model was based on the observation of Dowker and Critchley [106] that when the conformal anomaly of the energy-momentum tensor is taken into account, de Sitter space with energy density approaching the Planck density turns out to be a self-consistent solution of the Einstein equations with quantum corrections. Starobinsky showed that the corresponding solution is unstable; the curvature scalar starts to decrease slowly at some moment, and this decrease accelerates. Finally, after the oscillatory stage, the universe is reheated, and is then described by the standard hot universe model.

The formal description of the decay of the initial de Sitter space in the Starobinsky model bears a close resemblance to the theory of the decay of the unstable state $\varphi = 0$ in the new inflationary universe scenario. When this model was proposed, it elicited an enormous amount of interest from cosmologists [287]. But the origin of the unstable de Sitter state in the Starobinsky model remained somewhat enigmatic — the conventional wisdom was that either such a state came into being as a result of an asymmetric collapse of a previously existing universe [288], or that the universe appeared "from nothing" in an unstable quasivacuum state [289, 290]. These suggestions seemed rather more complicated than the principles underlying the new inflationary universe scenario. Furthermore, the original Starobinsky model, like the first versions of the new inflationary universe scenario, leaves us with density inhomogeneities $\frac{\delta\rho}{\rho}$ after inflation that are too large [107], and it fails to provide a solution to the primordial monopole problem.

Subsequently, however, it proved to be possible to modify this model and implement it in a manner that was similar in spirit to the chaotic inflation scenario [108–110]. The crux of this modification entailed replacing the study of one-loop corrections to the energy-momentum tensor T_μ^ν with an examination of a gravitational theory in which terms quadratic in the curvature tensor $R_{\mu\nu\alpha\beta}$ are added to the Einstein Lagrangian $\frac{R}{16\pi G}$.

In general, this is far from an innocuous procedure, inasmuch as metric perturbations then turn out to be described by fourth-order equations, and this frequently leads to particles having imaginary mass (tachyons) or negative energy (indefinite metric) [291]. Fortunately, these problems do not appear if just one term $\frac{R^2 M_P^2}{96\pi^2 M^2}$ is added, with $M^2 \ll M_P^2$. When the sign in front of R^2 is correctly chosen, this term leads to the emergence of a scalar excitation (scalaron) corresponding to a particle with positive energy and mass $M^2 > 0$. Taking the term $\sim R^2$ into consideration, the Einstein equations are then modified. Specifically, in a flat Friedmann space ($k = 0$), Eq. (1.7.12) for the universe filled by a uniform field φ is replaced by

$$H^2 = \frac{8\pi}{3M_P^2}\left[\frac{1}{2}\dot{\varphi}^2 + V(\varphi)\right] - \frac{H^2}{M^2}\left[\dot{H} + 2\frac{\ddot{H}}{H} - \left(\frac{\dot{H}}{H}\right)^2\right]. \tag{9.4.1}$$

Let us first neglect the contribution to (9.4.1) from the field φ, and consider solutions of the modified Einstein equations in the absence of matter fields. Equation (9.4.1) will then admit of a solution satisfying the conditions $|\dot{H}| \ll H^2$, $|\ddot{H}| \ll |\dot{H}H|$, or in other words, a solution that describes an inflationary universe with a slowly changing parameter H [108, 109]:

$$H = \frac{1}{6}M^2(t_1 - t), \tag{9.4.2}$$

$$a(t) = a_0 \exp\left(\frac{M^2}{12}(t_1 - t)^2\right). \tag{9.4.3}$$

These conditions prevail until H becomes smaller than M; after that, the stage of inflation ends, and H starts to oscillate about some mean value $H_0(t) \sim \frac{1}{t}$. The universe then heats up, and it can subsequently be described by the familiar hot universe theory.

Strictly speaking, Eqs. (9.4.2) and (9.4.3) are only applicable if R^2, $R_{\mu\nu}R^{\mu\nu} \ll M_P^4$. Moreover, depending on the initial value of H, inflation may start much later than the Planck time, at which R^2 and $R_{\mu\nu}R^{\mu\nu}$ become of the same order as M_P^4. We thus arrive once again at the problem of the evolution of a universe in which inflation takes place only in regions with suitable initial conditions (in no way associated with high-temperature phase transitions). In other words, we again wind up with the chaotic inflation scenario, where the curvature scalar R (equal to $12H^2$ at the time of inflation) takes on the role of the scalar inflaton field. In the more general case in which scalar fields φ (9.4.1) are also present, several different stages of inflation are possible, where the dominant effects are either associated with the scalar fields or they are purely gravitational, as described above [110].

The succession of different stages is governed by the relationship between the mass M of the scalaron and the mass m of the scalar field φ when $\varphi \sim M_P$ ($m \sim \sqrt{\lambda}\, M_P$ in the $\frac{\lambda}{4}\varphi^4$ theory). When $m \gg M$, the stage in which the field φ dominates rapidly comes to an end, and the next stage of inflation is associated with purely gravitational effects. During that stage, as usual, density inhomogeneities $\frac{\delta\rho}{\rho}$ are produced which on a galactic scale are given to order of magnitude by [107, 221]

$$\frac{\delta\rho}{\rho} \sim 10^3 \frac{M}{M_P}, \tag{9.4.4}$$

or in other words, $\frac{\delta\rho}{\rho} \sim 10^{-5}$ when

$$M \sim 10^{11}\ \mathrm{GeV}. \tag{9.4.5}$$

Reheating of the universe in this instance is also driven by purely gravitational effects [52, 134]. According to (7.9.17),

$$T_R \sim 10^{-1}\sqrt{\Gamma M_P} \sim 10^{-1}\sqrt{\frac{M^3}{M_P}} \sim 10^6\ \mathrm{GeV}. \tag{9.4.6}$$

To account for baryogenesis at a temperature $T \lesssim T_R \sim 10^6\,GeV$ as in the case of the Shafi–Vilenkin model and a number of other models based on supergravity, it is necessary to invoke the nonstandard mechanisms described in Section 7.10. Note, however, that terms like $\frac{R^2 M_P^2}{96\pi^2 M^2}$, if they do appear in elementary particle theories or superstring theories, will do so, as a rule, with $M \sim M_P$ rather than $M \sim 10^{-8} M_P$. It therefore seems more likely that realistic theories will yield $m \ll M$. The modified Starobinsky model may then turn out to be responsible for the description of the earliest stages of inflation, while the formation of the observable structure of the universe and its reheating take place during the stage when the scalar field φ dominates. Reference 110 describes a more detailed investigation of the combined model (9.4.1), effects related to the scalar field φ, and effects associated with quadratic corrections to the Einstein Lagrangian.

9.5 Inflation in Kaluza–Klein and superstring theories

It was noted in Chapter 1 that our fondest hopes for constructing a unified theory of all fundamental interactions have been tied in recent years to Kaluza–Klein and superstring theories. One feature common to both of these theories is the proposition that original space-time has a dimensionality $d \gg 4$. Theories with $d = 10$ [17], $d = 11$ [16], $d = 26$ [94], and even $d = 506$ [95, 96] have all been entertained. The assumption is that $d - 4$ dimensions are compactified, and that space takes on dimensions of order M_P^{-1} in the corresponding directions, so that we are actually able to move only in the one remaining time and three remaining space directions. It is usually assumed that the compactified directions are spatial, but the possibility of compactifying multidimensional time has also aroused some interest [292, 293]. The properties of a compactified space, in the final analysis, determine the basic properties of the elementary particle theory that it engenders.

Unfortunately, neither specific elementary particle models based on Kaluza–Klein and superstring theories nor associated cosmological models have yet come close to fruition. Nevertheless, it does make sense to examine the results that have been obtained in this field.

One of the most interesting and detailed models of inflation based on Kaluza–Klein models is that of Shafi and Wetterich [237]. The basis for this model is the Einstein action with corrections that are quadratic in the d-dimensional curvature:

$$S = -\frac{1}{V_{\mathcal{D}}}\int d^dx\,\sqrt{g_d}\left\{\alpha\hat{R}^2 + \beta\,\hat{R}_{\hat{\mu}\hat{\nu}}\hat{R}^{\hat{\mu}\hat{\nu}} + \gamma\,\hat{R}_{\hat{\mu}\hat{\nu}\hat{\sigma}\hat{\lambda}}\hat{R}^{\hat{\mu}\hat{\nu}\hat{\sigma}\hat{\lambda}} + \delta\cdot\hat{R} + \varepsilon\right\}. \qquad (9.5.1)$$

Here $\hat{\mu}, \hat{\nu}, \ldots = 0, 1, 2, \ldots, \quad d - 1$; $\hat{R}_{\hat{\mu}\hat{\nu}\hat{\sigma}\hat{\lambda}}$ is the curvature tensor in d-dimensional space; $V_{\mathcal{D}}$ is a volume in a $\mathcal{D}$-dimensional compactified space, $\mathcal{D} = d - 4$; and α, β, and γ are dimensionless parameters. With

$$\zeta = \mathcal{D}(\mathcal{D} - 1)\,\alpha + (\mathcal{D} - 1)\,\beta + 2\gamma > 0, \qquad (9.5.2)$$

$$\delta > 0, \tag{9.5.3}$$

$$\varepsilon = \frac{1}{4}\delta^2 \mathcal{D}(\mathcal{D}-1)\zeta^{-1}, \tag{9.5.4}$$

the equations for the d-dimensional metric have a solution of the form $M^4 \times S^{\mathcal{D}}$, where M^4 is a Minkowski space, and $S^{\mathcal{D}}$ is a sphere of radius

$$L_0^2 = \frac{2\zeta}{\delta}. \tag{9.5.5}$$

For

$$\chi = (\mathcal{D}-1)\beta + 2\gamma > 0\,, \tag{9.5.6}$$

the effective gravitational constant describing the interaction at large distances in M^4 space is positive:

$$G^{-1} = M_P^2 = 16\pi\frac{\chi}{\zeta}\delta. \tag{9.5.7}$$

A question remains concerning the stability of the solution $M^4 \times S^{\mathcal{D}}$, but it has at least been proven that with certain constraints on the parameters of the theory, compactification is stable against variations of the radius of the sphere $S^{\mathcal{D}}$ [294].

To describe cosmological evolution in this model, it is convenient to introduce the four-dimensional scalar field

$$\varphi(x) = \ln\frac{L(x)}{L_0}. \tag{9.5.8}$$

After a suitable change of scale of the metric $g_{\mu\nu}^{\sim}(x)$, the effective action in four-dimensional space can be expressed as

$$
\begin{aligned}
S = &-\int d^4x \sqrt{g_4} \\
&\Big[M_P^2 R/16\pi + \exp \mathcal{D}\varphi \cdot \left(\alpha R^2 + \beta R_{\mu\nu} R^{\mu\nu} + \gamma R_{\mu\nu\sigma\lambda} R^{\mu\nu\sigma\lambda}\right) \\
&-\frac{1}{2} f^2(\varphi)\, \partial_\mu\varphi\, \partial^\mu\varphi - \frac{1}{2} f_R(\varphi) \cdot R\, \partial_\mu\varphi\, \partial^\mu\varphi \\
&- \tilde{h}(\varphi)\, \partial_\mu\varphi\, \partial^\mu R + V(\varphi) + \Delta L_{kin}\Big].
\end{aligned}
\tag{9.5.9}
$$

Here $\mu, \nu, \ldots = 0, 1, 2, 3$, and ΔL_{kin} takes in terms that comprise many derivatives of the field φ, like $\partial_\mu \partial_\nu \varphi \cdot \partial^\mu \partial^\nu \varphi$, etc. The potential $V(\varphi)$ takes the form

$$
V(\varphi) = \left(\frac{M_P^2}{16\pi}\right)^2 \frac{\mathcal{D}(\mathcal{D}-1)}{4\zeta}\, e^{-\mathcal{D}\varphi} \left(\frac{1 - e^{-2\varphi}}{1 - \sigma\, e^{-2\varphi}}\right)^2, \tag{9.5.10}
$$

where $\sigma = \frac{\chi}{\zeta} - 1$. The functions $f^2(\varphi)$, $f_R(\varphi)$, and $\tilde{h}(\varphi)$ in (9.5.9) depend on α, β, γ, and $\mathcal{D}$. From (9.5.10), it follows that $V(\varphi) \geq 0$; $V(\varphi)$ tends to zero only when $\varphi = 0$ — that is, when $L(x) = L_0$. When $R_{\mu\nu\sigma\lambda} \neq 0$, however, the term

$$
e^{\mathcal{D}\varphi} \mathcal{K}_\varphi \equiv e^{\mathcal{D}\varphi}\left(\alpha R^2 + \beta R_{\mu\nu} R^{\mu\nu} + \gamma R_{\mu\nu\sigma\lambda} R^{\mu\nu\sigma\lambda}\right) \tag{9.5.11}
$$

also contributes to the equation of motion of the field φ, and the function

$$
W(\varphi) = V(\varphi) + e^{\mathcal{D}\varphi} \mathcal{K}_\varphi \tag{9.5.12}
$$

then plays the role of the potential energy of the field φ.

On the other hand, it is readily demonstrated that at the stage of inflation, the term (9.5.11) makes a contribution $\sim e^{\mathcal{D}\varphi} H^2 \dot{H}$ to the Einstein equations in four-dimensional space-time, and with $H =$ const, this contribution can be neglected. In that approximation, then, the rate of inflation of the universe does not depend on the fact that the additional term (9.5.11) is present, and is governed solely by the potential $V(\varphi)$,

$$H^2 = \frac{8\pi}{3M_P^2} V(\varphi), \tag{9.5.13}$$

while at the same time, the evolution of the field φ depends on the form of the potential $W(\varphi)$ (9.5.12):

$$3H h^2(\varphi)\dot{\varphi} = -\frac{\partial W}{\partial \varphi} = -\frac{\partial V}{\partial \varphi} - \mathcal{D} e^{\mathcal{D}\varphi} \mathcal{K}_\varphi(H(\varphi)). \tag{9.5.14}$$

The function $h^2(\varphi)$ appears in (9.5.14) by virtue of the non-minimal nature of the kinetic energy terms pertaining to the field φ in (9.5.9). This function varies slowly as φ changes, and goes to a constant as $\varphi \to \infty$. The function $\frac{\partial W}{\partial \varphi}$ at large φ behaves as follows:

$$\lim_{\varphi \to \infty} \frac{\partial W}{\partial \varphi} = \left(\frac{M_P^2}{16\pi}\right)^2 \frac{\mathcal{D}^2(\mathcal{D}-1)}{4\zeta} (\mu - 1)\, e^{-\mathcal{D}\varphi}, \tag{9.5.15}$$

where

$$\mu - 1 = \frac{\mathcal{D}-4}{12\zeta}\left[3(\mathcal{D}-1)\beta + 2(\mathcal{D}+3)\gamma\right]. \tag{9.5.16}$$

For $\mu > 1$, the potential $W(\varphi)$ approaches some constant from below, with the corresponding difference becoming exponentially small. This implies that the field φ rolls down to the minimum of $W(\varphi)$ at $\varphi = 0$ exponentially slowly. On the other hand, the potential $V(\varphi)$, which determines the rate of expansion of the universe, is also exponentially small at large φ. Nonetheless, with an initial value of $\varphi \gtrsim O(1)$ and a reasonable choice of the constants α, β, and γ, it is possible simultaneously to obtain both a high degree of inflation and small density perturbations. In particular, the duration of the inflationary stage in this model is given approximately [237] by

$$\Delta t \sim H^{-1} \frac{2\mathcal{K}_\infty}{\mu - 1} \varphi, \tag{9.5.17}$$

where $\mathcal{K}_\infty$ is defined by

$$\lim_{\varphi \to \infty} h^2(\varphi) = h_\infty^2 = \frac{M_P^2}{16\pi} \frac{\mathcal{D}}{4\zeta} \mathcal{K}_\infty . \tag{9.5.18}$$

The quantity $\mathcal{K}_\infty$ may be expressed in terms of α, β, and γ, and is usually of order unity. It can easily be shown that by choosing $\alpha, \beta, \gamma \sim 1$, which is quite natural, one can obtain $\Delta t \sim 70\ H^{-1}$, assuming an initial value $\varphi \sim 3$ [237].

In this model, the quantity $\frac{\delta\rho}{\rho}$ is given by an expression like (7.5.21), the only difference being that $\dot{\varphi}\, h(\varphi)$ appears instead of $\dot{\varphi}$:

$$\frac{\delta\rho}{\rho} \sim 0.2\, \frac{H^2}{\dot{\varphi}\, h(\varphi)} \sim 0.2\, \frac{H^2 \Delta t}{h_\infty \varphi} \sim \frac{2H}{M_P} \left(\frac{\pi\varphi}{\mathcal{D}\mathcal{K}_\infty} \right)^{1/2} \frac{H \Delta t}{\varphi} . \tag{9.5.19}$$

At the epoch of interest, $H\Delta t \sim 70$ and $\varphi \sim 3$, so

$$\frac{\delta\rho}{\rho} \sim C\, \frac{H}{M_P} , \tag{9.5.20}$$

where $C = O(1)$. In particular, $\frac{\delta\rho}{\rho} \sim 10^{-5}$ when

$$\frac{H}{M_P} \sim \frac{1}{8} \left(\frac{\mathcal{D}(\mathcal{D}-1)}{6\pi\zeta} \right)^{1/2} \exp\left(-\frac{\mathcal{D}}{2}\varphi\right) \sim 10^{-5} . \tag{9.5.21}$$

For $\varphi \sim 3$, (9.5.20) is satisfied in theories with $d = \mathcal{D} + 4 = O(10)$. One interesting feature of the perturbation spectrum obtained under these circumstances is its decline at large φ — that is, at long wavelengths. This is related to the fact that the behavior of the field and the rate of expansion of the universe in this model are determined by the two different functions $W(\varphi)$ and $V(\varphi)$, respectively, rather than the single function $V(\varphi)$.

The Shafi–Wetterich model is also interesting in that the curvature of the effective potential $W(\varphi)$ is quite large for $\varphi \ll 1$. After inflation, φ oscillates in the vicinity of $\varphi = 0$ at close to the Planck frequency, and the universe is reheated very rapidly and efficiently. In this model, the temperature can reach $T_R \sim 10^{17}$ GeV after reheating [295].

The main problem here is related to the initial conditions required for inflation. Indeed, it would be unnatural to assume, within the framework of Kaluza–Klein theories, that three-dimensional space has been infinite from the very beginning, since that would mean that the distinction between compactified and non-compactified dimensions would have to have been inserted into the theory from the outset, rather than arising spontaneously. It would be more natural to suppose that the universe has been compact since its birth, but that it has expanded at different rates in different directions: in three dimensions, it has grown exponentially, while in $d-4$ dimensions, it has gradually acquired a size of approximately $L_0 \sim M_P^{-1}$ (9.5.5), (9.5.7). To phrase it differently, we are dealing with a compact (closed, for example) universe governed by an expansion law that is asymmetric in different directions.

In Chapter 1, it was noted that a closed universe has a typical total lifetime of order M_P^{-1}, and the only thing that can rescue it from collapse is inflation that begins immediately after it has emerged from a state of Planck energy density. In the Shafi–Wetterich model, however, inflation ought to begin when $\varphi \gtrsim 3$, $H \lesssim 10^{-5} M_P$ (9.5.21), that is, when $V(\varphi) \ll M_P^4$. In that event, inflation cannot save the universe from a premature death. In order to circumvent this problem, it was suggested in [237] that the entire universe came into being as a result of a quantum jump from the space-time foam (from “nothing”) into a state with $\varphi \gtrsim 3$, $H \lesssim 10^{-5} M_P$, a possibility we shall discuss in the next chapter. Unfortunately, however, estimates of the probability for such processes [296] lead to an expression of the form $P \sim \exp\left(-\frac{M_P^4}{V(\varphi)}\right)$, giving $P \sim \exp\left(-10^{10}\right)$ in the present case. Thus, the outlook for a natural implementation of the inflationary universe scenario in the context of the Shafi–Wetterich model is not very good. In fact, we are the victims here of difficulties of the same type as those that prevent a successful implementation of the new inflationary universe scenario.

It might be hoped that these problems will all go away when we make the transition to a superstring theory. Such theories present several different candidates for the inflaton field responsible for the inflation of the universe — it may be some combination of the dilaton field that appears in superstring theory and the logarithm of the compactification radius. Regrettably, our current understanding of the phenomenological and cosmological aspects of superstring theory are still not entirely satisfactory. Existing models of inflation based on superstring theories [297] rest on various assumptions about the structure of these theories, and these assumptions are not always well-founded. But it is the initial conditions, as before, that are the main problem. Our view is that the initial conditions prerequisite to the onset of inflation in most of the models based on superstring theories that have been proposed thus far are unnatural.

Does this mean that we are headed down the wrong road? At the moment, that is a very difficult question to answer. It is quite possible that with the future development of superstring theory, the inflationary universe scenario will be implemented in the context of the latter in some nontrivial way (see, for example, Ref. 353). On the other hand, one should recall that over the past decade, three palace revolts have taken place in the land of elementary particles. In place of grand unified theories came theories based on supergravity, followed by Kaluza–Klein theories, and finally the presently reigning superstring theory. The inflationary universe scenario can be successfully implemented in some of these theories; in some, this has not yet been accomplished, but there are no "no-go" theorems that say it is impossible. In our opinion, what we have encountered here is a somewhat nonstandard aspect of a standard situation. A theory should be constructed in such a way that it describes experimental data, but this cannot always be done, and the theory must then be changed. Until recently, however, cosmological data have not been counted among the most important experimental facts. This situation has now been radically altered, and it might just be that models in which inflation of the universe cannot be implemented in a natural way will be rejected as being inconsistent with the experimental data (if, of course, we find no alternative solution for all of the cosmological problems outlined in Chapter 1 that is not based on the inflationary universe scenario). In analyzing the current state of affairs in this field, it must also be borne in mind that our understanding of the inflationary universe scenario, and in particular the most important question of initial conditions, is far from complete. In recent years, our conception of the initial-condition problem in cosmology and our ideas about the global structure of the inflationary universe have undergone a significant change. Progress in this field depends primarily on the development of quantum cosmology, the topic to which we now turn.

10

Inflation and Quantum Cosmology

If a man will begin with certainties, he shall end in doubts; but if he will be content to begin with doubts he shall end in certainties.

Francis Bacon (1561–1626)
The Advancement of Learning, Book V

10.1 The wave function of the universe

Quantum cosmology is conceptually one of the most difficult branches of theoretical physics. This is due not just to such difficulties as the ultraviolet divergences encountered in the quantum theory of gravitation, but also in large measure to the fact that the very formulation of the problems studied in quantum cosmology is not at all a trivial matter. The results of research often appear paradoxical, and it requires an especially open mind not to dismiss them outright.

The foundations of quantum cosmology were laid at the end of the 1960's by Wheeler and DeWitt [298, 299]. But prior to the advent of the inflationary universe scenario, a description of the universe as a whole within the framework of quantum mechanics seemed to most scientists to be an unnecessary luxury. When one describes macroscopic objects using quantum mechanics, the results are usually the same as those given by classical mechanics. If the universe is indeed the largest macroscopic entity that exists, then why bother to describe it using quantum theory?

In the standard hot universe theory, this was a perfectly legitimate question, since according to that theory the observable part of the universe resulted from the expansion of a region containing a total of perhaps 10^{87} elementary particles. But in the inflationary universe scenario, the entire observable part of the universe (and possibly the entire universe itself) was formed by virtue of the rapid expansion of a region of size $l \lesssim M_P^{-1} \sim 10^{-33}$ cm, containing perhaps not a single elementary particle! Quantum effects could thus have played a pivotal role in events during the earliest stages of expansion of the universe.

Until recently, the fundamental working tool in quantum cosmology has been the Wheeler–DeWitt equation for the wave function of the universe $\Psi(h_{ij}, \varphi)$, where h_{ij} is the three-dimensional spatial metric, and φ is the matter field. The Wheeler–DeWitt equation is essentially the Schrödinger equation for the wave function in the stationary state given by $\frac{\partial \Psi}{\partial t} = 0$ (see below). It describes the behavior of the quantity Ψ in so-called superspace — the space of all three-dimensional metrics h_{ij} (not to be confused with the superspace used to describe supersymmetric theories!). A detailed exposition of the corresponding theory may be found in [298-301]. But the most interesting results in this sphere were obtained using a simplified approach in which only a portion of the full superspace, known as a minisuperspace, was considered, giving a description of a homogeneous Friedmann universe; the scale factor of the universe a took up the role of all the quantities h_{ij}. In this section, we will therefore illustrate the basic problems relating to the calculation and interpretation of the wave function of the universe with an example of the minisuperspace approach. In subsequent sections, we will discuss the limits of applicability of this approach, the results obtained via recently developed stochastic methods for describing an inflationary universe [134, 135, 57, 132, 133], and a number of other questions with a bearing on quantum cosmology.

Let us consider, then, the theory of a scalar field φ with the Lagrangian

$$\mathcal{L}(g_{\mu\nu}, \varphi) = -\frac{R M_P^2}{16\pi} + \frac{1}{2}\partial_\mu \varphi\, \partial^\mu \varphi - V(\varphi) \qquad (10.1.1)$$

in a closed Friedmann universe whose metric can be represented in the form

$$ds^2 = N^2(t)\,dt^2 - a^2(t)\,d\Omega_3^2, \tag{10.1.2}$$

where $N(t)$ is an auxiliary function that defines the scale on which the time t is measured, and $d\Omega_3^2 = d\chi^2 + \sin^2\chi\,(d\theta^2 + \sin^2\theta\,d\varphi^2)$ is the element of length on a three-dimensional sphere of unit radius. To obtain an effective Lagrangian that depends on $a(t)$ and $\varphi(t)$, one must integrate over angular variables in the expression for the action $S(g,\varphi)$, which with the factor $\sqrt{g}$ taken into account gives $2\pi^2 a^3$. Then, making use of the fact that the universe is closed (has no boundaries), one obtains the effective Lagrangian in a form that depends only on a and $\dot{a}$, but not on $\ddot{a}$:

$$L(a,\varphi) = -\frac{3M_P^2\pi}{4}\left(\frac{\dot{a}^2 a}{N} - Na\right) + 2\pi^2 a^3 N\left(\frac{\dot{\varphi}^2}{2N^2} - V(\varphi)\right). \tag{10.1.3}$$

The canonical momenta are

$$\pi_\varphi = \frac{\partial L}{\partial\dot{\varphi}} = \frac{2\pi^2 a^3}{N}\dot{\varphi}, \tag{10.1.4}$$

$$\pi_a = \frac{\partial L}{\partial\dot{a}} = -\frac{3M_P^2\pi}{2N}\dot{a}a, \tag{10.1.5}$$

$$\pi_N = \frac{\partial L}{\partial\dot{N}} = 0, \tag{10.1.6}$$

and the Hamiltonian is

$$\begin{aligned}\mathcal{H} &= \pi_\varphi\dot{\varphi} + \pi_a\dot{a} - L(a,\varphi)\\ &= -\frac{N}{a}\left(\frac{\pi_a^2}{3\pi M_P^2} + \frac{3\pi M_P^2}{4}a^2\right) + \frac{N}{a}\left(\frac{\pi_\varphi^2}{4\pi^2 a^2} + 2\pi^2 a^4 V(\varphi)\right)\\ &= \mathcal{H}_a + \mathcal{H}_\varphi.\end{aligned} \tag{10.1.7}$$

Here $\mathcal{H}_a$ and $\mathcal{H}_\varphi$ are the effective Hamiltonians of the scale factor a and scalar field φ in the Friedmann universe. The equation relating the canon-

$$0=\frac{\partial\mathcal{H}}{\partial N}=\frac{\mathcal{H}}{N}=-\frac{1}{a}\left(\frac{\pi_a^2}{3\pi M_P^2}+\frac{3\pi M_P^2}{4}a^2\right)+\frac{1}{a}\left(\frac{\pi_\varphi^2}{4\pi^2a^2}+2\pi^2a^2V(\varphi)\right). \tag{10.1.8}$$

Upon quantization, Eq. (10.1.8) gives the relation that governs the wave function of the universe:

$$i\frac{\partial\Psi(a,\varphi)}{\partial t}=\mathcal{H}\Psi=0. \tag{10.1.9}$$

In the usual fashion, the canonical variables are replaced with the operators

$$\begin{aligned}\varphi\to\varphi,\quad \pi_\varphi\to\frac{1}{i}\frac{\partial}{\partial\varphi},\\ a\to a,\quad \pi_a\to\frac{1}{i}\frac{\partial}{\partial a},\end{aligned} \tag{10.1.10}$$

and Eq. (10.1.9) takes the form

$$\left(-\frac{1}{3\pi M_P^2}\frac{\partial^2}{\partial a^2}+\frac{3\pi M_P^2}{4}a^2+\frac{1}{4\pi^2a^2}\frac{\partial^2}{\partial\varphi^2}-2\pi^2a^4V(\varphi)\right)\Psi(a,\varphi)=0. \tag{10.1.11}$$

This then is the Wheeler–DeWitt equation in minisuperspace.

Strictly speaking, it should be pointed out that ambiguities relating to the commutation properties of a and π_a can arise in the derivation of Eq. (10.1.11). Instead of the term $-\frac{\partial^2}{\partial a^2}$ in (10.1.11), one sometimes writes $-\frac{1}{a^p}\frac{\partial}{\partial a}a^p\frac{\partial}{\partial a}$, where the parameter p can take on various values. In the semiclassical approximation, which will be of particular interest to us later on, the actual value of this parameter is unimportant, and in particular, one can take $p=0$ and seek a solution of (10.1.11).

Clearly, however, Eq. (10.1.11) has many different solutions, and one of the most fundamental questions facing us is which of these solutions actually describes our universe. Before launching into a discussion of this question, we make several remarks of a general nature that bear upon the interpretation of the wave function of the universe.

First of all, we point out that the wave function of the universe depends on the scale factor a but, according to (10.1.9), *it is time-inde-*

pendent. How can this be reconciled with the fact that our observable universe does depend on time?

Here we encounter one of the principal paradoxes of quantum cosmology, a proper understanding of which is exceedingly important. The universe *as a whole* does not depend on time because the very concept of such a change presumes the existence of some immutable reference that does not appertain to the universe, but relative to which the latter evolves. If by "the universe" we mean "everything," then there remains no *external observer* according to whose clocks the universe could develop. But in actuality, we are not asking why the universe is developing, we are inquiring as to why we *perceive* it to be developing. We have thereby separated the universe into two parts: a macroscopic observer with clocks, and all the rest. The latter can perfectly well evolve in time (according to the clocks of the observer), despite the fact that the wave function of the *entire* universe is time-independent [299].

In other words, we arrive at our customary picture of a world that evolves in time only after the universe has been divided into two macroscopic parts, each of which develops semiclassically. The situation that ensues is reminiscent of the theory of tunnelling through a barrier: the wave function is defined inside the barrier, but it yields the probability amplitude of finding a particle propagating in real time only outside the barrier, where classical motion is allowed. By analogy, the universe too exists in its own right, in a certain sense, but one can only speak of its *temporal* existence in the context of the semiclassical evolution of the part that remains after a macroscopic observer with clocks has emerged.

Thus, by the very fact of his existence, an observer somehow reduces the overall wave function of the universe to that part which describes the world that is observable to him. This is exactly the point of view espoused in the standard Copenhagen interpretation of quantum mechanics — the observer becomes not just a passive viewer, but something more like a participant in the creation of the universe [302].

The situation is somewhat different in the many-worlds interpretation of quantum mechanics [303–309], which presently enjoys a sizable following among quantum cosmologists. In this interpretation, the wave function $\Psi(h_{ij}, \varphi)$ simultaneously describes all possible universes together with the observers (of all possible kinds) that inhabit them. In performing a measurement, rather than reducing the wave function of all of these universes to the wave function of one of them (or a part of one of them), an observer simply refines the issue of who he is and in which of these universes he resides. The same results are then obtained as in the standard approach, but without recourse to the somewhat ill-founded assumption of the reduction of the wave function at the instant of measurement.

We shall not engage here in a detailed discussion of the interpretation of quantum mechanics, a problem which becomes particularly acute in the context of quantum cosmology [302, 309]; instead, we return to our discussion of the evolution of the universe.

Another manifestation of the fact that the universe as a whole does not change in time is that the wave function $\Psi(a, \varphi)$ depends only on the quantities a and φ, and not on whether the universe is contracting or expanding. One might interpret this to mean that at the point of maximum expansion of a closed universe, the arrow of time is somehow reversed, the total entropy of the universe begins to decrease — and observers are rejuvenated [310]. However, to determine the direction of the arrow of time, one must first divide the universe into two quasiclassical subsystems, one of which contains an observer with clocks. In general, the wave function of each such subsystem will certainly not be symmetric under a change in the sign of $\dot{a}$. After the division of the universe into two semiclassical subsystems, one can make use of the usual classical description of the universe, according to which the total entropy of the universe can only increase with time, and there is no way in which the direction of the arrow of time can ever be reversed at the instant of maximum expansion [311].

We have discussed all these problems in such detail here in order to demonstrate that not just the solution but even the formulation of problems in the context of quantum cosmology is a nontrivial matter. The question of whether entropy can decrease in a contracting universe, whether the arrow of time can be reversed in a singularity or at the point of maximum expansion of a closed universe, whether the universe can oscillate, has thus far bothered many experts in quantum cosmology; see, for example, [312, 313]. Above, we enunciated our own viewpoint in this regard, but it should be understood that the comprehensive investigation of these questions is only just beginning.

The Wheeler–DeWitt equation (10.1.11) has many different solutions, and it is very difficult to ascertain which of them actually describes our universe. One of the most interesting suggestions here was advanced by Hartle and Hawking [314], who proposed that the universe possesses a ground state, or a state of least excitation, similar to the vacuum state in quantum field theory in Minkowski space. By carrying out short-time measurements in Minkowski space, one can see that the vacuum is not empty, but is instead filled with virtual particles. Similarly, the universe that we observe might be a virtual state (but with a very long lifetime, due to inflation), and the probability of winding up in such a state might be determinable if the wave function of the ground state of the universe were known. According to the Hartle and Hawking hypothesis, the wave function $\Psi(a,\varphi)$ of the ground state of the universe with scale factor a which is filled with a homogeneous field φ is given in the semiclassical approximation by

$$\Psi(a,\varphi) \sim \mathrm{N}\, e^{-S_E(a,\varphi)} . \qquad (10.1.12)$$

Here N is a normalizing factor, and $S_E(a,\varphi)$ is the Euclidean action corresponding to solutions of the equations of motion for $a(\varphi(\tau),\tau)$ and $\varphi(\tau)$ with boundary conditions $a(\varphi(0),0) = a(\varphi)$, $\varphi(0) = \varphi$ in space with a metric that has Euclidean signature.

The reason for choosing this particular solution of the Wheeler–DeWitt equation was explained as follows. Consider the Green's function of a particle that moves from the point $(0,t')$ to $(\boldsymbol{x},0)$:

$$\langle \boldsymbol{x},0|0,t'\rangle = \sum_n \Psi_n(\boldsymbol{x})\Psi_n(0)\, e^{iE_n t'} = \int d\boldsymbol{x}(t) \exp\{iS[\boldsymbol{x}(t)]\}, \qquad (10.1.13)$$

where $\Psi_n(\boldsymbol{x})$ is a time-independent eigenfunction of the energy operator with eigenvalue $E_n \geq 0$. Let us now perform a rotation $t \to -i\tau$ and take the limit as $\tau' \to -\infty$. The only term that survives in the sum (10.1.13) is the one corresponding to the smallest eigenvalue E_n (normalized to zero). This implies that

$$\Psi_0(\boldsymbol{x}) \sim \mathrm{N} \int d\boldsymbol{x}(\tau) \exp\{-S_E[\boldsymbol{x}(\tau)]\} . \qquad (10.1.14)$$

Hartle and Hawking have argued that the generalization of this result to quantum cosmology in the semiclassical approximation will yield (10.1.13). For a slowly varying field φ (and this is precisely the most interesting case from the standpoint of implementing the inflationary universe scenario), the solution of the Euclidean version of the Einstein equations for $a(\varphi,\tau)$ is

$$a(\varphi,\tau) \sim H^{-1}(\varphi)\cos[H(\varphi)\tau] \equiv a(\varphi)\cos[H(\varphi)\tau], \quad (10.1.15)$$

where $H(\varphi) = \sqrt{\dfrac{8\pi V(\varphi)}{3M_P^2}}$, and the corresponding Euclidean action is

$$S_E(a,\varphi) = -\frac{3M_P^4}{16V(\varphi)}, \quad (10.1.16)$$

whereupon

$$\begin{aligned}\Psi[a(\varphi),\varphi] &\sim N\exp\left(\frac{3M_P^4}{16V(\varphi)}\right) = N\exp\left(\frac{\pi M_P^2}{2H^2(\varphi)}\right)\\ &= N\exp\left(\frac{\pi M_P^2 a^2(\varphi)}{2}\right).\end{aligned} \quad (10.1.17)$$

Hence, it should follow that the probability of detecting a closed universe in a state with field φ and scale factor $a(\varphi) = H^{-1}(\varphi)$ is

$$\begin{aligned}P[a(\varphi),\varphi] &\sim N^2|\Psi[a(\varphi),\varphi]|^2 \sim N^2\exp\left(\frac{3M_P^4}{8V(\varphi)}\right)\\ &= N^2\exp[\pi M_P^2 a^2(\varphi)].\end{aligned} \quad (10.1.18)$$

If the ground state of the universe were a state with $\varphi = \varphi_0$ and $0 < V(\varphi_0) \ll M_P^4$, then the normalization factor N^2 that ensures a total probability of all realizations being unity would have to be

$$N \sim \exp[-\pi M_P^2 a_0^2] = \exp\left(-\frac{3M_P^4}{8V(\varphi_0)}\right), \quad (10.1.19)$$

where $a_0 = H^{-1}(\varphi_0)$. From Eqs. (10.1.18) and (10.1.19), it follows that

$$P[a(\varphi),\varphi] \sim \exp\left[\frac{3M_P^4}{8}\left(\frac{1}{V(\varphi)} - \frac{1}{V(\varphi_0)}\right)\right]. \quad (10.1.20)$$

To calculate the probability that $a \ll a_0 = H^{-1}(\varphi_0)$ or $a \gg a_0 = H^{-1}(\varphi_0)$ when $\varphi = \varphi_0$, we must venture outside the confines of the quasiclassical approximation (10.1.12) or solve Eq. (10.1.11) directly in the WKB approximation. According to [314],

$$\Psi(a \ll a_0) \sim \exp\left[\frac{\pi}{2} M_P^2 (a^2 - a_0^2)\right], \tag{10.1.21}$$

$$\Psi(a \gg a_0) \sim \exp\left[\frac{i\, H(\varphi_0)\, M_P^2 a^3}{3}\right] + \exp\left[-\frac{i\, H(\varphi_0)\, M_P^2 a^3}{3}\right]. \tag{10.1.22}$$

Unfortunately, the arguments used by Hartle and Hawking to justify (10.1.12) are far from universally applicable. In fact, the Euclidean rotation alluded to above can be used to eliminate all but the zeroth-order term from (10.1.13) only if $E_n > 0$ for all $n > 0$. While the energy of excitations of the scalar field φ is positive, the energy of the scale factor a is negative, so that these sum to zero; see (10.1.7) and (10.1.9). In such a situation, there is no general prescription for extracting the ground state Ψ_0 from the sum (10.1.13) by rotation to Euclidean space. To investigate the properties of the field φ on scales much smaller than the size of the closed universe, this is an unimportant issue, and we can simply quantize the field φ against the background of the classical gravitational field and perform the standard Euclidean rotation $t \to -i\tau$. This is exactly the reason why the probability density function (10.1.20) is the same as the distribution (7.4.7), which was derived using more conventional methods. On the other hand, in those situations where the scale factor a itself must be quantized (for instance, in a description of the quantum creation of the universe from a state with $a = 0$, i.e., from "nothing" [315–317, 289, 290, 318]), the corresponding problem becomes much more serious.

Fortunately, this can be avoided if the quantum properties of the field φ are unimportant for our purposes at the epoch of interest — for example, if φ is a classical slowly varying field whose sole role is to produce a nonzero vacuum energy $V(\varphi)$ (cosmological term). One can then neglect quantum effects associated with the scalar field, and to isolate the ground state $\Psi(a, \varphi)$, corresponding to the lowest excitation state of the scale factor a, one need only carry out the rotation $t \to +i\tau$ in order to suppress the contribution to (10.1.13) from negative-energy excitations.[1] This gives

contribution to (10.1.13) from negative-energy excitations.[1] This gives

$$\Psi(a,\varphi) \sim \mathrm{N}\ e^{S_E(a,\varphi)} \sim \mathrm{N}\ \exp\left(-\frac{3\mathrm{M_P^4}}{16\mathrm{V}(\varphi)}\right), \qquad (10.1.23)$$

and the probability that the universe appears in a state with field φ is

$$\mathrm{P}[a(\varphi),\varphi)] \sim |\Psi|^2 \sim \mathrm{N}^2 \exp\left(-\frac{3\mathrm{M_P^4}}{8\mathrm{V}(\varphi)}\right). \qquad (10.1.24)$$

Equation (10.1.23) was first obtained using the method described above [319]; it was later derived by Zeldovich and Starobinsky [320], Rubakov [321], and Vilenkin [322] using a different method. For the reasons to be discussed soon, we will call (10.1.23) *tunneling wave function.*

The obvious difference between Eqs. (10.1.24) and (10.1.18)–(10.1.21) is in the sign of the argument of the exponential. This difference is extremely important, since according to (10.1.18) and (10.1.20), the probability of *detecting* the universe in a state with a large value of $\mathrm{V}(\varphi)$ is exponentially small. In contrast, Eq. (10.1.24) tells us that the universe is most likely *created* in a state with $\mathrm{V}(\varphi) \sim \mathrm{M_P}^4$. This is consistent with our previous expectations, and leads to a natural implementation of the chaotic inflation scenario [319].

In order to comprehend the physical meaning of the Hartle–Hawking wave function (10.1.12), let us compare the solutions of Eqs. (10.1.21) and (10.1.22) with solutions for the scalar field (1.1.3). One possible interpretation of the solution (10.1.21) is that the wave function $\exp\left[\frac{i\,\mathrm{H}(\varphi_0)\,\mathrm{M_P^2}\,a^3}{3}\right]$ describes a universe with decreasing scale factor a (compare with the wave function of a particle with momentum p, $\psi \sim e^{-ipx}$), while the wave function $\exp\left[-\frac{i\,\mathrm{H}(\varphi_0)\,\mathrm{M_P^2}\,a^3}{3}\right]$ corresponds to a universe with increasing scale factor. Bearing in mind, then, that the corresponding motion takes place, according to (10.1.11), in a theory for

[1] It should be borne in mind that there are no *physical* excitations of the gravitational field with negative energy. Therefore, for a consistent quantization of the scale factor a one should introduce Faddeev-Popov ghosts. However, as usual, ghosts do not contribute to $\Psi(a,\varphi)$ in the semiclassical approximation.

$$V(a) = \frac{3\pi M_P^2}{4} a^2 - 2\pi^2 a^4 V(\varphi), \qquad (10.1.25)$$

the interpretation of the solutions (10.1.21) and (10.1.22) becomes quite straightforward (although different authors are still not in complete agreement on this point). The wave function (10.1.22) describes a wave incident upon the barrier $V(a)$ from the large-a side and a wave reflected from the barrier; when $a < H^{-1}$ (i.e., incidence below the barrier height), the wave is exponentially damped in accordance with (10.1.21) (see Fig. 35). The physical meaning of this solution is most easily grasped if one recalls that a closed de Sitter space with $V(\varphi_0) > 0$ first contracts and then expands: $a(t) = H^{-1} \cosh(Ht)$. The Hartle–Hawking wave function (10.1.21) accounts for the "broadening" of this quasiclassical trajectory, and allows for the fact that at the quantum level, the scale factor can become less than H^{-1} at the point of maximum contraction. The absence of exponential suppression for $a > H^{-1}$ (10.1.22) is related to the fact that values $a > H^{-1}$ are classically allowed [314]. Observational cosmological data put the present-day energy density of the vacuum $V(\varphi_0)$ at no more than 10^{-29} g·cm^{-3}, which corresponds to $H^{-1} \gtrsim 10^{28}$ cm. The evolution of a de Sitter space whose minimal size exceeds 10^{28} cm has nothing in common with the evolution of the universe in which we now live. Therefore, within the scope of the foregoing interpretation, the Hartle–Hawking wave function does not give a proper description of our universe in the minisuperspace approximation studied above. We face a similar difficulty if we attempt (without justification) to use this wave function instead of the function (10.1.23) to account for the very earliest stages in the evolution of the universe, since according to (10.1.18) and (10.1.20), the likelihood of a prolonged inflationary stage would in that case be exponentially small.

One might suggest another possible interpretation of the Hartle–Hawking wave function, namely that its square gives the probability density function for an observer to detect that he is in a universe of a given type not at the instant of its creation, but at the instant of his first measurement, prior to which he cannot say anything about the evolution of the universe.[2] Such an interpretation may turn out to be eminently reason-

[2] From this standpoint, one might tentatively say that the wave function (10.1.23) is associated with the creation of the universe, while (10.1.12) is associated with the

able (and, in the final analysis, independent of the choice of observer) if, as originally supposed by Hartle and Hawking, a ground state actually exists for the system in question, so that the probability distribution under consideration turns out to be stationary, like the vacuum state or ground state of an equilibrium thermodynamic system. And in fact, as we have already noted, the Hartle–Hawking wave function provides a good description of the quasistationary distribution of the field φ in an intermediate metastable state (see (10.1.20) and (7.4.7)).

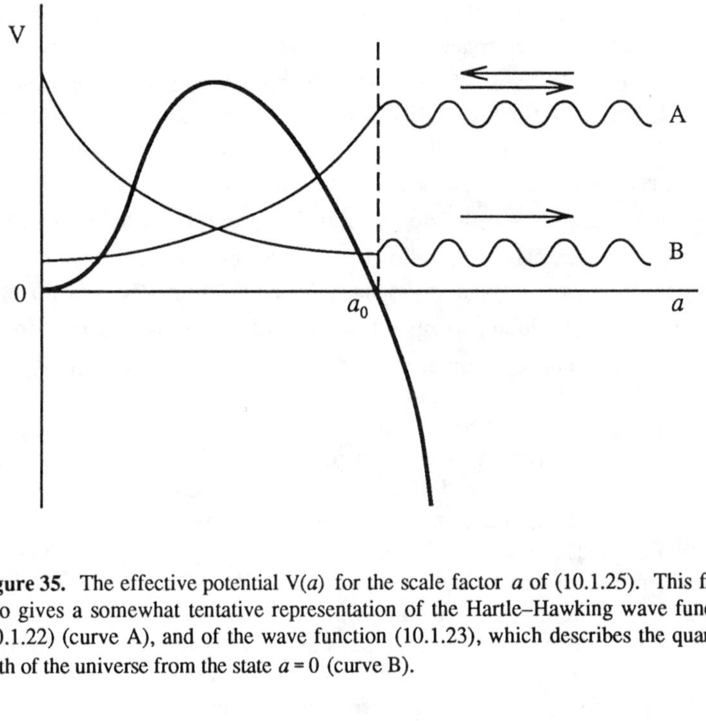

Figure 35. The effective potential V(a) for the scale factor a of (10.1.25). This figure also gives a somewhat tentative representation of the Hartle–Hawking wave function (10.1.22) (curve A), and of the wave function (10.1.23), which describes the quantum birth of the universe from the state $a = 0$ (curve B).

On the other hand, a stationary distribution (10.1.20) of the field φ is only possible if $m^2 = \frac{d^2V}{d\varphi^2} \ll H^2 \sim \frac{V(\varphi)}{M_P^2}$ in the vicinity of the absolute minimum of V(φ). Not a single realistic model of the inflationary uni-

creation of an observer.

verse satisfies this requirement. The only stationary distribution of a (quasi)classical field φ in realistic models that we are presently aware of (see the discussion of this question in Sections 7.4, 10.2, and 10.3) is the trivial delta-function distribution, with the field totally concentrated at the minimum of $V(\varphi)$. But this is not at all the result that researchers in quantum cosmology are trying to obtain when they discuss the Hartle–Hawking wave function and assume that the appropriate probability distribution is given by Eq. (10.1.20).

All these caveats notwithstanding, we would rather not draw any hasty conclusions, which would be a particularly dangerous thing to do in a science whose ultimate foundations have yet to be laid. The mathematical structure proposed by Hartle and Hawking is quite elegant in and of itself, and quite possibly we may yet find a way to take advantage of it. Our fundamental objection to the possibility of a stationary distribution of the field φ in an inflationary universe is based upon a study of a (typical) situation in which the field possesses one (or a few) absolutely stable vacuum states. But instances are known in which theories are characterized by the value of some time-independent field, topological invariant, or other parameter characterizing properties of the vacuum state which might govern, say, the strength of CP violation, the energy of the vacuum, and so forth.

One such parameter is the angle θ which characterizes the vacuum properties in quantum chromodynamics [183]. It is possible that the cosmological term and many coupling constants in elementary particle theory [345, 346, 349] are also vacuum parameters of this type. Their time-independence may be guaranteed by superselection rules of some sort [346]. But in the context of the many-worlds interpretation of quantum mechanics, the question of what the properties of the world (of the vacuum state) are in which the observer finds himself at the instant of his first observation is a perfectly reasonable one. The suggestion that the appropriate probability distribution will be given by the square of the Hartle–Hawking wave function [346] seems to us worthy of serious consideration. At the same time, the question of choosing between the Hartle–Hawking function and the function (10.1.23) under these circumstances becomes especially important. As we pointed out earlier, the Hartle–Hawking wave function actually gives the proper results when one considers the (quasi)stationary distribution of a scalar field φ with positive energy density in a classical de Sitter background (see (10.1.20)). On the other hand, if the evolution of matter fields (and vacuum parameters) is insignificant, then the wave function may possibly be determined by an expression like (10.1.23). We will return to the discussion of this question in Section 10.7.

A possible interpretation (and an alternative derivation) of the wave function (10.1.23) can also be obtained by studying tunneling through the barrier (10.1.25), but from the direction of small a rather than large a [320–322]. Indeed, one can readily show that (with $\varphi \sim$ const) a solution of Eq. (10.1.11) exists, which behaves as $\exp\left(-\frac{\pi}{2}M_P^2a^2\right)$ for

$$a < a_0 = H^{-1}(\varphi) \gg M_P^{-1}$$

(cf. (10.1.21)), while for $a \gg H^{-1}(\varphi)$, it is represented by a wave

$$\sim \exp\left(-\frac{\pi M_P^2 a_0^2}{2} - \frac{iHM_P^2 a^3}{3}\right)$$

emerging from the barrier and moving off toward large a; see Fig. 35. Damping of the wave as it emerges from the barrier is of order

$$\exp\left(-\frac{\pi M_P^2 a_0^2}{2}\right) \sim \exp\left(-\frac{3M_P^4}{16V(\varphi)}\right),$$

which exactly corresponds to Eq. (10.1.23) above. The wave function (10.1.23) thus describes the process of quantum creation of a closed inflationary universe filled with a homogeneous field φ due to tunneling from a state with scale factor $a = 0$, or in other words, from "nothing" [319–322].

We now attempt to provide a plausible interpretation of this result, and to clarify the reason why the probability of the quantum creation of a closed universe only becomes large when $V(\varphi) \sim M_P^4$. To this end we examine a closed de Sitter space with energy density $V(\varphi)$. Its volume at the epoch of maximum contraction ($t = 0$) is of order $H^{-3}(\varphi) \sim M_P^3[V(\varphi)]^{-3/2}$, and the total energy of the scalar field contained in de Sitter space at that instant is approximately $E \sim V(\varphi)H^{-3}(\varphi) \sim \frac{M_P^3}{\sqrt{V(\varphi)}}$. When $V(\varphi) \sim M_P^4$, the total energy of the scalar field is $E \sim M_P$. By the uncertainty principle, one cannot rule out the possibility of quantum fluctuations of energy E lasting for a time $\Delta t \sim E^{-1} \sim M_P^{-1}$. However, within a time of this order (or slightly longer), de Sitter space of initial size $\sim H^{-1} \sim M_P^{-1}$ becomes exponentially large, and one can consider it to be an inflationary universe emerging from "nothing" (or from the space-time foam). For small $V(\varphi)$, the probability that this process will come to pass should be extremely low, since as $V(\varphi)$ decreases, the minimum energy E of a scalar field in de Sitter space increases, rather than falling off, and the typical lifetime of the corresponding quantum fluctuation becomes much shorter than the Planck time.

It is important to note here that we are dealing with the creation of a compact universe with no boundaries, so that no supplementary conditions such as $\varphi = 0$ are required at the boundary of the bubble that is formed (compare this with the discussion of the "budding" of the universe from Minkowski space in Section 10.3). If the effective potential $V(\varphi)$ is flat enough, so that the field φ rolls down to its minimum in a time much longer than H^{-1}, then at the instant of its creation, the universe will "know nothing" of the location of the minimum of $V(\varphi)$ or how far the initial field φ is displaced from it. To a first approximation, the probability of creation of the universe is given solely by the magnitude of $V(\varphi)$, in accordance with (10.1.24).

Generally speaking, the extent to which the probability of creation of a universe with $V(\varphi) \ll M_P^4$ is suppressed may be lessened somewhat if particle creation at the time of tunneling is taken into account [321]. Furthermore, exponential suppression may be absent, by and large, during the creation of a compact (flat) universe with nontrivial topology [320]. For us, the only important thing is that, as expected, there is no exponential suppression of the probability for creation of an inflationary universe with $V(\varphi) \sim M_P^4$ — that is, from the present point of view, the initial conditions for implementation of the chaotic inflation scenario are also found to be quite natural.

Note that the distinction between the creation of the universe from a singularity and quantum creation from "nothing" at Planck density is rather tentative. In either case, one is dealing with the emergence of a region of classical space-time from the space-time foam. The terminological distinction consists of the fact that by creation from "nothing," one usually means that a description of the evolution of the universe according to the classical equations of motion begins only at large enough a. But due to large quantum fluctuations of the metric at $\rho \gtrsim M_P^4$, it also turns out to be impossible to describe the universe near the singularity (i.e., at small a). An important feature of either case is that as $a \to 0$, only quantum cosmology can provide a description of the evolution of the universe, a circumstance that can lead to surprising consequences.

Consider, for example, a possible model for the evolution of a closed inflationary universe. This model will be incomplete, and aspects of its interpretation will be open to argument, but on the whole it furnishes a good illustration of some novel possibilities being discussed within the context of quantum cosmology.

Thus, suppose that the universe was originally in a state $a = 0$. Quantum fluctuations of the metric at that time were extremely large, and there were neither clocks nor rulers. Any observations made by an imaginary observer at that epoch would have been uncorrelated with one another, and one could not even have said which of those observations came first or last. The results of measurements could not be remembered, which implies that with each new measurement the observer would effectively find himself in a completely new space. If during one of these measurements he found himself inside a hot universe that was not passing through an inflationary stage, then the characteristic lifetime of such a universe would turn out to be of order M_P^{-1} and its total energy $E \sim M_P$; it would therefore be essentially indistinguishable from a quantum fluctuation. But if the observer detected that he was in an inflationary universe, he would then be able to make clocks and rulers, and over an exponentially long period of time he could describe the evolution of the universe with the aid of the classical Einstein equations. After a certain time, the universe would be reheated, following which it would proceed through a state of maximum expansion and begin to contract. When it had reached a state of Planck density (which would occur when $a \gg M_P^{-1}$), it would become impossible to use clocks and rulers and thereby introduce any meaningful concept of time, entropy density, and so forth, due to large quantum fluctuations of the metric. One could say that quantum fluctuations near the singularity in effect erase from the memory of the universe any information about the properties it had during its period of quasiclassical evolution. Consequently, after Planck densities have been attained, subsequent observations again become disordered; it is even dubious, strictly speaking, that one could say they are *subsequent*. At some point, the observer detects that he is in an inflationary universe, and everything begins anew. Here the parameters of the inflationary universe depend only on the value of the wave function $\Psi(a, \varphi)$, and not on its history, which has been "forgotten" during the passage through the purgatory of Planck densities. We thus obtain a model for an oscillating universe in which there is no increase of entropy during each successive cycle [298, 323]. Other versions of this model exist, and are based on a hypothesized limiting density $\rho \sim M_P^4$ [313] or gravitational confinement at $\rho \gtrsim M_P^4$ [116].

The examples considered in this section indicate how interesting the investigation of solutions of the Wheeler–DeWitt equation is, and by the same token, how difficult it is to choose and interpret a satisfactory solution. Studies of this question are only just beginning [324]. Some of the problems encountered are related to the minisuperspace approximation employed, and some to the fact that we wish to derive (or guess) the correct solution to the full quantum mechanical problem without an adequate understanding of the properties of the global structure of the inflationary universe at a more elementary level. To fill this gap, it will prove very useful to study the properties of the inflationary universe via the stochastic approach to inflation, which occupies an intermediate position between the classical description of the inflationary universe and an approach based on solving the Wheeler–DeWitt equation.

10.2 Quantum cosmology and the global structure of the inflationary universe

One of the principal shortcomings of an approach based on minisuperspace is the initial presumption of global homogeneity of the universe. The only explanation for the homogeneity of the universe which is known at present is based on the inflationary universe scenario. However, as we showed in Section 1.8, effects related to long-wave fluctuations of the scalar field prevent the geometry of an inflationary universe from having anything in common with that of a homogeneous Friedmann space on much larger scales (at $l \gtrsim l^*$). Instead of a homogeneous universe that comes into being as a whole at some instant of time $t = 0$, we must deal with a globally inhomogeneous self-regenerating inflationary universe, whose evolution has no end and quite possibly may not have had a unique beginning. Thus, many of the most important properties of the inflationary universe cannot be understood or studied within the context of the minisuperspace approach.

In Chapter 1, we presented the simplest description of the mechanism of self-regeneration of inflationary domains of the universe in the chaotic inflation scenario [57]. Below, we engage in a more detailed investigation of this problem [132, 133].

One could carry out this investigation in the coordinate system (7.5.8), which is especially convenient for analyzing density inhomogeneities in an inflationary universe [218, 220]. If, however, one is interested in describing the evolution of the universe from the point of view of a co-moving observer, then it is more convenient to go to a synchronous coordinate system, which can be so chosen that the metric of an inflationary universe on scales much larger than H^{-1} may be written in the form [135, 133]

$$ds^2 \sim dt^2 - a^2(\mathbf{x}, t)\, d\mathbf{x}^2, \tag{10.2.1}$$

where

$$a(\mathbf{x}, t) \sim \exp\left\{ \int_0^t H[\varphi(\mathbf{x}, t)]\, dt \right\}. \tag{10.2.2}$$

What these expressions mean is that an inflationary universe in a neighborhood of size $l \gtrsim H^{-1}$ about every point x looks like a homogeneous inflationary universe with Hubble parameter $H[\varphi(\mathbf{x}, t)]$. To study the global structure of the inflationary universe in this approximation, it turns out to be sufficient to study the independent local evolution (in accordance with the "no hair" theorem in de Sitter space) of the field φ within each individual region of the inflationary universe of size $l \sim H^{-1}$ (or with *initial* size $l_0 \sim H^{-1}$), and then attempt to discern the overall picture using Eqs. (10.2.1) and (10.2.2). The local evolution of the field φ in regions whose size is of order H^{-1} is governed by the diffusion equations (7.3.22), (7.4.4), and (7.4.5), taking into consideration the dependence of the diffusion and mobility coefficients $D = \frac{H^3}{8\pi^2}$ on the magnitude of the field φ [135, 133].

The simplest possibility would be to study stationary solutions of Eqs. (7.4.4) and (7.4.5). The corresponding solution in the general case might also depend on the stationary probability flux j_c = const, and for $V(\varphi) \ll M_P^4$, it is given by [135]

$$P_c(\varphi) \sim \text{const}\cdot \exp\left[\frac{3M_P^4}{8V(\varphi)}\right] - 2j_c\frac{\sqrt{6\pi V(\varphi)}}{M_P V'(\varphi)}. \qquad (10.2.3)$$

Unfortunately, attempts to provide a physical interpretation of this solution meet with all sorts of difficulties. As in Section 7.4, let us first consider the case $j_c = 0$. One can readily show that Eq. (10.2.3) is identical to the square of the Hartle–Hawking wave function (10.1.17), namely (10.1.18). But since the effective potential $V(\varphi)$ vanishes at its minimum, which corresponds to the vacuum state in the observable part of the universe, the distribution (10.2.3) is unnormalizable. This is an especially easy problem to comprehend in the chaotic inflation scenario for theories with $V(\varphi) \sim \varphi^{2n}$; in these theories, inflation takes place only when $\varphi \gtrsim M_P$. In such theories, then, there is no diffusion flux out of the region with $\varphi \lesssim M_P$ and into the region with $\varphi \gtrsim M_P$. But such a flux would provide the only way to compensate for the classical rolling down of the field to the minimum of $V(\varphi)$, which is necessary for the existence of a stationary distribution $P_c(\varphi)$ when $j_c = 0$.

The issue of the interpretation of the second term in (10.2.3) is even more complicated. As we noted in Section 7.4, formally, in theories with $V(\varphi) \sim \varphi^{2n}$, this solution does not exist in general, as it is odd in φ while $P_c(\varphi)$ must always be positive. We can avert this problem to a certain extent by recalling that in these theories, Eq. (7.4.5) itself holds only on the segment $M_p \lesssim \varphi \lesssim \varphi_P$, where $V(\varphi_P) \sim M_P^4$. But this is not an entirely satisfactory response. It is actually straightforward to show that the second term in (10.2.3) is a solution of Eq. (7.3.22) in which the first (diffusion) term is omitted. Thus, we are simply dealing with a classical "rolling" of the field φ away from a region with $V(\varphi) \gg M_P^4$, where the diffusion equation is not valid. In this instance, a stationary distribution $P_c(\varphi, t)$ can only be sustained through a constant flux j_c out of a region with $V(\varphi_P) \gg M_P^4$. One could attempt to interpret this flux as a current corresponding to the probability of quantum creation of new domains of the universe with $V(\varphi_P) \gtrsim M_P^4$, per unit initial coordinate volume. But as Starobinsky has already emphasized in [135], where a solution of the type (10.2.3) was first derived, at present we can neither give a rigorous existence proof for such a solution, nor can we say anything definite about the magnitude of j_c if it is nonzero. The feasibility of the interpretation suggested above does not follow from the derivation of Eqs. (7.4.4) and (7.4.5) given in [132–135]. Moreover, since the vast preponderance of the initial coordinate volume of the inflationary universe eventually transforms into a state with $\varphi \lesssim M_P$, $V(\varphi_P) \ll M_P^4$, the validity of the assumed constancy of the probability current for quantum creation of the new domains of the universe per unit *initial* volume seems not to be well-founded. It is not entirely clear, for example, why it is necessary to require a stationary distribution $P_c(\varphi, t)$ rather than a probability distribution for finding the field φ at time *t*, per unit *physical* volume, and taking into account the increase in volume due to the quasiexponential expansion of the universe (10.2.2), which proceeds at different rates in regions filled with different fields φ.

One can attain a proper understanding of the situation for stationary solutions only through a comprehensive analysis of nonstationary solutions of the diffusion equation under the most general initial conditions for the distribution $P_c(\varphi, t)$. As stated above, we are interested in the probability $P_c(\varphi, t)$ of finding the field φ at time t in a region of initial size $O(H^{-1})$. Due to inflation, the initial inhomogeneities of the field φ on this scale become exponentially small, while the amplitude of quasiclassical perturbations $\delta\varphi$ with wavelengths $l \gtrsim H^{-1}$ on this scale do not exceed H; see (7.3.12). Bearing in mind that inflation occurs in theories with $V(\varphi) \sim \varphi^{2n}$ when $\varphi \gtrsim M_P$, the condition $V(\varphi) \ll M_P^4$ implies that $\delta\varphi \sim H \ll \varphi$, or in other words the field φ is, to a very good approximation, homogeneous on a scale $l \sim H^{-1}$.

We may therefore assume without loss of generality that at time $t = 0$, the field φ is equal to some constant φ_0 in the region under consideration — the size of which is $O(H^{-1})$ — or in other words, $P_c(\varphi, t = 0) = \delta(\varphi - \varphi_0)$. Solutions of Eq. (7.3.22) with such initial conditions were investigated in [132, 133], and there it was found that these solutions are all nonstationary. The distribution $P_c(\varphi, t)$ first broadens, and its center is then displaced toward small φ, being governed by the same law as the classical field $\varphi(t)$. Meanwhile, the distribution $P_p(\varphi, t)$ of the physical volume occupied by the field φ behaves differently depending on the value of the initial field $\varphi = \varphi_0$. For small φ_0, $P_p(\varphi, t)$ behaves in almost the same way as $P_c(\varphi, t)$, but for sufficiently large φ_0, the distribution $P_p(\varphi, t)$ starts to edge toward large values of the field φ as t increases, which leads to the onset of the self-regenerating inflationary regime discussed in Section 1.8.

Referring the reader to [132, 133] for details, let us elucidate the behavior of the distributions $P_c(\varphi, t)$ and $P_p(\varphi, t)$ using the theory $V(\varphi) = \frac{\lambda}{4}\varphi^4$ as an example. In order to do so, we divide the quasiclassical field φ into a homogeneous classical field $\varphi(t)$ and inhomogeneities $\delta\varphi(\boldsymbol{x}, t)$ with wavelengths $l \gtrsim H^{-1}$ (see (7.5.7)):

$$\varphi(\boldsymbol{x}, t) = \varphi(t) + \delta\varphi(\boldsymbol{x}, t). \tag{10.2.4}$$

It can readily be shown that during the inflationary stage, the equations of motion for $\varphi(t)$ and $\delta\varphi$ in the metric (10.2.1), (10.2.2) are given, to terms linear in $\delta\varphi$, by

$$3H\dot{\varphi} = -\frac{dV}{d\varphi} = -\lambda\varphi^3, \tag{10.2.5}$$

$$3H\delta\dot{\varphi} - \frac{1}{a^2}\Delta\delta\varphi = -\left[V'' - \frac{(V')^2}{2V}\right]\delta\varphi = -\frac{5}{2}\lambda\varphi^2\delta\varphi. \tag{10.2.6}$$

The term $\frac{(V')^2}{2V}\delta\varphi$ in (10.2.6) appears by virtue of the dependence of the Hubble parameter H on φ. Equations (10.2.5) and (10.2.6) make it clear that to lowest order in $\delta\varphi$, an investigation of the evolution of the field $\varphi(\boldsymbol{x}, t)$ reduces to an investigation of the motion of the homogeneous field $\varphi(t)$ as governed by the classical equation of motion (10.2.5), and to a subsequent investigation of the evolution of the distribution $P_c(\delta\varphi, t)$ subject to the initial condition $P_c(\delta\varphi, 0) \sim \delta(\delta\varphi)$.

It is important that when $\varphi \gg M_P$ (that is, during inflation), the effective mass squared of the field $\delta\varphi$,

$$m_{\delta\varphi}^2 = V'' - \frac{(V')^2}{2V} = \frac{5}{2}\lambda\varphi^2, \tag{10.2.7}$$

be much less than the square of the Hubble parameter: $m_{\delta\varphi}^2 \ll H^2$. This means that during the first stage of "broadening" of the delta functional distribution $P_c(\delta\varphi, 0) \sim \delta(\delta\varphi)$, right up to the time

$$t_1 \sim \frac{3\mathrm{H}}{2m_{\delta\varphi}^2} \sim (2\sqrt{\lambda}\,\mathrm{M_P})^{-1},$$

the dispersion squared of fluctuations $\delta\varphi$ grows linearly with time, (7.3.12):

$$\left\langle \delta\varphi^2 \right\rangle = \frac{\mathrm{H}^3(\varphi)\, t}{4\pi^2} = \frac{\sqrt{\lambda}\,\varphi^6}{3\sqrt{6\pi}\,\mathrm{M_P^3}}\, t\,. \tag{10.2.8}$$

The growth of $\left\langle \delta\varphi^2 \right\rangle$ then slows down (see (7.3.13)), and by

$$t_2 \sim \frac{\sqrt{6\pi}}{\sqrt{\lambda}\,\mathrm{M_P}} \sim 10\, t_1,$$

the dispersion of fluctuations $\delta\varphi$ will have essentially reached its asymptotic value (7.3.3):

$$\Delta_0 = \sqrt{\left\langle \delta\varphi^2 \right\rangle} = \mathrm{C}\sqrt{\frac{3\mathrm{H}^4}{8\pi^2 m_{\delta\varphi}^2}} \sim \sqrt{\frac{\lambda}{15}}\,\frac{\varphi^3}{\mathrm{M_P^2}}, \tag{10.2.9}$$

where $\mathrm{C} \sim 1$. Equation (1.7.22) tells us that at this stage (with $t \ll t_2$), the mean field $\varphi(t)$ hardly decreases at all. For $t > t_2$, both the field $\varphi(t)$ and the quantity $\mathrm{H}(\varphi)$ start to fall off rapidly, and that is why fluctuations produced at $t \gg t_2$ make a negligible contribution to the total dispersion $\Delta(t) = \sqrt{\left\langle \delta\varphi^2 \right\rangle}$. The latter quantity is basically determined by fluctuations that make their appearance when $t \lesssim t_2$. To analyze the behavior of $\Delta(t)$ for $t > t_2$, it is sufficient to note that the amplitude of fluctuations $\delta\varphi$ that arise when $t < t_2$ subsequently behaves in the same way as the magnitude of $\dot{\varphi}$ [114] (the reason being that, as one can readily prove, $\dot{\varphi} = \dfrac{d\varphi}{dt}$ obeys the same equation of motion (10.2.6) as $\delta\varphi$). This implies that for $t \gg t_2$,

$$\Delta(t) \sim \Delta_0 \frac{\dot{\varphi}(t)}{\dot{\varphi}(t_2)} \sim \Delta_0 \frac{\dot{\varphi}(t)}{\dot{\varphi}(0)}\,. \tag{10.2.10}$$

In the theory with $\mathrm{V}(\varphi) = \dfrac{\lambda}{4}\varphi^4$, Eq. (1.7.22) implies that $\dot{\varphi} \sim \varphi(t)$; that is, for $t \gg t_2$,

$$\Delta(t) = C\sqrt{\frac{\lambda}{15}}\,\frac{\varphi(t)\varphi_0^2}{M_P^2}, \tag{10.2.11}$$

with $C \sim 1$.

The foregoing is an elementary derivation of the expressions (10.2.8)–(10.2.11) for the dispersion $\Delta(t)$, as an attempt to elucidate the physical nature of the phenomena taking place [57, 78]. The same results can be obtained in a more formal manner by solving the diffusion equation (7.3.22) directly for the distribution $P_c(\varphi, t)$ with the initial condition $P_c(\varphi, t) \sim \delta(\varphi - \varphi_0)$. This problem has been solved in [132, 133]; here we simply present the final result for $\Delta(t)$ in theories with $V(\varphi) = \dfrac{\lambda\varphi^n}{nM_P^{n-4}}$:

$$\Delta^2(t) = \frac{4\lambda\,\varphi^{n-2}(t)}{3n^2 M_P^n}\left[\varphi_0^4 - \varphi^4(t)\right]. \tag{10.2.12}$$

In particular, in a theory with $V(\varphi) = \dfrac{\lambda}{4}\varphi^4$,

$$\Delta(t) = \frac{1}{2}\sqrt{\frac{\lambda}{3}}\,\frac{\varphi(t)}{M_P^2}\left[\varphi_0^4 - \varphi^4(t)\right]^{1/2}. \tag{10.2.13}$$

Using (1.7.21), one can easily show that this result is consistent with Eqs. (10.2.8)–(10.2.11), which were obtained by a simpler method.

In what follows, it will be especially important to analyze the evolution of the scalar field φ during the initial stage of the process (for a time $t \lesssim t_2$), during which time the field $\varphi(t)$ changes by an amount $\Delta\varphi \lesssim \varphi_0$. We already know from (10.2.12) that at that stage $\Delta(t) \ll \varphi(t)$ if $V(\varphi_0) \ll M_P^4$. Thus, if the initial energy density is much less than the Planck density, the dispersion of the scalar field distribution during the stage in question will always be much less than the mean value of the field $\varphi(t)$ — that is, we are justified in investigating the evolution of the field $\varphi(x, t)$ to first order in $\delta\varphi(x, t)$ (10.2.5), (10.2.6). On the other hand, when $\Delta(t) \ll \varphi(t)$, $P_c(\varphi, t)$ is a Gaussian distribution in the vicinity of its maximum at $\varphi = \varphi(t)$, i.e.,

$$P_c(\varphi, t) \sim \exp\left(-\frac{[\varphi - \varphi(t)]^2}{2\Delta^2}\right)$$
$$= \exp\left(-\frac{3n^2\,[\varphi - \varphi(t)]^2\, M_P^n}{8\lambda\, \varphi^{n-2}(t)\left[\varphi_0^4 - \varphi^4(t)\right]}\right), \tag{10.2.14}$$

where $\varphi(t)$ is a solution of Eq. (10.2.5); see (1.7.21), (1.7.22). In particular,

$$\varphi(t) = \varphi_0 \exp\left(-\sqrt{\frac{\lambda}{6\pi}}\, M_P\, t\right) \tag{10.2.15}$$

in the theory with $V(\varphi) = \frac{\lambda}{4}\varphi^4$, and

$$\varphi^{2-\frac{n}{2}}(t) = \varphi_0^{2-\frac{n}{2}} - t\left(2 - \frac{n}{2}\right)\sqrt{\frac{n\lambda}{24\pi}}\, M_P^{3-\frac{n}{2}} \tag{10.2.16}$$

in the theory with $V(\varphi) = \dfrac{\lambda\varphi^n}{n M_P^{n-4}}$ for $n \neq 4$.

In any given domain of the universe, then, the distribution $P_c(\varphi, t)$ is nonstationary, and in the course of time, the probability of detecting a large field φ at any given point of space becomes exponentially low.

If, however, one would like to know the distribution $P_P(\varphi, t)$ of the physical volume of the universe, taking account of expansion proportional to

$$\exp\left(\int_0^t H(x, t)\, dt\right),$$

that contains the field φ at time t, the answer will be completely different. To attack this question, let us consider for definiteness the evolution of the distribution $P_c(\varphi, t)$ for a time Δt, during which the mean field $\varphi(t)$ decreases by $\Delta\varphi = \dfrac{\varphi_0}{N} \ll \varphi_0$, where N is some number satisfying $N \gg 1$. According to (10.2.14),

$$\mathrm{P}_c(\varphi, \Delta t) \sim \exp\left(-\frac{3n^2 \mathrm{N}\left[\varphi - \varphi_0\left(1 - \frac{1}{\mathrm{N}}\right)\right]^2 \mathrm{M}_{\mathrm{P}}^n}{32\lambda\, \varphi_0^{n+2}}\right). \tag{10.2.17}$$

We see from this result that the fraction of the original coordinate volume remaining in a state with $\varphi = \varphi_0$ after a time Δt is

$$\mathrm{P}_c(\varphi_0, \Delta t) \sim \exp\left(-\frac{3n^2 \mathrm{M}_{\mathrm{P}}^n}{32\lambda\, \mathrm{N}\, \varphi_0^n}\right) = \exp\left(-\frac{3n\, \mathrm{M}_{\mathrm{P}}^4}{32\mathrm{N}\mathrm{V}(\varphi_0)}\right). \tag{10.2.18}$$

Notice that $\mathrm{P}_c(\varphi_0, \Delta t) \ll 1$ when $\mathrm{V}(\varphi_0) \ll \mathrm{M}_{\mathrm{P}}^4$. In other words, the dispersion Δ is much less than the difference $\varphi_0 - \varphi(t)$. In a time Δt, the volume of a region with $\varphi = \varphi_0$ expands by a factor $e^{3\mathrm{H}(\varphi_0)\Delta t}$, on the average. It then follows from (10.2.15) and (10.2.16) that

$$\Delta t = \frac{2}{\mathrm{N}}\sqrt{\frac{6\pi}{n\lambda}}\left(\mathrm{M}_{\mathrm{P}}^{\frac{n}{2}-3}/\varphi^{\frac{n}{2}-2}\right). \tag{10.2.19}$$

The original volume occupied by the field φ_0 thus changes in a time Δt given by (10.2.19) by a factor $\mathrm{P}_p(\varphi_0, \Delta t)$, where

$$\begin{aligned}\mathrm{P}_{\mathrm{P}}(\varphi_0, \Delta t) &\sim \mathrm{P}_c(\varphi_0, \Delta t)\exp\left[3\mathrm{H}(\varphi_0)\Delta t\right] \\ &= \exp\left(-\frac{3n^2}{32\lambda \mathrm{N}}\frac{\mathrm{M}_{\mathrm{P}}^n}{\varphi_0^n} + \frac{24\pi}{\mathrm{N}n}\frac{\varphi_0^2}{\mathrm{M}_{\mathrm{P}}^2}\right).\end{aligned} \tag{10.2.20}$$

Clearly, then, when $\varphi_0 \gg \alpha\varphi^*$, where

$$\varphi^* = \lambda^{-\frac{1}{n+2}}\mathrm{M}_{\mathrm{P}}, \quad \alpha = \left(\frac{n^3}{2^8\pi}\right)^{\left(\frac{1}{n+2}\right)} = \mathrm{O}(1), \tag{10.2.21}$$

the volume occupied by the field φ_0 will grow during the time Δt rather than decrease. The same thing will be repeated during the next time interval Δt, and so on. This means that during inflation, regions of the inflationary universe with $\varphi > \varphi^*$ reproduce themselves endlessly: the process of inflation, once having started, continues forever unabated, and the volume of the inflationary part of the universe grows without bound.

havior of the distribution $P_p(\varphi, \Delta t)$ for $\varphi - \varphi_0 \gg \Delta\varphi = \frac{\varphi_0}{N}$. In point of fact, a field φ that occasionally jumps to values with $\varphi - \varphi_0 \gg \frac{\varphi_0}{N}$ due to quantum fluctuations in some domain of the inflationary universe cannot be significantly reduced in a time $\sim\Delta t$, either by classical rolling (by $\Delta\varphi \ll \varphi - \varphi_0$) or by diffusion (by $\sim\Delta \ll \Delta\varphi$). The volume of all regions occupied by the field φ increases in a time Δt (10.2.19) by a factor $\exp[3H(\varphi)\Delta t]$, whereupon

$$P_P(\varphi, \Delta t) \sim P_c(\varphi, \Delta t) \exp[3H(\varphi)\Delta t]$$
$$= \exp\left\{-\frac{3n^2 N[\varphi - \varphi(t)]^2}{32\lambda}\frac{M_P^n}{\varphi_0^{n+2}} + \frac{24\pi}{N n}\frac{\varphi_0^2}{M_P^2}\left(\frac{\varphi}{\varphi_0}\right)^{\frac{n}{2}}\right\}. \tag{10.2.22}$$

It can readily be shown that when $\varphi > \beta\varphi^*$, where $\beta = \left(\frac{n^2 N}{2^6}\right)^{\frac{1}{n+2}} = O(1)$, the maximum of the distribution $P_p(\varphi, \Delta t)$ is shifted not towards $\varphi < \varphi_0$, like the maximum of $P_c(\varphi, \Delta t)$, but towards $\varphi > \varphi_0$.

This means that when $\varphi \gg \varphi^*$, the universe not only continually regenerates itself, but in the process, most of the physical volume of the universe gradually fills with a larger and larger field φ [132, 133]. This result is entirely consistent with the result obtained in Section 1.8 by more elementary methods [57].

10.3 The self-regenerating inflationary universe and quantum cosmology

The possibility of an eternally existing, self-regenerating universe is one of the most important and surprising consequences of the theory of the inflationary universe, and it merits detailed discussion (see also Section 1.8). Let us first of all provide a more accurate interpretation of the results obtained in Section 10.2.

The physical meaning of the distributions $P_c(\varphi, t)$ and $P_p(\varphi, t)$ is as follows. Consider a domain of the inflationary universe having initial size $l \gtrsim H^{-1}$, and let us assume that initially (at $t = 0$) it is uniformly filled throughout its entire volume by observers with identical clocks, which are synchronized at $t = 0$. In that event, the quantity $P_c(\varphi, t)$ determines the fraction of all observers who at time t as measured by their own clocks (i.e., in a synchronous coordinate system) are located in a region filled with a practically homogeneous (on a scale $l \gtrsim H^{-1}(\varphi)$) semiclassical field φ. The distribution $P_p(\varphi, t)$ determines how much of the physical volume of the universe is occupied by observers who, at time t as measured by their own clocks, live in regions filled with the field φ that is homogeneous on a scale larger than $H^{-1}(\varphi)$.

The results obtained in the preceding section imply that in no particular region of the inflationary universe can the distribution $P_c(\varphi, t)$ be stationary. It can be quasistationary during tunneling from a metastable vacuum state to a stable state, as in the case of Hawking–Moss tunneling in the new inflationary universe scenario. But in any model in which the universe becomes hot after inflation and the field φ rolls down to its minimum at $V(\varphi) = 0$, the function $P_c(\varphi, t)$ cannot (and should not) be a nontrivial stationary distribution (at least within the range of validity of the approximation that we have employed; see below). In other words, the fraction of all observers initially situated in an unstable state away from the absolute minimum of the effective potential $V(\varphi)$ should decrease with time. This conclusion is confirmed by results obtained above — for example, see Eq. (10.2.14), which shows that the probability of remaining in an unstable state $\varphi \gtrsim M_P$ in a theory with $V(\varphi) = \frac{\lambda}{4}\varphi^4$ after a time $t \gtrsim \frac{\sqrt{6\pi}}{\sqrt{\lambda} M_P} \ln \frac{\varphi_0}{M}$ becomes exponentially small.

At the same time, when $\varphi_0 \gg \varphi^*$ (see (10.2.21)), the distribution $P_p(\varphi, t)$ begins to increase with increasing φ in a region with $\varphi \gtrsim \varphi^*$, i.e., the fraction of the volume of the inflationary universe occupied by observers who at time t by their clocks find themselves located in an unstable state $\varphi \gtrsim \varphi^*$ increases at large φ and t, and consequently the total volume of the inflationary regions of the universe continues to grow without bound. It follows from (10.2.22) that at large t, most of the volume of the universe should be in a state with the very largest possible value of the field φ, such that $V(\varphi) \sim M_P^4$.

Here, to be sure, we must state an important reservation. The fraction of the volume of the universe in a state with a given field φ *at a given time* depends on what we mean by the word *time*. The results obtained above refer to the proper time t of comoving observers whose clocks were synchronized at some time $t = 0$, when they were all quite close to one another. The same phenomena can be described using another coordinate system, namely the coordinates (7.5.8), which are especially convenient in investigating the density inhomogeneities produced during inflation. In order to distinguish between the proper time t in the synchronous coordinates and the "time" in the coordinates (7.5.8), we denote the latter here as τ. Investigation of diffusion in the coordinate system (7.5.8) also shows that the total volume of the universe filled with the field $\varphi > \varphi^*$ increases exponentially with time τ [133]. But due to the specific way in which the "time" τ is defined, the rate of exponential expansion of the universe, which is $\sim e^{H\tau}$ in the coordinates (7.5.8), is the same everywhere, regardless of any local increases or decreases in the field φ. The fraction of the physical volume of the universe filled with a large field φ on a hypersurface of constant τ therefore falls off in almost the same way as $P_c(\varphi, t)$, and thus the fraction of the physical volume of a self-regenerating universe that is transformed over time into a state with the largest possible value of the field φ depends on what exactly one means by the word *time*. It is for just this reason that we have engaged here in a more detailed discussion of this question, the answer to which turns out to depend on exactly how the question is formulated. Fortunately, however, our basic conclusion about self-regeneration and exponential expansion of regions of the universe filled with a field $\varphi > \varphi^*$ is coordinate-independent [133].

It is worthwhile to examine these results from yet another standpoint. If the universe is self-regenerating, then the standard question about the initial conditions *over the whole universe* may be irrelevant, since the universe may turn out not to have had a global initial spacelike singular hypersurface to play the role of a global Cauchy hypersurface. At present, we do not have sufficient reason to believe that the universe as a whole was created approximately 10^{10} years ago in a singular state, prior to which classical space-time did not exist at all. Inflation could begin and end at different times in different domains of the universe, and this would be completely consistent with the existing observational data.

Accordingly, the matter density in different regions of the universe will drop to $\rho_0 \sim 10^{-29}\ g \cdot cm^{-3}$ at different times, approximately 10^{10} years after inflation ends in each of these regions. It is just after this point in each of these regions that the conditions required for the emergence of observers like ourselves will first appear. The number of such observers should plainly be proportional to the volume of the universe at the density hypersurface(s) with $\rho = \rho_0 \sim 10^{-29}$ g·cm^{-3}. Therefore, having investigated the question of what processes take part in the creation of most of the volume of the universe at the density hypersurface $\rho = \rho_0 \sim 10^{-29}$ g·cm^{-3} (i.e., 10^{10} years after inflation ceases in each particular domain), we have prepared ourselves to assess the most likely history of the part of the universe that we are able to observe.

In order to look into this question, one should take into account that the universe expands by a factor of approximately 10^{30} over the 10^{10} years after the end of inflation, and that during inflation in the theory with $V(\varphi) = \frac{\lambda}{4}\varphi^4$, the universe typically expands by a factor $\exp\left(\frac{\pi\varphi_0^2}{M_P^2}\right)$, where φ_0 is the original value of the field φ. However, for $\varphi_0 \gtrsim \varphi^* \sim \lambda^{-1/6} M_P$, this result is modified.

Indeed, as at the end of Section 10.2, let us consider a domain of the inflationary universe in which the scalar field, due to its long-wave quantum fluctuations during the time Δt, jumps from $\varphi = \varphi_0$ up to some value φ such that $\varphi - \varphi_0 \gg |\Delta\varphi|$, where $\Delta\varphi$ is the value of the classical decrease in the field φ during the time Δt. If the jump in the field φ is large enough, its mean value in this domain will return to the original value $\varphi = \varphi_0$, mainly due to classical rolling. According to (1.7.25), during classical rolling, the domain under consideration inflates by an additional factor of $\exp\left[\frac{\pi}{M_P^2}(\varphi^2 - \varphi_0^2)\right]$.

The probability of a large jump in the field φ is exponentially suppressed (see (10.2.14), (10.2.22)), but it is not hard to prove that when $\varphi \gg \varphi^*$, this suppression pays us back with interest on account of the aforementioned additional inflation of the region filled with the field φ making the jump. This means that most of the volume of the universe after inflation (for example, on the hypersurface $\rho = \rho_0$) results from the evolution of those relatively rare but additionally inflated regions in which the field φ has jumped upward as a result of long-wavelength quantum fluctuations. To continue this line of reasoning, it can be shown that the overwhelming preponderance of the physical volume of the universe in a state with given density $\rho = \rho_0$ is formed as a result of the inflation of regions in which the field φ, over the longest possible times, has been fluctuating about its maximum possible values, such that $V(\varphi) \sim M_P^4$. In that sense, a state with potential energy density close to the Planck value (i.e., the space-time foam) can be considered to be a source, continuously producing the greater part of the physical volume of the universe. We shall return to this point subsequently, but for the moment we wish to compare our conclusions with the basic expectations and assumptions that have been made in analyses of the wave function of the universe.

In deriving the expression for the wave function (10.1.12), (10.1.17) proposed by Hartle and Hawking, it was assumed that the universe has a stationary ground state, or a state of least excitation (vacuum), the wave function $\Psi(a, \varphi)$ of which they attempted to determine; see Section 10.1. The square of this wave function $|\Psi(a, \varphi)|^2$ should then give the stationary distribution for the probability of detecting the universe in a state with the homogeneous scalar field φ and scale factor a (10.1.18), (10.1.20). The fact that the quasistationary distribution $P_c(\varphi)$ (10.2.3) is proportional to the square of the Hartle–Hawking wave function could be taken as a important indication of the validity of this assumption. The results obtained in the preceding section, however, show that with fairly general initial conditions, the distribution $P_c(\varphi, t)$ in the inflationary universe scenario does not wind up in the stationary regime (10.2.3). Nevertheless, another type of stationary regime is possible, described in part by the distribution $P_p(\varphi, t)$. In this regime, the universe continually produces exponentially expanding regions (mini-universes) containing the large field φ (with $\varphi^* \lesssim \varphi \lesssim \varphi_P$), where $V(\varphi_P) \sim M_P^4$, and the properties of the universe within such regions do not depend on the properties of neighboring regions (by virtue of the "no hair" theorem for de Sitter space), nor do they depend on the history or epoch of formation of those regions. Here we can speak of stationarity in the sense that regions of the inflationary universe containing a field $\varphi \gtrsim \varphi^*$ constantly come into being, and in an exponentially large neighborhood of each such region, the average properties of the universe are the same and do not depend on the epoch at which the region was formed. This implies that the inflationary universe has a fractal structure [133, 325].

Thus, the Hartle–Hawking wave function may be useful in describing the intermediate stages of inflation, during which the distribution $P_c(\varphi, t)$ can sometimes (in the presence of metastable vacuum states) turn out to be quasistationary. This function may also turn out to be quite useful in the study of certain other important problems of quantum cosmology — see, for example, a discussion of this possibility in Section

10.7. At the present time, however, we are unable to ascribe to this wave function the fundamental significance sometimes assigned to it in the literature.

What then can we say about the tunneling wave function (10.1.23) used to describe the quantum creation of the universe?

To answer this question, let us study in somewhat more detail the first (diffusive) stage of spreading of the initial distribution $P_c(\varphi, 0) = \delta(\varphi - \varphi_0)$ when the magnitude of the classical field φ is almost constant, $\varphi(t) \sim \varphi_0$. Equation (10.2.14) holds when the dispersion Δ and the difference between φ and φ_0 are much less than φ_0 itself. At the same time, when $\varphi - \varphi_0 \sim \varphi_0$ the distribution $P_c(\varphi, t)$ is far from Gaussian. In order to calculate $P_c(\varphi, t)$ in that case, one should bear in mind that the classical rolling of the field φ, i.e., the last term in the diffusion equation (7.3.22), can be neglected during the first stage:

$$\frac{\partial P_c(\varphi,t)}{\partial t} = \frac{2\sqrt{2}}{3\sqrt{3\pi}\, M_P^3} \frac{\partial^2}{\partial \varphi^2}\left[V(\varphi) P_c(\varphi,t)\right]. \qquad (10.3.1)$$

It is convenient to seek a solution of this equation in the form

$$P_c(\varphi,t) \sim A(\varphi,t) \cdot \exp\left[-\frac{S(\varphi)}{t}\right],$$

where $A(\varphi, t)$ and $S(\varphi)$ are relatively slowly varying functions of φ and t.

It can readily be shown that in the theory $V(\varphi) = \dfrac{\lambda \varphi^n}{n M_P^{n-4}}$ with $\varphi \ll \varphi_0$, the corresponding solution is

$$P_c(\varphi,t) = A \exp\left[-\frac{3\sqrt{6\pi}}{t\lambda\sqrt{\lambda}\,(3n-4)^2}\left(\frac{M_P}{\varphi}\right)^{\frac{3n}{2}-1}\right], \qquad (10.3.2)$$

and neglect of the last term in (7.3.22) is warranted when

$$t \lesssim \Delta t(\varphi) = \sqrt{\frac{6\pi}{n\lambda}}\, M_P^{-1}\left(\frac{M_P}{\varphi}\right)^{\frac{n}{2}-2} \qquad (10.3.3)$$

(compare with (10.2.19)). If the effective potential $V(\varphi)$ is not too steep ($n \le 4$), then the diffusion approximation first ceases to work at small φ, and subsequently at $\varphi \sim \varphi_0$. One can say in that event that regions of space with a small field φ, in which classical motion prevails over quantum fluctuations, are formed by virtue of a quantum diffusion process that operates for a time $t \lesssim \Delta t(\varphi)$, and the probability distribution for the creation of a region (mini-universe) with a given field φ when quantum diffusion ceases to dominate ($t \sim \Delta t(\varphi)$) is, according to (10.3.2) and (10.3.3),

$$P_c(\varphi, \Delta t(\varphi)) \sim \exp\left[-C\frac{M_P^4}{V(\varphi)}\right], \tag{10.3.4}$$

where $C = O(1)$. This formula holds for $n \leq 4, \varphi \ll \varphi_0$, regardless of the initial value of the field φ_0. In particular, when $V(\varphi_0) \gtrsim M_P^4$, it can be interpreted as the probability for the quantum creation of a (mini-)universe from the space-time foam with $V(\varphi) \gtrsim M_P^4$. It is easily seen that up to a factor $C = O(1)$, (10.3.4) and the probability (10.1.24) of quantum creation of a universe from "nothing" are identical.

Is this merely formal consistency between these equations, or is there more to it than that? An answer requires additional investigation, but there are some ideas on this score that may be enunciated right now. First of all, note that rather than describing the creation of the entire universe from "nothing," Eq. (10.3.4) describes the creation of only a part of the universe of size greater than $H^{-1}(\varphi)$ due to quantum diffusion from a previously existing region of the inflationary universe. Moreover, Eq. (10.3.4) holds in theories with $V(\varphi) \sim \varphi^n$ only when $n \leq 4$; for $n > 4$, it can be shown that

$$P_c(\varphi, \Delta t(\varphi)) = P_c(\varphi, \Delta t(\varphi_0)) \sim \exp\left[-\frac{C M_P^4}{V(\varphi)}\left(\frac{\varphi_0}{\varphi}\right)^{\frac{n}{2}-2}\right]. \tag{10.3.5}$$

In theories in which the interval between φ and φ_0 contains segments where the field φ rolls down rapidly and the universe is not undergoing inflation, equations like (10.3.4) and (10.3.5) will generally not be valid; that is, the diffusion equation that we have used will not be applicable to such segments. More specifically, one cannot derive equations like (10.3.4) for the probability of diffusion from a space-time foam with $V(\varphi_0) \sim M_P^4$ onto the top of the effective potential at $\varphi = 0$ in the new inflationary universe scenario. Meanwhile, it is usually assumed that Eq. (10.1.23) (perhaps somewhat modified to take account of the effects of quantum creation of particles during tunneling [321]) can describe the quantum creation of the universe as a whole, even if a continuous diffusive transition between φ_0 and φ is not possible.

This indicates that we are dealing here with two distinct complementary or competing processes described by Eqs. (10.3.4) and (10.1.24), respectively. However, experience with the Hawking–Moss theory of tunneling (7.4.1) engenders a certain amount of caution in this regard. Recall that Eq. (7.4.1), which was originally derived via the Euclidean approach to tunneling theory, was interpreted as giving the probability of uniform tunneling over the entire universe [121]. However, a rigorous derivation of Eq. (7.4.1) and a justification for this interpretation were lacking. In our opinion, a rigorous derivation of Eq. (7.4.1) was first provided by solving the diffusion equation (7.3.22), and its interpretation was different from the original one based on the Euclidean approach [134, 135], though consistent with the interpretation proposed in [209]. Likewise, neither approach to deriving Eq. (10.1.24) (using the (anti-)Wick rotation $t \to i\tau$ or considering tunneling from the point $a = 0$) is sufficiently rigorous, and the interpretation of (10.1.24) as the probability of tunneling from "nothing" also falls somewhere on the borderline between physics and poetry. One of the fundamental questions to emerge here had to do what *exactly* was tunneling, if there were no *incoming* wave. A plausible response is that one simply cannot identify the incoming wave within the framework of the minisuperspace approach. Indeed, by solving the diffusion equation in the theory of chaotic inflation, it was shown that during inflation there is a steady process of creation of inflationary domains whose density is close to the Planck density, and whose size is $l \sim l_P \sim M_P^{-1}$. Tunneling (or diffusion) involving an increase in the size of each such region and a change in the magnitude of the scalar field within each region can be (approximately) associated with the process of quantum creation of the universe. The process of formation of such Planck-size inflationary domains (the "incoming wave") cannot be described in the context of the minisuperspace approach, but it has a simple interpretation within the scope of the stochastic approach to inflation.

We have thus come closer to substantiating the validity of Eq. (10.1.24) as a probability of quantum creation of the universe from "nothing." It is nevertheless still not entirely clear whether in this expression there is anything that is both true and at the same time different from Eq. (10.3.4), which was derived with the aid of the stochastic approach to inflation, and which has a much more definite physical meaning. This is a particularly important question, as it pertains to the theory of the quantum creation of the universe in the state $\varphi = 0$, corresponding to a local maximum of $V(\varphi)$ located at $V(\varphi) \ll M_P^4$, since diffusion into this state from a space-time foam with $V(\varphi) \sim M_P^4$ is impossible.

To conclude this section, let us examine one more question, relating to the possibility of creating an inflationary universe from Minkowski space. The issue here is that quantum fluctuations in the latter can bring into being an inflationary domain of size $l \gtrsim H^{-1}(\varphi)$, where φ is a scalar field produced by quantum fluctuations in this domain. The "no hair" theorem for de Sitter space implies that such a domain inflates in an entirely self-contained manner, independent of what occurs in the surrounding space. We could then conceive of a ceaseless process of creation of inflationary mini-universes that could take place even at the very latest stages of development of the part of the universe that surrounds us.

A description of the process whereby a region of the inflationary universe is produced as a result of quantum fluctuations could proceed in a manner similar to that for the formation of regions of the inflationary universe with a large field φ through the buildup of long-wave quantum fluctuations $\delta\varphi$. The basic difference here is that long-wave fluctuations $\delta\varphi$ of the massive scalar field φ at the time of inflation with $m \ll H$ are "frozen" in amplitude, while there is no such effect in Minkowski space. But if the buildup of quantum fluctuations in some region of Minkowski space were to engender the creation of a fairly large and homogeneous field φ, then that region in and of itself could start to inflate, and such a process could stabilize ("freeze in") the fluctuations $\delta\varphi$ that led to its onset. In that event, one could sensibly speak of a self-consistent process of formation of inflationary domains of the universe due to quantum fluctuations in Minkowski space.

Without pretending to provide a complete description of such a process, let us attempt to estimate its probability in theories with $V(\varphi) = \dfrac{\lambda\varphi^n}{n M_P^{n-4}}$. A domain formed with a large field φ will only be a part of de Sitter space if in its interior $(\partial_\mu\varphi)^2 \ll V(\varphi)$. This means that the size of the domain must exceed $l \sim \varphi\, V(\varphi)^{-1/2}$, and the field inside must be greater than M_P. Such a domain could arise through the buildup of quantum fluctuations $\delta\varphi$ with a wavelength

$$k^{-1} \gtrsim l \sim \varphi\, V^{-1/2}(\varphi) \sim m^{-1}(\varphi)$$

One can estimate the dispersion $<\varphi^2>_{k<m}$ of such fluctuations using the simple formula

$$<\varphi^2>_{k<m} \sim \frac{1}{2\pi^2}\int_0^{m(\varphi)} \frac{k^2\,dk}{\sqrt{k^2+m^2(\varphi)}} \sim \frac{m^2}{\pi^2} \sim \frac{V(\varphi)}{\pi^2\varphi^2}, \qquad (10.3.6)$$

and for a Gaussian distribution $P(\varphi)$ for the appearance of a field φ which is sufficiently homogeneous on a scale l, one has [133]

$$P(\varphi) \sim \exp\left[-C\,\frac{\pi^2\varphi^4}{V(\varphi)}\right], \qquad (10.3.7)$$

where $C = O(1)$. In particular, for a theory with $V(\varphi) = \frac{\lambda}{4}\varphi^4$,

$$P(\varphi) \sim \exp\left[-C\,\frac{4\pi^2}{\lambda}\right]. \qquad (10.3.8)$$

Naturally, this method is rather crude; nevertheless, the estimates that it provides are quite reasonable, to order of magnitude. For example, practically the same lines of reasoning could be employed in assessing the probability of tunneling from the point $\varphi = 0$ in a theory with $V(\varphi) = -\frac{\lambda}{4}\varphi^4$; see Chapter 5. The estimate that one obtains for the formation of a bubble of the field φ is also given by Eq. (10.3.8). This result is in complete accord with the equation $P \sim \exp\left(-\frac{8\pi^2}{3\lambda}\right)$ (5.3.12), which was derived using Euclidean methods. In fact, one can easily verify that *all* results concerning tunneling which were obtained in Chapter 5 can be reproduced (up to a numerical factor $C = O(1)$ in the exponent) by using the simple method suggested above. This makes the validity of the estimates (10.3.7), (10.3.8) quite plausible.

The main objection to the possibility of quantum creation of an inflationary universe in Minkowski space is that energy conservation forbids the production of an object with positive energy out of the vacuum in this space. Within the scope of classical field theory, in which the energy is everywhere positive, such a process would therefore be impossible (a related problem is discussed in [213, 326]). But at the quantum level, the energy density of the vacuum is zero by virtue of the cancellation between the positive energy density of classical scalar fields, along with their quantum fluctuations, and the negative energy associated with quantum fluctuations of fermions, or the bare negative energy of the vacuum. The creation of a positive energy-density domain through the buildup of long-wave fluctuations of the field φ is inevitably accompanied by the creation of a region surrounding that domain in which the long-wave fluctuations of the field φ are suppressed, and the vacuum energy density is consequently negative. Here we are dealing with the familiar quantum fluctuations of the vacuum energy density about its zero point. It is important here that from the point of view of an external observer, the total energy of the inflationary region of the universe (and indeed the total energy of the closed inflationary universe) does not grow exponentially; the region that emerges forms a universe distinct from ours, to which it is joined only by a connecting throat (wormhole) which, like a black hole, can disappear by virtue of the Hawking effect [327, 213]. At the same time, the shortfall of long-wave fluctuations of the field φ surrounding that region is quickly replenished by fluctuations arriving from neighboring regions, so the negative energy of the region near the throat can be be rapidly spread over a large volume around the inflationary domain.

Our discussion of the creation of an inflationary universe in Minkowski space is highly speculative, and is only intended to illustrate the basic feasibility of such a process; this is clearly a problem that requires closer study. If this process can actually transpire, and if it is accompanied by burnout of the wormhole connecting the parent (Minkowski) space with the inflationary universe that is its offspring, then the theory will have one more mode for the stationary production of regions of an inflationary universe. We wish to emphasize, however, that in our approach, the likelihood of this regime being realized is in no way related to the distribution $P_c(\varphi)$, which is proportional to the square of the wave function (10.1.17). The Euclidean approach to the theory of baby-universe formation has been developed in [350–352], and will be discussed in Section 10.7.

10.4 The global structure of the inflationary universe and the problem of the general cosmological singularity

One of the most important consequences of the inflationary universe scenario is that under certain conditions, once a universe has come into being it can never again collapse as a whole and disappear completely. Even if it initially resembles a homogeneous closed Friedmann universe, it will most likely cease to be locally homogeneous and become markedly inhomogeneous on the largest scales, and there will then be no global end of the world such as that which occurs in a homogeneous closed Friedmann universe.

There are versions of the theory of the inflationary universe in which self-regeneration of the universe does not take place, such as the Shafi–Wetterich model [237], which is based on a study of inflation in a particular version of Kaluza–Klein theory — see Section 9.5. But for most of the inflationary models studied thus far, the evolution of the universe and the process of inflation has no end. In the old Guth scenario, for example, when the probability of forming bubbles of a new phase with $\varphi \neq 0$ is low enough, these bubbles will never fill the entire physical volume of the universe, since the distance separating any two of them increases exponentially, and the resulting increase in the volume of the universe in the state $\varphi = 0$ is greater than the decrease of this volume due to the creation of new bubbles [53, 113, 327, 328]. One encounters a similar effect in the new inflationary universe scenario [266, 267], and a detailed theory of this process [204] is quite similar to the corresponding theory in the chaotic inflation scenario [57, 133], the basic difference being that in both the old and new inflationary universe scenarios one is dealing with the production of regions containing a field φ close to zero and with $V(\varphi) \ll M_P^4$, while in the chaotic inflation scenario there can be a steady output of regions with very high values of $V(\varphi)$, right up to $V(\varphi) \sim M_P^4$. We shall subsequently find this to be a very important circumstance.

The possibility of the ceaseless regeneration of inflationary regions of the universe, which implies the absence of a general cosmological singularity (i.e., a global spacelike singular hypersurface) *in the future,* compels us to reconsider the problem of the *initial* cosmological singularity as well. At present, it seems unnecessary to assume that there was a unique beginning to this endless production of inflationary regions. Models of a nonsingular universe based on this idea have been proposed in the context of both the old [327, 328] and new [267] inflationary universe scenarios. According to these models, most of the physical volume of the universe remains forever in a state of exponential expansion with $\varphi \sim 0$, engendering ever newer exemplars of our type of mini-universe.

Unfortunately, it is not yet entirely clear how one would go about implementing this possibility. To understand where the main difficulty lies, recall that exponentially expanding (flat) de Sitter space is not geodesically complete — it comprises only a part of the closed de Sitter space, which originally contracts (at $t < 0$) rather than expands:

$$a(t) = H^{-1} \cosh Ht$$

(see Section 7.2). In de Sitter space contracting at an exponential rate, a phase transition from the state $\varphi = 0$ can in principle take place in a finite time over the entire volume, and there would then remain no regions that could lead to an infinite expansion of the universe for $t > 0$. This question, and the problem of the geodesic completeness of a self-regenerating inflationary universe, requires further study, both because the theory of phase transitions in an exponentially contracting space is not well understood, and because the *global* geometry of a self-regenerating universe differs from the geometry of de Sitter space. At present, therefore, we cannot say with absolute certainty that the new inflationary universe scenario with no initial singularities is impossible. That there be *no* singularities in the past, however, is probably too strong a requirement.

A more natural possibility is suggested by the chaotic inflation scenario, where most of the physical volume of the universe at a hypersurface of given density is comprised of regions which, by virtue of fluctuations in the field φ, passed through a stage with $V(\varphi) \sim M_P^4$. In this scenario, classical space-time behaves as if it were in a state of dynamic equilibrium with the space-time foam: regions of classical space are continually created out of the space-time foam, and some of these are reconverted to the foam with $V(\varphi) \gtrsim M_P^4$. In that sense, the occurrence of spatial "singularities" is part and parcel of this scenario. At the same time, what is graphically clear about this scenario is that instead of dealing with the problem of creation of an *entire* universe from a singularity, prior to which *nothing* existed, and its subsequent dénouement into *nothingness*, we are simply concerned with an endless process of interconversion of phases in which quantum fluctuations of the metric are either large or small. Our results imply that once it has arisen, classical space-time — a phase in which quantum fluctuations of the metric are small — will never again disappear. Even more than in the new inflationary universe scenario, the global geometric properties of a region filled with this phase[3] differ from those of de Sitter space. If this region turns out to be geodesically complete, then one can plausibly discuss a model in which the universe has neither a unique beginning nor a unique end.

Actually, however, as we have already pointed out, this possibility arises in the chaotic inflation scenario without even taking the self-regeneration process into consideration. Specifically, if the universe is finite and initially no larger than the Planck size, $l \lesssim M_P^{-1}$, then it is not unreasonable to suppose that at some initial time $t = 0$ (to within perhaps $\Delta t \sim M_P^{-1}$) the entire universe came into being as a whole out of the space-time foam (in classical language, it appeared from a singularity). If, however, the universe is infinite, then the possibility that an infinity of causally disconnected regions of classical space will simultaneously appear out of the space-time foam seems totally unlikely.[4]

To avoid confusion, it should again be emphasized that the existence of an initial general (global) spacelike cosmological singularity is not, in and of itself, a necessary consequence of the general topological theorems on singularities. This conclusion is based primarily on the assumption of global homogeneity of the universe. Within the framework of the hot universe theory, such an assumption, even though it had no fundamental justification to back it up, nevertheless seemed unavoidable, since in that theory the observable part of the universe arose by virtue of the expansion of an enormous number of causally disconnected regions in which for some unknown reason the matter density was virtually the same (see the discussion of the homogeneity and horizon problems in Chapter 1). On

3 We traditionally refer to this particular region as the universe, although strictly speaking, *the universe* (i.e., all that exists) includes those regions occupied by the space-time foam as well.

4 From this standpoint, open and flat Friedmann models, which are quite useful in the description of local properties of our universe, cannot correctly describe the global structure of the inflationary universe at any stage of its existence. At the same time, the model of a closed Friedmann universe *can* describe the global properties of the universe, but only during the earliest stages of its evolution, until diffusion of the field φ results in a large distortion of the original metric.

the other hand, in the inflationary universe scenario, the assumption that the universe is globally homogeneous is unnecessary, and in many cases it is simply wrong. Therefore, in the context of inflationary cosmology, the conventional statement that in the very early stages of evolution of the universe there was some instant of time before which there was no time at all (see Section 1.5) is, at the very least, not well-founded.

10.5 Inflation and the Anthropic Principle

One of the main desires of physicists is to construct a theory that unambiguously predicts the observed values for all parameters of all the elementary particles that populate our universe. The noble idealism of the researcher compels many to believe that the correct theory describing our world should be both beautiful and unique. This does not at all imply, however, that all parameters of elementary particles in such a theory must be uniquely calculable. For example, in supersymmetric SU(5) theory, the effective potential $V(\Phi, H)$ for the Higgs fields Φ and H that figure in this theory has several different minima, and without taking gravitational effects into account, the vacuum energy $V(\Phi, H)$ would be the same at all of these minima. Each of the minima corresponds to a different type of symmetry breaking in this theory, i.e., to different properties of elementary particles. Gravitational interactions remove the energy degeneracy between these minima. But the lifetime of the universe in a state corresponding to any such minimum turns out either to be infinite or at least many orders of magnitude greater than 10^{10} years [329]. This means that prescribing a specific grand unified theory will not always enable one to uniquely determine properties of elementary particles in our universe. An even richer spectrum of possibilities comes to the fore in Kaluza–Klein and superstring theories, where an exponentially large variety of compactification schemes is available for the original multidimensional space; the type of compactification determines the coupling constants, the vacuum energy, the symmetry breaking properties in low-energy elementary particle physics, and finally, the effective dimensionality of the space we live in (see Chapter 1). Under these circumstances, the most varied sets of elementary-particle parameters (mass, charge, etc.) can appear. It is conceivable that this is the very reason why we have not yet been able to identify any particular regularity in comparisons of the electron, muon, proton, W-meson, and Planck masses. Most of the parameters of elementary particles look more like a collection of random numbers than a unique manifestation of some hidden harmony of Nature. Meanwhile, it was pointed out long ago that a minor change (by a factor of two or three) in the mass of the electron, the fine-structure constant α_e, the strong-interaction constant α_s, or the gravitational constant would lead to a universe in which life as we know it could never have arisen. For example, increasing the mass of the electron by a factor of two and one-half would make it impossible for atoms to exist; multiplying α_e by one and one-half would cause protons and nuclei to be unstable; and more than a ten percent increase in α_s would lead to a universe devoid of hydrogen. Adding or subtracting even a single spatial dimension would make planetary systems impossible, since in space-time with dimensionality $d > 4$, gravitational forces between distant bodies fall off faster than r^{-2} [330], and in space-time with $d < 4$, the general theory of relativity tells us that such forces are absent altogether.

Furthermore, in order for life as we know it to exist, it is necessary that the universe be sufficiently large, flat, homogeneous, and isotropic. These facts, as well as a number of other observations and remarks, lie at the foundation of the so-called Anthropic Principle in cosmology [77]. According to this principle, we observe the universe to be as it is because only in such a universe could observers like ourselves exist. There are presently several versions of this principle extant (the Weak Anthropic Principle, the Strong Anthropic Principle, the Final Anthropic Principle, etc.) — see [331]. All versions, formulated in markedly different ways, in one way or another interrelate the properties of the universe, the properties of elementary particles, and the very fact that mankind exists in this universe.

At first glance, this formulation of the problem looks to be faulty, inasmuch as mankind, having appeared 10^{10} years after the basic features of our universe were laid down, could in no way influence either the structure of the universe or the properties of the elementary particles within it. In reality, however, the issue is not one of cause and effect, but just of correlation between the properties of the observer and the properties of the universe that he observes (in the same sense as in the Einstein–Podolsky–Rosen experiment [332], where there is a correlation between the states of two different particles but no interaction between them). In other words, what is at issue is the *conditional probability* that the universe will have the properties that we observe, with the obvious and apparently trivial condition that observers like ourselves, who take an interest in the structure of the universe, do indeed exist.

All this discussion can make sense only if one can actually compare the probabilities of winding up in different universes having different properties of space and matter, but that is possible only if such universes do in fact exist. If it is not so, any talk of altering the mass of the electron, the fine structure constant, and so forth is perfectly meaningless.

One possible response to this objection is that the wave function of the universe describes both the observer and the rest of the universe in all its possible states, including all feasible variants of compactification and spontaneous symmetry breaking (see Section 10.1). By making a measurement that improves our knowledge of the properties of the observer, one simultaneously obtains information about the rest of the universe, just as by measuring the spin of one particle in the Einstein–Podolsky–Rosen experiment one promptly obtains information about the spin of the other [302, 304, 359].

This answer, in our view, is correct and entirely sufficient. Nevertheless, it would be more satisfying to have an alternative reply to the foregoing objection, one that is conceptually simpler and that does not require an analysis of the somewhat obscure foundations of quantum cosmology for its justification. Moreover, we would like to obtain an answer to another (and in our opinion, the most important) objection to the Anthropic Principle, namely that it does not seem at all necessary for the existence of life as we know it to have identical conditions (homogeneity, isotropy, ratios $\frac{n_B}{n_\gamma} \sim 10^{-9}$, $\frac{\delta\rho}{\rho} \sim 10^{-5}$, etc.) over the whole observable part of the universe. The random occurrence of such uniformity seems completely unlikely.

As noted in Chapter 1, both of these objections can be dispatched particularly simply by the theory of the self-regenerating inflationary universe. Specifically, long-wave fluctuations are generated during inflation not only of the inflaton field φ, which drives inflation, but of all other scalar fields Φ with mass $m_\Phi \ll H$ as well (and having a small coupling constant ξ in possible interactions of the type $\xi R\Phi^2$). In the chaotic inflation scenario, this means that in certain regions of the universe where the potential energy density of the scalar field φ permanently fluctuates near $V(\varphi) \sim M_P^4$ (by virtue of the self-regeneration process taking place in such regions), long-wave fluctuations of practically every scalar field Φ grow until the mean value of the potential energy density of each of these fields approaches M_P^4. This follows simply from an analysis of the Hawking–Moss distribution (7.4.1) for the field Φ with $V(\Phi = 0, \varphi) \sim M_P^4$).

The upshot of this process is that a distribution of the scalar fields φ and Φ is set up that is quite uniform on exponentially large scales because of inflation; but on the scale of the universe as a whole, on the other hand, the fields can take on practically any value for which their potential energy density does not exceed the Planck value. In those regions of the universe where inflation has ended, the fields φ and Φ roll down to different *local* minima of $V(\varphi, \Phi)$, and since all feasible initial conditions for rolling are realized in different regions of the universe, the universe becomes divided into different exponentially large domains containing the fields φ and Φ corresponding to all local minima of $V(\varphi, \Phi)$, i.e., all possible types of symmetry breaking in the theory.

At the stage of strong fluctuations with $V(\varphi, \Phi) \sim M_P^4$, not only can the magnitudes of the scalar fields vary, but large metric fluctuations can also be generated, leading to local compactification or decompactification of space in Kaluza–Klein or superstring theories. If a region of space with changing compactification is inflating and has an initial size $H^{-1} \sim M_P^{-1}$ (at Planck densities, the probability of this happening should not be too low), then as a result of inflation this region will turn into an exponentially large domain with a new type of compactification (for example, a different dimensionality) [78].

Thus, the universe will be partitioned into enormous regions (mini-universes), which will manifest all possible types of compactification and all possible types of spontaneous symmetry breaking compatible with the process of inflation, leading to an exponential growth in the size of these regions. Reference [333] contains an implementation of this scenario in the context of certain specific Kaluza–Klein theories.

It should be emphasized that due to the unlimited temporal extent of the process of inflation in a self-regenerating universe, such a universe will support an unbounded collection of mini-universes of all types, even if the probability of creation of some of them is extremely small. But this is just what is needed in support of the so-called Weak Anthropic Principle: we reside in regions with certain space-time and matter properties not because other regions are impossible, but because regions of the required type exist, and life as we know it (or, to be more precise, the carbon based life of human observers of our type) would not be possible in others.[5] It is

[5] Zelmanov proposed a similar formulation of the Anthropic Principle [77], saying that

important here that the total volume of regions in which we could live be unbounded, so that life as we know it will come into being even if the probability of its spontaneous appearance is vanishingly small. This does not mean, of course, that we can choose the laws of physics at random. The issue simply involves choosing one type of compactification and symmetry breaking or another, as allowed by the theory at hand. The search for theories in which *the part of the world that surrounds us* can have the properties that we observe is still a difficult problem, but it is much easier than a search for theories in which *the whole world* is not permitted to have properties different from those in the part where we now live.

Naturally, most of what we have said would remain true if we simply considered an infinitely large universe with chaotic initial conditions. But when inflation is not taken into account, the Anthropic Principle is incapable of explaining the uniformity of properties of the observable part of the universe (see Chapter 1). Furthermore, the mechanism for self-regeneration of the inflationary universe enables one to put the Anthropic Principle on a firm footing, given the most natural initial conditions in the universe, and regardless of whether it is finite or infinite.

We now consider several examples that demonstrate the various ways in which the Anthropic Principle can be applied in inflationary cosmology.

1) Consider first the symmetry breaking process in supersymmetric SU(5) theory. After inflation, the universe breaks up into exponentially large domains containing the fields Φ and H, corresponding to all possible types of symmetry breaking. Among these domains, there will be some in the SU(5)-symmetric phase and some in the SU(3) × U(1)-symmetric phase, corresponding to the type of symmetry breaking that we observe. Within each of these domains, the vacuum state will have a lifetime many orders of magnitude longer than the 10^{10} years which have passed since the end of inflation in our part of the universe. We live inside a domain with SU(3) × U(1) symmetry, within which there are strong, weak, and electromagnetic interactions of the type actually observed. This occurs not because there are no other regions whose properties differ from ours, and not because life is totally impossible in other regions, but because life *as we know it* is only possible in a region with SU(3) × U(1) symmetry.

2) Consider next the theory of the axion field θ with a potential of

---

we are witnesses only to certain definite kinds of physical processes because other processes take place without witnesses.

2) Consider next the theory of the axion field θ with a potential of the form (7.7.22):

$$V(\theta) \sim m_\pi^4 \left(1 - \cos \frac{\theta}{\sqrt{2}\,\Phi_0}\right). \qquad (10.5.1)$$

The field θ can take any value in the range $-\sqrt{2}\,\pi\Phi_0$ to $\sqrt{2}\,\pi\Phi_0$. A natural estimate for the initial value of the axion field would therefore be $\theta = O(\Phi_0)$, and the initial value of $V(\theta)$ should be of order m_π^4. An investigation of the rate at which the energy of the axion field falls off as the universe expands shows that for $\Phi_0 \gtrsim 10^{12}$ GeV, most of the energy density would presently be contributed by axions, while the baryon energy density would be considerably lower than its presently observed value of $\rho_B \gtrsim 2\cdot 10^{-31}$ g$\cdot$cm^{-3}. (Since the universe becomes almost flat after inflation, the overall matter density in the universe should now be $\rho_c \sim 2\cdot 10^{-29}$ g$\cdot$cm^{-3}, independent of the value of the parameter Φ_0.) This information was used to derive the strong constraint $\Phi_0 \lesssim 10^{11}$–10^{12} GeV [49]. This is not a terribly felicitous result, as axion fields with $\Phi_0 \sim 10^{15}$–10^{17} GeV show up in a natural way in many models based on superstring theory [50].

Let us now take a somewhat closer look at whether one can actually obtain the constraint $\Phi_0 \lesssim 10^{11}$–10^{12} GeV in the context of inflationary cosmology.

As we remarked in Section 7.7, long-wave fluctuations of the axion field θ are generated at the time of inflation (if Peccei–Quinn symmetry breaking, resulting in the potential (10.5.1), takes place before the end of inflation). By the end of inflation, therefore, a quasihomogeneous distribution of the field θ will have appeared in the universe, with the field taking on all values from $-\sqrt{2}\,\pi\Phi_0$ to $\sqrt{2}\,\pi\Phi_0$ at different points in space with a probability that is almost independent of θ [276, 224]. This means that one can always find exponentially large regions of space within which $\theta \ll \Phi_0$. The energy of the axion field always remains relatively low in such regions, and there is no conflict with the observational data.

In and of itself, this fact does not suffice to remove the constraint $\Phi_0 \lesssim 10^{12}$ GeV, since when $\Phi_0 \gg 10^{12}$ GeV, only within a very small fraction of the volume of the universe is the axion field energy density small enough by comparison with the baryon density. It might therefore seem unlikely that we just so happened make our appearance in one of these particular regions. Consider, for example, those regions initially containing a field $\theta_0 \ll \Phi_0$, for which the present-day ratio of the energy density of the axion field to the baryon density is consistent with the observational data. It can be shown that the total number of baryons in regions with $\theta \sim 10\,\theta_0$ should be ten times the number in regions with $\theta \sim \theta_0$. One might therefore expect the probability of randomly winding up in a region with $\theta \sim 10\,\theta_0$ (contradicting the observational data) to be ten times that of winding up in a region with $\theta \sim \theta_0$. Closer examination of this problem indicates, however, that the mean matter density in galaxies at time $t \sim 10^{10}$ years is proportional to θ^8, and in regions with $\theta \sim 10\,\theta_0$ it should be 10^8 times higher than in regions like our own, with $\theta \sim \theta_0$ [334]. A preliminary study of the star formation process in galaxies with $\theta \sim 10\,\theta_0$ indicates that solar-type stars are most likely not formed in such galaxies. If that is really the case, then the conditions required for the appearance of life as we know it can only be realized when $\theta \sim \theta_0$, and that is exactly why we find ourselves to be in one such region rather than in a typical region with $\theta \gg \theta_0$. Generally speaking, then, the observational data do not imply that $\Phi_0 \lesssim 10^{12}$ GeV. In any event, since regions with $\theta \sim \theta_0$ most assuredly exist, a derivation of the constraint $\Phi_0 \lesssim 10^{12}$ GeV would require that one show that it is much more improbable for life as we know it to arise in regions with $\theta \sim \theta_0$ than in regions with $\theta \gg \theta_0$. As we have already said, an investigation of this question indicates just the opposite.

The preceding discussions are quite general in nature, and can be applied not just to the theory of axions, but to the theory of any other light, weakly interacting scalar fields as well — dilatons, for example [335]. In principle, by applying the Anthropic Principle to axion cosmology, one might attempt to explain why the present-day baryon density ρ_B is 10^{-1}–10^{-2} times the total matter density $\rho_0 \sim \rho_c$ in the universe. Actually, for $\theta \ll \theta_0$, the energy density of the axion field would be low ($\rho_a \sim \theta^2$), so the main contribution to ρ_0 would come from baryons, $\rho_0 \sim \rho_c \sim \rho_B$. But only a small fraction of the baryons in the universe (proportional to $\frac{\theta}{\Phi_0}$) are to be found in regions with $\theta \ll \theta_0$. At the same time, for $\theta \gg \theta_0$, the conditions of life would be markedly different from our own, and it is most unlikely that one would find himself inside such a region of the universe. The location of the maximum probability for the existence of life as we know it, as a function of θ, depends on the value of Φ_0, and for a particular value $\Phi_0 \gg 10^{12}$ GeV, the maximum may be attained precisely in a state with initial value $\theta \sim \theta_0$, in which now $\rho_B \sim 10^{-1}$–$10^{-2} \rho_0$. Consequently, a study of the theory of star and galaxy formation together with a detailed study of the conditions necessary for the existence of life as we know it may actually enable us to determine the most likely value of the parameter Φ_0 in the theory of axions.

3) The preceding results can be employed practically unchanged to avoid one of the principal difficulties encountered in using the mechanism for generating the baryon asymmetry of the universe proposed by Affleck and Dine [97, 98]. Recall that the baryon asymmetry produced by this mechanism is typically too large: the value of $\frac{n_B}{n_\gamma}$ ranges from $-O(1)$ to $+O(1)$, depending on the magnitude of the angle θ in isotopic spin space between the initial values of two different scalar fields. According to the inflationary universe scenario, one can always find exponentially large regions in which this angle is small and $\frac{n_B}{n_\gamma} \sim 10^{-9}$. Such regions occupy a very small fraction of the total volume of the universe. But in regions, say, with $\frac{n_B}{n_\gamma} \sim 10^{-7}$, the density of matter in galaxies will be eight orders of magnitude greater than in our own, and life as we know it will either be impossible or extremely unlikely. Naturally, there are a number of other ways to get rid of the excess baryon asymmetry of the universe (see Section 7.10), but interestingly enough, the Anthropic Principle as applied within the scope of the theory of the inflationary universe may turn out to be sufficient to resolve this problem all by itself.

4) The last example that we present here is somewhat different from its predecessors. We know that in the standard Friedmann cosmology, the universe, if it is closed, will spend approximately half its life in a state of expansion, and the other half in a state of contraction. A similar phenomenon can also occur locally in an inflationary universe on scales at which density inhomogeneities that came into being during the inflationary stage become large, with $\frac{\delta\rho}{\rho} \sim 1$ [336]. The question that arises is then 'Why is the observable part of the universe expanding?' Do we live in an expanding part of the universe by accident, or is there some special reason for this circumstance?

The answer to this question is related to the fact that in the simplest $\frac{\lambda}{4}\varphi^4$ theory with $\lambda \sim 10^{-14}$, for example, the size of a homogeneous locally

Friedmann part of the universe is of order $l \sim M_P^{-1} \exp(\pi\lambda^{-1/3}) \sim M_P^{-1} \cdot 10^{6.10^4}$ (1.8.8), and a typical mass concentrated in such a region is of order $M \sim M_P \cdot 10^{2 \cdot 10^5}$, so according to (1.3.15), the typical time until the beginning of contraction within such a region is $t \sim 10^{2 \cdot 10^5}$ years [336]. In a self-regenerating universe, inasmuch as it exists without end, there should be regions that are much older and regions that are much younger. We happen to live in a relatively young region, which has existed a total of 10^{10} years since the end of inflation (in this region). This is related to the fact that life *as we know it* exists near solar-type stars, whose maximum lifetime is 10^{10}–10^{11} years. This is precisely why the part of the universe that surrounds us is still in the initial stage of its expansion, and that expansion (within the framework of the simple model considered here) should last at least another $10^{2 \cdot 10^5}$ years.

Taken by itself, the foregoing certainly does not mean that no life at all is possible during the contraction stage [336]; the issue is simply that at the current rate of evolution of living organisms (and also taking into account the probable decay of baryons after 10^{35}–10^{40} years), observers $10^{2 \cdot 10^5}$ years hence will scarcely resemble the way we are now.

We wish to emphasize once again that the so-called Weak Anthropic Principle, as formulated and used above, is conceptually quite simple. It involves an assessment of the probability of observing a region of the universe with given properties, under the condition that the fundamental properties of the observer himself also be known. The preceding discussions require no philosophical sophistication, and their import is of a trivial, mundane nature — for instance, we live on the surface of the Earth not because there is more room here than in interstellar space, but simply because in interstellar space we could never breathe.

Furthermore, the richness and heuristic value of the results obtained via the Weak Anthropic Principle have impelled many authors to try to expand and generalize it as much as possible, even if such a generalization is presently not entirely justified (see [331]). The possibility of such a generalization is suggested by the unusually important role played by the concept of the observer in the construction and interpretation of quantum cosmology. Most of the time, when discussing quantum cosmology, one can remain entirely within the bounds set by purely physical categories, regarding the observer simply as an automaton, and not dealing with questions of whether he has consciousness or feels anything during the process of observation [305]. This we have done in all of the preceding discussions. But we cannot rule out the possibility *a priori* that carefully avoiding the concept of consciousness in quantum cosmology constitutes an artificial narrowing of one's outlook. A number of authors have underscored the complexity of the situation, replacing the word *observer* with the word *participant,* and introducing such terms as "self-observing universe" (see, for example, [302, 323]). In fact, the question may come down to whether standard physical theory is actually a closed system with regard to its description of the universe as a whole at the quantum level: is it really possible to fully understand what the universe is without first understanding what life is?

Leaving aside the question of how well motivated such a statement of the problem is, let us note that similar problems often appear in the development of science. We know, for example, that classical electrodynamics is incomplete, an example being the problem of the self-acceleration of an electron, requiring the use of quantum theory for a solution [65]. Quantum electrodynamics likely suffers from the zero-charge problem [156, 157], which can be circumvented by including electrodynamics in a unified nonabelian gauge theory [3]. The quantum theory of gravitation presents even greater conceptual difficulties, and attempts have been made to overcome these by significantly broadening and generalizing the original theory [14–17]. We do not know whether it is possible to assign an exact meaning to many of the concepts employed in quantum cosmology (probability of creation of the universe from "nothing," splitting of the universe in the many-worlds interpretation of quantum mechanics, etc.) without stepping outside the confines of the existing theory, and possible ways of generalizing this theory are still far from clear. The only thing we can do at this point is to attempt to draw upon analogies from the history of science which may prove to be instructive.

Prior to the advent of the special theory of relativity, space, time, and matter seemed to be three fundamentally different entities. In fact, space was thought to be a kind of three-dimensional coordinate grid which, when supplemented by clocks, could be used to describe the motion of matter. Special relativity did away with the insuperable distinction between space and time, combining them into a unified whole. But space-time nevertheless remained something of a fixed arena in which the properties of matter became manifest. As before, space itself possessed no intrinsic degrees of freedom, and it continued to play a secondary, subservient role as a backdrop for the description of the truly substantial material world.

The general theory of relativity brought with it a decisive change in this point of view. Space-time and matter were found to be interdependent, and there was no longer any question of which was the more fundamental of the two. Space-time was also found to have its own inherent degrees of freedom, associated with perturbations of the metric — gravitational waves. Thus, space can exist and change with time in the absence of electrons, protons, photons, etc.; in other words, in the absence of anything that had *previously* (i.e., prior to general relativity) been subsumed by the term *matter*. (Note that because of the weakness with which they interact, gravitational waves are exceedingly difficult to detect experimentally, an as-yet unsolved problem.)

A more recent trend, finally, has been toward a unified geometric theory of all fundamental interactions, including gravitation. Prior to the end of the 1970's, such a program — a dream of Einstein's — seemed unrealizable; rigorous theorems were proven on the impossibility of unifying spatial symmetries with the internal symmetries of elementary particle theory [337]. Fortunately, these theorems were sidestepped after the discovery of supersymmetric theories [85]. In principle, with the help of supergravity, Kaluza–Klein, and superstring theories, one may hope to construct a theory in which all matter fields will be interpreted in terms of the geometric properties of some multidimensional superspace [13–17]. Space would then cease to be simply a requisite mathematical adjunct for the description of the real world, and would instead take on greater and greater independent significance, gradually encompassing all the material particles under the guise of its own intrinsic degrees of freedom. Of course, this doesn't at all mean that the concept of matter becomes useless. The issue at hand is simply the revelation of the fundamental unity of space, time, and matter, which is hidden from us in much the same way that the unity of the weak and electromagnetic interactions was hidden until recently.

According to standard materialistic doctrine, consciousness, like space-time before the invention of general relativity, plays a secondary, subservient role, being considered just a function of matter and a tool for the description of the truly existing material world. It is certainly possible that nothing similar to the modification and generalization of the concept of space-time will occur with the concept of consciousness in the coming decades. But the thrust of research in quantum cosmology has taught us that the mere statement of a problem which might at first glance seem entirely metaphysical can sometimes, upon further reflection, take on real meaning and become highly significant for the further development of science. We should therefore like to take a certain risk and formulate several questions to which we do not yet have the answers.

Is it not possible that consciousness, like space-time, has its own intrinsic degrees of freedom, and that neglecting these will lead to a description of the universe that is fundamentally incomplete? Will it not turn out, with the further development of science, that the study of the universe and the study of consciousness will be inseparably linked, and that ultimate progress in the one will be impossible without progress in the other? After the development of a unified geometrical description of the weak, strong, electromagnetic, and gravitational interactions, will the next important step not be the development of a unified approach to our entire world, including the world of consciousness?

All of these questions might seem somewhat naïve and out of place in a serious scientific publication, but to work in the field of quantum cosmology without an answer to these, and without even trying to discuss them, gradually becomes as difficult as working on the hot universe theory without knowing why there are so many different things in the universe, why nobody has ever seen parallel lines intersect, why the universe is almost homogeneous and looks approximately the same at different locations, why space-time is four-dimensional, and so on (see Section 1.5). Now, with plausible answers to these questions in hand, one can only be surprised that prior to the 1980's, it was sometimes taken to be bad form even to discuss them. The reason is really very simple: by asking such questions, one confesses one's own ignorance of the simplest facts of daily life, and moreover encroaches upon a realm which may seem not to belong to the world of positive knowledge. It is much easier to convince oneself that such questions do not exist, that they are somehow not legitimate, or that someone answered them long ago.

It would probably be best then not to repeat old mistakes, but instead to forthrightly acknowledge that the problem of consciousness and the related problem of human life and death are not only unsolved, but at a fundamental level they are virtually completely unexamined. It is tempting to seek connections and analogies of some kind, even if they are shallow and superficial ones at first, in studying one more great problem — that of the birth, life, and death of the universe. It may conceivably become clear at some future time that these two problems are not so disparate as they might seem.

10.6 Quantum cosmology and the signature of space-time

The most significant modification to the concept of four-dimensional space-time that we have discussed so far is that of a space with one temporal and $d-1$ spatial coordinates, some of the latter being compactified. This construction, however, is clearly not the most general. Our intuitive ideas about space-time are linked to our study of the dynamics of objects whose dimensions may be arbitrarily small, but in the quantum theory of gravitation it is difficult (or impossible) to consider objects smaller than M_P^{-1}. If the theory is to be based on the study of extended objects like strings or membranes, many of our intuitive ideas about the geometrical objects associated with them (points, straight lines, etc.) will be found to be largely inadequate [17].

Unanswered questions arise, however, even at a simpler level. For example, why are there many spatial coordinates, but only one temporal coordinate? That is, why does our space have the signature (+ − − − ... −)? Why could it not be Euclidean, with signature (+ + ... +) or (− − ... −)? Why are just the spatial dimensions compactified, and not the temporal? Are transformations which change the signature of the metric possible [292]?

In the context of a model of the universe consisting of large regions with differing properties, these may all prove to be sensible questions. It is therefore worth considering, if only briefly, how the properties of the universe would be altered under a change in the signature of the metric. There are many aspects to this question, some of which stand out particularly clearly in supergravity and the theory of superstrings. For instance, the 16-component Majorana–Weyl spinors required for a formulation of supergravity in a space with $d=10$ only exist for three different signatures of the metric: 1 + 9 (one temporal and nine spatial dimensions), 5 + 5, and 9 + 1 [338]; a supersymmetric theory has only been formulated in the first case.

There exists one additional more general problem that arises in a very broad class of theories when space contains more than one time dimension. The problem is most readily understood by studying a scalar field in a flat space with signature (+ + − −) as an example. The usual dispersion relation for the field φ, which takes the form $k_0^2 = \boldsymbol{k}^2 + m^2$ in Minkowski space, then becomes

$$k_0^2 = k_2^2 + k_3^2 + m^2 - k_1^2 . \tag{10.6.1}$$

Here the momentum k_1 can clearly change the sign of the effective mass squared in (10.6.1), or in other words, it can induce exponentially rapid growth of fluctuations of the field φ when $k_1^2 > k_2^2 + k_3^2 + m^2$:

$$\delta\varphi \sim \exp\left(\sqrt{k_1^2 - k_2^2 - k_3^2 - m^2}\, t\right). \tag{10.6.2}$$

This effect is analogous to the instability of the vacuum state with $\varphi = 0$ in the theory of a scalar field with negative mass squared (see (1.1.5), (1.1.6)). In the theory (1.1.5), however, the development of the instability came to a halt when the sign of the effective squared mass $m^2(\varphi)$ changed as the field φ increased. But in the present example, the instability grows without bound, since there are exponentially growing modes with arbitrary values of m^2 for sufficiently large momenta k_1. Since the instability is associated precisely with the very largest momenta (shortest wavelengths), the existence of such an instability will most likely be a general feature of theories in a space with more than one time dimension, regardless of either the topology of the space or whether the additional time dimensions are compactified. In some theories it proves to be possible to avoid instability in modes corresponding to particles that have relatively low mass after compactification [293], but there still remains an instability due to heavy particles with masses m of the order of the reciprocal of the compactification radius R_c^{-1}. From (10.6.2), it follows that this instability is not the least bit less dangerous. One might hold out hope that for some reason there might be a cutoff at k_0, $k_1 \sim R_c^{-1}$ in this theory; modes with $m \gtrsim R_c^{-1}$ might then not appear. But if there were a cutoff at momenta of order R_c^{-1}, it would become impossible even to discuss compactification in conventional semiclassical language. To put it differently, until such time as we are capable of describing a classical space containing more than one time dimension, instability seems unavoidable.

In Euclidean space there is no instability, but there is also no evolution over the course of time, which is necessary for the existence of life as we know it. Furthermore, Euclidean space also lacks the requisite instability with respect to exponential growth of the universe, which leads to inflation and makes the universe so large.

To summarize this section, we might say that where there is no time, there is neither evolution nor life, and where there is too much time, instability is rampant and life is short. From this perspective, the familiar sig-

place within a relatively orderly framework.

10.7 The cosmological constant, the Anthropic Principle, reduplication of the universe and life after inflation

As noted in Section 1.5, one of the most difficult problems in modern physics is the problem of the vacuum energy, or the cosmological constant. There have been a great many interesting suggestions as to how this problem might be solved — for example, see [17, 78, 116, 292, 335, 339–359]. This multitude of proposals can be divided into two main groups. The most attractive possibility is that due to some mechanism related to a symmetry of the theory, for example, the vacuum energy must be exactly zero. The second possibility, presently an active topic of discussion among experts in the theory of the formation of the large-scale structure of the universe, is that there is some sort of mechanism that may engender a vacuum energy density ρ_v at the present time of the same order of magnitude as the present total matter density $\rho_0 \sim \rho_c \sim 2 \cdot 10^{-29}$ g$\cdot$cm^{-3}. While deep-seated reasons may exist for the vanishing of the vacuum energy, however, ensuring the equality of ρ_v and ρ_0 at the present epoch (if only to order of magnitude) is difficult without placing unnatural constraints on the parameters of the theory.

The Anthropic Principle suggests one possible escape from this situation. To illustrate the basic idea behind this approach to the problem of the cosmological constant, consider the theory of a scalar field Φ with effective potential $V(\Phi, \varphi) = \alpha\, M_P^3\, \Phi + V(\varphi)$ [78, 341]. Here $V(\varphi)$ is the potential of the field φ responsible for inflation, with a minimum at the point φ_0. We shall assume that the constant α is very small, $\alpha \lesssim 10^{-120}$. Fluctuations of the field Φ that set in at the time of inflation result in space being partitioned into regions with all possible values of $V(\Phi, \varphi_0)$, ranging from $-M_P^4$ to $+M_P^4$. In those regions where $V(\Phi, \varphi_0) \ll -10^{-29}$ g$\cdot$cm^{-3} at present, the universe looks locally like de Sitter space with negative vacuum energy. In such regions, all structures come into being and pass out of existence in much less than 10^{10} years, and life as we know it cannot emerge. In regions with $V(\Phi, \varphi) > 2 \cdot 10^{-29}$ g$\cdot$cm^{-3}, inflation is still ongoing, and if the potential $V(\Phi, \varphi_0)$ is very flat ($\alpha \lesssim 10^{-120}$) the field Φ will vary quite slowly; the time needed for $V(\Phi, \varphi_0)$ to decrease to

10^{-29} g·cm^{-3} will then be more than 10^{10} years. In regions with $V(\Phi, \varphi_0) \gtrsim 10^{-27}$ g·cm^{-3}, the standard mechanism for galaxy formation is altered significantly, and for $V(\Phi, \varphi_0) \gg 10^{-27}$ g·cm^{-3}, galaxies and stars like our own are hardly formed at all [348]. This still does not tell us why presently $V(\Phi, \varphi_0) \lesssim 10^{-29}$ g·cm^{-3}, but the fact that the observational constraints on the vacuum energy density must be satisfied in at least a few percent of the "habitable" reaches of the universe makes the problem of the cosmological constant much less acute. An even better model would be one in which the spectrum of possible values of the vacuum energy ρ_v is discrete rather than continuous, and includes states with $\rho_v = 0$ but not those with energy density less than 10^{-27} g·cm^{-3}. With the enormous number of possible types of compactification in Kaluza–Klein theories, such a possibility can actually be realized [292], and a similar possibility emerges in superstring theory [353]. In any case, the very fact that the Anthropic Principle may enable us to narrow the range of possible values of ρ_v in the observable part of the universe from -10^{94} g·cm$^{-3} \lesssim \rho_v \lesssim 10^{94}$ g·cm^{-3} to -10^{-29} g·cm$^{-3} \lesssim \rho_v \lesssim 10^{-27}$ g·cm^{-3} (i.e., to reduce the range by a factor of 10^{121}) seems worthy of note.

We will soon return to a discussion of using the Anthropic Principle to solve the problem of the cosmological constant. Here we consider the possibility that the cosmological constant vanishes because of some hidden symmetry. Several suggestions have been made on this score. One of the most interesting and promising ideas has to do with the application of supersymmetric theories, and superstring theories in particular [17]. In certain versions of such theories, the vacuum energy in the absence of supersymmetry breaking vanishes to all orders of perturbation theory [17, 353]. In the real world, however, supersymmetry is broken, and it is still unclear whether the vacuum energy remains at zero after supersymmetry breaking. Another possibility has to do with so-called dilatation invariance, which (if certain rather strong assumptions are made) might be of some help despite the fact that it is also broken [335] (see also [354]). Below, we discuss one more possibility with a direct bearing on quantum cosmology that shows how many surprises this science may hold in store [344].

Consider a model that simultaneously describes two different universes X and $\overline{X}$, with coordinates x_μ and $\overline{x}_\alpha$ respectively (μ, $\alpha = 0, 1, 2, 3$) and metrics $g_{\mu\nu}(x)$ and $\overline{g}_{\alpha\beta}(\overline{x})$, containing the fields $\varphi(x)$ and $\overline{\varphi}(\overline{x})$ with action of the following unusual type [344]:[6]

$$S = N \int d^4x\, d^4\overline{x}\, \sqrt{g(x)}\, \sqrt{\overline{g}(\overline{x})} \times \left\{ \frac{M_P^2}{16\pi} R(x) + L[\varphi(x)] - \frac{M_P^2}{16\pi} R(\overline{x}) - L[\overline{\varphi}(\overline{x})] \right\}. \tag{10.7.1}$$

Here N is a normalizing constant. The action (10.7.1) is invariant under general covariant transformations in each of the universes individually. A novel symmetry of the action (10.7.1) is the symmetry under the transformations $\varphi(x) \to \overline{\varphi}(x)$, $g_{\mu\nu}(x) \to \overline{g}_{\alpha\beta}(x)$, $\overline{\varphi}(\overline{x}) \to \varphi(\overline{x})$, $\overline{g}_{\alpha\beta}(\overline{x}) \to g_{\mu\nu}(\overline{x})$ with a subsequent change of sign, $S \to -S$. For reasons that will soon become clear, we call this antipodal symmetry. (In principle, other terms that do not violate this symmetry could be added to the integrand of Eq. (10.7.1), such as any odd function of $\varphi(x) - \overline{\varphi}(\overline{x})$; this would not affect our basic result.)

One immediate consequence of antipodal symmetry is invariance under a shift in the values of the effective potentials $V(\varphi) \to V(\varphi) + C$, $V(\overline{\varphi}) \to V(\overline{\varphi}) + C$, where C is an arbitrary constant. Thus, nothing in the theory depends on the value of the potentials $V(\varphi)$ and $V(\overline{\varphi})$ at their absolute minima φ_0 and $\overline{\varphi}_0$ (note that $\varphi_0 = \overline{\varphi}_0$ and $V(\varphi_0) = V(\overline{\varphi}_0)$ by virtue of the same symmetry). This is precisely why it proves possible to solve the cosmological constant problem in the theory (10.7.1).

However, the main reason for invoking this new symmetry was not just to solve the cosmological constant problem. Just as the theory of mirror particles was originally proposed in order to make the theory CP-symmetric while maintaining CP-asymmetry in its observable sector, the theory (10.7.1) is proposed in order to make the theory symmetric with respect to the choice of the sign of energy. This removes the old prejudice that even though the overall change of sign of the Lagrangian (i.e., of both

[6] Somewhat different (but similar) models have also been considered elsewhere [116, 293].

its kinetic and potential terms) does not change the solutions of the theory, one *must say* that the energy of all particles is positive. This prejudice was so strong that several years ago people preferred to quantize *particles* with *negative energy* as *antiparticles* with *positive energy*, which resulted in the appearance of such meaningless concepts as negative probability. We wish to emphasize that there is no problem with performing a consistent quantization of theories which describe particles with negative energy. Difficulties appear only when there exist interacting species with both signs of energy. (As noted in Section 10.1, this is one of the main problems of quantum cosmology, where one must quantize fields with positive energy and the scalar factor a with negative energy.) In the present case no such problem exists, just as there is no problem of antipodes falling off the opposite side of the Earth. The reason is that the fields $\bar{\varphi}(\bar{x})$ do not interact with the fields $\varphi(x)$, and the equations of motion for the fields $\bar{\varphi}(\bar{x})$ are the same as for the fields $\varphi(x)$ (the overall minus sign in front of $L[\bar{\varphi}(\bar{x})]$ does not change the Lagrange equations). In other words, in spite of the fact that from the standpoint of the sign of the energy of matter, universe $\bar{X}$ is an antipodal world where everything is upside-down, there is no instability there, and particles of the field $\bar{\varphi}(\bar{x})$ are completely unaware that they have energy of the "wrong" sign, just as our antipodal counterparts living on the other side of the globe are completely unruffled by the fact that they are upside-down from our point of view.

Similarly, gravitons from different universes do not interact with each other. However, some interaction between the two universes does exist. In the theory (10.7.1), the Einstein equations take the form

$$\begin{aligned} R_{\mu\nu}(x) - \frac{1}{2} g_{\mu\nu}(x)\, R(x) = & -8\pi G\, T_{\mu\nu}(x) \\ & - g_{\mu\nu}(x) < \frac{R(\bar{x})}{2} + 8\pi G\, L[\bar{\varphi}(\bar{x})] >, \end{aligned} \tag{10.7.2}$$

$$\begin{aligned} R_{\alpha\beta}(\bar{x}) - \frac{1}{2} g_{\alpha\beta}(\bar{x})\, R(\bar{x}) = & -8\pi G\, T_{\alpha\beta}(\bar{x}) \\ & - \bar{g}_{\alpha\beta}(\bar{x}) < \frac{R(x)}{2} + 8\pi G\, L[\varphi(x)] >. \end{aligned} \tag{10.7.3}$$

Here $G = M_P^{-2}$, $T_{\mu\nu}$ is the energy-momentum tensor of the field $\varphi(x)$, $T_{\alpha\beta}$ is the energy–momentum tensor of the field $\bar{\varphi}(\bar{x})$, and the meaning of the angle-bracket notation is

$$\langle R(x)\rangle = \frac{\int d^4x \sqrt{g(x)}\, R(x)}{\int d^4x \sqrt{g(x)}}, \tag{10.7.4}$$

$$\langle R(\bar{x})\rangle = \frac{\int d^4\bar{x} \sqrt{\bar{g}(\bar{x})}\, R(\bar{x})}{\int d^4\bar{x} \sqrt{\bar{g}(\bar{x})}}, \tag{10.7.5}$$

and similarly for $\langle L[\varphi(x)]\rangle$ and $\langle L[\bar{\varphi}(\bar{x})]\rangle$. Thus, although particles in universes X and $\bar{X}$ do not interact with each other, the universes themselves *do* interact, but only globally: each one makes a time-independent contribution to the average vacuum energy density of the other, with averaging taking place over the entire history of the universe. At the quantum cosmological level, when one writes down the equations for universe X, for example, averaging should take place over *all* possible states of universe $\bar{X}$ — that is, the result should not depend on the initial conditions in each of the two universes.

Generally speaking, it is extremely difficult to calculate the averages (10.7.4) and (10.7.5), but in the inflationary universe scenario (at least at the classical level), everything turns out to be quite simple. Indeed, after inflation the universe becomes almost flat, and its lifetime becomes exponentially long (or even infinite, if it is open or flat). In that event, the dominant contribution to the mean values $\langle R\rangle$ and $\langle L\rangle$comes from the late stages of the evolution of the universe, when the fields $\varphi(x)$ and $\bar{\varphi}(\bar{x})$ relax near the global minima of $V(\varphi)$ and $V(\bar{\varphi})$. As a consequence, the mean value of $-L[\varphi(x)]$ is the same, to exponentially high accuracy, as the value of the potential $V(\varphi)$ at its global minimum at $\varphi = \varphi_0$, and the mean value of the curvature scalar $R(x)$ is identical to its value during the late stages of evolution of the universe X, when the universe transforms to the state $\varphi = \varphi_0$, corresponding to the global minimum of $V(\varphi)$. The analogous statement also holds true for $\langle L[\bar{\varphi}(\bar{x})]\rangle$and $\langle R(\bar{x})\rangle$. For that reason, Eqs. (10.7.2) and (10.7.3) take the following form in the late stages of evolution of the universes X and $\bar{X}$:

$$R_{\mu\nu}(x) - \frac{1}{2} g_{\mu\nu}(x)\, R(x) = 8\,\pi\, G\, g_{\mu\nu}(x) \left[V(\overline{\varphi}_0) - V(\varphi_0) \right] - \frac{1}{2} g_{\mu\nu}(x)\, R(\overline{x}), \tag{10.7.6}$$

$$R_{\alpha\beta}(\overline{x}) - \frac{1}{2} g_{\alpha\beta}(\overline{x})\, R(\overline{x}) = 8\,\pi\, G\, g_{\alpha\beta}(\overline{x}) \left[V(\varphi_0) - V(\overline{\varphi}_0) \right] - \frac{1}{2} g_{\alpha\beta}(\overline{x})\, R(x), \tag{10.7.7}$$

yielding

$$R(x) = 2\, R(\overline{x}) + 32\,\pi\, G \left[V(\varphi_0) - V(\overline{\varphi}_0) \right], \tag{10.7.8}$$

$$R(\overline{x}) = 2\, R(x) + 32\,\pi\, G \left[V(\overline{\varphi}_0) - V(\varphi_0) \right]. \tag{10.7.9}$$

Recall from our earlier discussion that $\varphi_0 = \overline{\varphi}_0$ and $V(\varphi_0) = V(\overline{\varphi}_0)$ by virtue of antipodal symmetry. This implies that in the late stages of evolution of the universe X,

$$R(x) = -R(\overline{x}) = \frac{32\pi}{3} G \left[V(\varphi_0) - V(\overline{\varphi}_0) \right] = 0\,. \tag{10.7.10}$$

We emphasize that the contribution made by universe $\overline{X}$ to the effective vacuum energy of universe X does not depend on the time t in the latter. The cancellation represented by (10.7.10) therefore takes place only in the late stages of evolution of universe X, and its sole effect is to add a constant term to $V(\varphi)$ in such a way as to obtain $V(\varphi_0) = 0$. Thus, the mechanism considered here does not alter the standard inflationary scenario at all.

Note that this model differs from the conventional Kaluza–Klein theory in which, as we have already stated, the introduction of two time coordinates immediately leads to instability. If would be straightforward to generalize the theory (10.7.1), for example, by writing the action as an integral over universes X_1, X_2, ..., and taking the Lagrangian to be a sum of Lagrangians from the various fields $\varphi_1(x_1)$, $\varphi_2(x_2)$, ..., each of which resides in only one of these universes. With such a scheme, our world would consist of arbitrarily many different universes interacting with each other only globally, the inhabitants of each having their own time and their own physical laws. This approach would provide a basis for the Anthropic

Principle in its strongest form.

Of course there are shortcomings to be dealt with. This scheme could be generalized to supersymmetric theories, but it would be difficult to do so in a way that ensures that the cosmological constant vanishes automatically. If the universe is self-regenerating, one could encounter difficulties in calculating the averages (10.7.4), (10.7.5), since they may become infrared-divergent and the result may depend on the cutoff. This question has yet to be thoroughly examined, due to the very complicated large-scale structure of the self-regenerating universe. However, one can easily avoid such questions in theories in which $V(\varphi)$ grows rapidly enough at $\varphi \geq \varphi^*$, since there will be no self-regeneration of the universe in such theories. Another problem is that the integral over $d^4\bar{x}$ in (10.7.1) renormalizes the effective Planck constant in the universe X, and one should take a very small normalization factor ($N \sim \exp\left(-\lambda^{1/3}\right)$ in the $\lambda\varphi^4$ theory) in order to compensate this renormalization. A further possibility is that in constructing the quantum theory in a reduplicated universe, one should only do so in each of the noninteracting universes individually, without taking the foregoing renormalization of N into consideration. Note also that the mechanism for cancellation of the cosmological term that was suggested above works independently of the value of N.

In any case, the very fact that (at least at the classical level) there is a large class of models within which the cosmological constant automatically vanishes, regardless of the detailed structure of the theory, seems noteworthy. Moreover, the possibility of building a consistent theory of many universes that interact with one another only globally may be of interest in its own right.

A very interesting and nontrivial generalization of the ideas we have discussed here has been proposed quite recently in papers by Coleman [345, 346], Giddings and Strominger [349], and Banks [347]; these are based on earlier work by Hawking [350], Lavrelashvili, Rubakov, and Tinyakov [351], and Giddings and Strominger [352] dealing with wormholes and the loss of coherence in quantum gravitation, as well as on a paper by Hawking [340] that treated a possible mechanism for the vanishing of the cosmological constant in the context of quantum cosmology.

The basic idea in Refs. 345-347 and 349 is that because of quantum effects, the universe can split into several topologically disjoint but globally interacting parts. Such processes can take place at any location in our universe (see [350–352], [133], and Section 10.3). The baby universes can carry off electron–positron pairs or some other combination of particles and fields if they are not prevented from doing so by conservation laws. The simplest way to describe this effect is to say that the existence of baby universes leads to a modification of the effective Hamiltonian that describes particles and fields in our own universe [345, 349]:

$$\mathcal{H}(x) = \mathcal{H}_0(\varphi(x), \psi(x), \dots) + \sum \mathcal{H}_i(\varphi(x), \psi(x), \dots) A_i . \quad (10.7.11)$$

This Hamiltonian describes the fields φ, ψ, ... in our universe on scales exceeding M_P^{-1}. In (10.7.11), $\mathcal{H}_0$ is the part of the Hamiltonian unrelated to topological fluctuations; the $\mathcal{H}_i$ are certain local functions of the fields φ, ψ, ...; the A_i represent a combination of creation and annihilation operators in the baby universes. Thus, for example, a term like $\mathcal{H}_1 A_1$, where $\mathcal{H}_1$ is constant, is associated with the possibility of a change in the vacuum energy density due to an interaction with baby universes, the term $\bar{e}(x)\, e(x)\, A_2$ is associated with possible electron–positron pair exchange, and so on. The operators A_i are independent of the position x, since baby universes cannot carry off energy or momentum. According to [345, 346], the requirement that the Hamiltonian in our universe be local,

$$[\mathcal{H}(x), \mathcal{H}(y)] = 0 \quad (10.7.12)$$

for spacelike $x - y$, implies that all of the operators A_i commute with each other. They can therefore all be simultaneously diagonalized by the "α–states" $|\alpha_i\rangle$, so that

$$A_i |\alpha_i\rangle = \alpha_i |\alpha_i\rangle . \quad (10.7.13)$$

If the quantum state of the universe is an eigenstate of the operators A_i, then one consequence of the complicated structure of the vacuum (10.7.13) will be the introduction of an infinitude of *a priori* undetermined parameters α_i into the effective Hamiltonian: in (10.7.11), one simply replaces the operators A_i by their eigenvalues α_i in the given state. If the universe is not originally in an eigenstate of the A_i, its wave function, after a series of measurements, will nonetheless quickly be reduced to such an eigenstate [345].

This makes it possible to consider anew many of the fundamental problems of physics. It is often supposed that the basic goal of theoretical physics is to find exactly what Lagrangian or Hamiltonian correctly describes our entire world. But one could well ask the following question: if we assume that there was a time when our universe (or the part in which we live) did not exist (at least as a classical space-time), in what sense can one speak of the existence of laws *at that time* which would have governed its birth and evolution? We know, for example, that the laws that control biological evolution are recorded in our genetic code. But where were the laws of physics recorded if there was no universe?

One possible answer is that the final structure of the effective Hamiltonian, including whatever values are taken on by the constants α_i, is not fixed until a series of measurements have been made, determining with finite accuracy which of the possible quantum states of the universe $|\alpha_i\rangle$ we live in [355]. This implies that the concept of an observer may play an important role not just in discussions of the various characteristics of our universe, but in the very laws by which it is governed.

In general, the wave function of the universe can depend on the parameters α_i. This possibility lies at the root of Coleman's explanation for the vanishing of the cosmological constant

$$\Lambda = \frac{8\pi V(\varphi_0)}{M_P^2},$$

where $V(\varphi_0)$ is the present value of the vacuum energy density. The basic idea goes back to a paper by Hawking [340], who made use of the Hartle–Hawking wave function (10.1.12), (10.1.17). According to Hawking, if the cosmological constant could for some reason take on arbitrary values, the probability of winding up in a universe with a given value of Λ is

$$P(\Lambda) \sim \exp[-S_E(\Lambda)] = \exp\frac{3M_P^4}{8V} = \exp\frac{3\pi M_P^2}{\Lambda} \qquad (10.7.14)$$

(compare with (10.1.18)). In the context of the approach that we are considering, which is based on the theory (10.7.11), (10.7.13), the cosmological constant, like any other constant, could actually take on different values, depending on exactly what quantum state we happen to be in. In calculating $P(\Lambda)$ for this case, however, one must also sum over all topologically disconnected configurations of the baby universes, which leads to a modified expression for $P(\Lambda)$ [346]:

$$P(\Lambda) \sim \exp\left(\exp\frac{3\pi M_P^2}{\Lambda}\right). \tag{10.7.15}$$

From (10.7.14) and (10.7.15), we see that $P(\Lambda)$ is sharply peaked at $\Lambda = 0$, i.e among all possible universes, the most probable are those with a vanishingly small cosmological constant.

Just how reliable is this conclusion? At the moment, it is difficult to say. In fact, the probability of formation of baby universes, resulting in a gravitational vacuum with highly complicated structure, has yet to be firmly established. The description of this process using Euclidean methods [350–352] differs from that obtained via the stochastic approach [133] (see Section 10.3, Eq. (10.3.7)). Furthermore, as noted in Section 10.2, the use of the Hartle–Hawking wave function in the inflationary cosmology is justified only when there exists a stationary distribution of the field φ, and therefore of $V(\varphi)$ and $\Lambda(\varphi)$. Thus far, it has not been possible to find any inflationary models in which such a stationary distribution might actually exist. The probability distribution for the quantity $\Lambda(\alpha_i)$ ought to have been stationary, but this is the probability distribution for finding a cosmological constant equal to Λ *in different universes* (or to be more precise, in different quantum states of the universe), rather than in different parts of a single universe. The stochastic approach is not capable of justifying expressions like (10.7.14) and (10.7.15) under these circumstances, and validity of Euclidean methods used for the derivation of eq. (10.7.15) is not quite clear.

Some authors argued that a correct distribution of probability for the universe to be in a given quantum state $|\alpha_i\rangle$ does not have a peak at $\Lambda = 0$, in contrast with Eq. (10.7.15) [356, 357]. It may be important also that actually we are not asking why the universe lives in a given quantum state $|\alpha_i\rangle$. We are just trying to explain the experimental fact that *we* live in the universe in the quantum state $|\alpha_i\rangle$ corresponding to $|\rho| = |V(\varphi_0)| \lesssim 10^{-29} g\cdot cm^{-3}$ [358, 359].

In this regard, recall that the application of the Anthropic Principle based on an analysis of the galaxy formation process enables one to place constraints on the vacuum energy density [348],

$$-10^{-29}\ g\cdot cm^{-3} \lesssim V(\varphi_0) \lesssim 10^{-27}\ g\cdot cm^{-3},$$

and these constraints are quite close to the experimental figure,

$$|V(\varphi_0)| \lesssim 10^{-29}\ \text{g}\cdot\text{cm}^{-3}.$$

This is an especially interesting result when viewed in the context of the baby universe theory which makes it possible to choose among the various Λ [345].

Can the Anthropic constraint on the vacuum energy density be made stronger, so as to make the inequality $V(\varphi_0) \lesssim 10^{-29}\ \text{g}\cdot\text{cm}^{-3}$ a necessary consequence of the Anthropic Principle? There is still no final answer, but there are a few inklings of how one might go about solving this problem.

As follows from the results obtained in Sections 10.2 - 10.4, life in all its possible forms will appear again and again in different domains of the self-regenerating inflationary universe. This does not mean, however, that one can be very optimistic about the fate of mankind. An investigation of this question reveals that within the presently observable part of the universe, life as we know it cannot endure indefinitely, due to the decay of baryons and the local collapse of matter [336]. The only possibility that we are presently aware of for the perpetual promulgation of life is that in the scenario under consideration, for example in the $\frac{\lambda\varphi^4}{4}$ theory, there must now exist a large number of domains in every region of size

$$\begin{aligned} l \gtrsim l^* &\sim 10^{30}\, M_P^{-1} \exp\left(\frac{\pi(\varphi^*)^2}{M_P^2}\right) \\ &\sim 10^{30}\, M_P^{-1} \exp\left(\frac{\pi(\varphi^*)^2}{M_P^2}\right) \end{aligned} \tag{10.7.16}$$

in which the process of inflation continues unabated, and will do so forever. There will always be sufficiently dense regions (like our own) near such domains, in which inflation came to an end relatively recently and in which baryons have not yet had a chance to decay. One possible survival strategy for mankind might consist of continual spaceflight bound for such regions. In the worst case, if we will be unable to travel to such distant places ourselves, we can try to send some information about us, our life and our knowledge, and maybe even stimulate development of such kinds of life there, which would be able to receive and use this information. In such a case one would have a comforting thought that even though life in our part of the universe will disappear, we will have some inheritors, and in this sense our existence is not entirely meaningless. (At least it would be not worse than what we have here now.)

Leaving aside the question of the optimal strategy for the survival of mankind, we wish to note that the appropriate process is necessarily impossible if the vacuum energy density $V(\varphi_0)$ is greater than

$$V^* \sim \rho_0 \cdot 10^{200} \exp\left(-6\pi\lambda^{-1/3}\right). \tag{10.7.17}$$

In the theory with $\lambda \sim 10^{-14}$, V^* is vanishingly small:

$$V^* \sim 10^{-5\cdot 10^6}\ \mathrm{g \cdot cm^{-3}} \ll \rho_0 . \tag{10.7.18}$$

The reason there is a critical value V^* is that when $V(\varphi_0) > V^*$, the size $H^{-1}(\varphi_0)$ of the event horizon in a world with vacuum energy density $V(\varphi_0)$ turns out to be less than the typical distance between domains in which the process of self-regeneration of the universe is taking place. (This distance is presently l^* of (10.7.16); by the time the vacuum energy $V(\varphi_0)$ begins to dominate, the distance will have grown by a factor of approximately $10^{-60} \exp\left(2\pi\lambda^{-1/3}\right)$.) Under such circumstances, it would be impossible in principle either to fly or to send a signal from our region of the universe to regions in the vicinity of self-regenerating domains; see Section 1.4.

Thus, in the present model, any quantum state of the universe $|a_i\rangle$ with vacuum energy density $V(\varphi_0) \gtrsim 10^{-5\cdot 10^6}\ \mathrm{g \cdot cm^{-3}}$ is a sort of cosmic prison, and life within the universe in such a state is inescapably condemned to extinction as a result of proton decay and the exponential fall off of the density of matter when the vacuum energy $V(\varphi_0)$ becomes dominant. There is still no consensus on the probability of the spontaneous emergence of complex life forms on Earth through just a single evolutionary chain. If, as some believe, this probability is extremely low, and if some mechanism of indefinitely long reproduction of life at $V(\varphi_0) < V^*$ does actually exist (and this is not at all obvious *a priori* [336]), then the existence of such a mechanism can drastically increase the fraction of the "habitable" universe in a quantum state with $V(\varphi_0) > 10^{-5\cdot 10^6}\ \mathrm{g \cdot cm^{-3}}$ as compared with the fraction in a state with $V(\varphi_0) > 10^{-5\cdot 10^6}\ \mathrm{g \cdot cm^{-3}}$ The net result could then be [359] that any observer like ourselves, who is capable of inquiring about the vacuum energy density, would very likely find himself in a universe corresponding to a quantum state $|\alpha_\iota>$ with

$$V(\varphi_0) \ll 10^{-29}\ \text{g}\cdot\text{cm}^{-3}.$$

Our treatment of this problem has merely provided a sketch of the course of future research, illustrating the novel possibilities which have emerged in recent years in elementary particle theory and cosmology. If we are to develop a successful strategy for the survival of mankind (if such a strategy exists), we will need to undertake a much deeper study of the global structure of the inflationary universe and the conditions required for the emergence and/or propagation of life therein. In any event, however, the possibility that there is a correlation between the value of the vacuum energy density and the existence of a mechanism of eternal reproduction of life in the universe seems to us to be noteworthy.

Conclusion

Elementary particle theorists and cosmologists might be compared to two teams tunneling toward each other through the enormous mountain of the unknown. The analogy, however, is not entirely accurate. If the two teams of construction workers miss each other, they will simply have built two tunnels instead of one. But in our case, if the particle theorists fail to meet the cosmologists, we wind up without any complete theory at all. Furthermore, even if they do meet, and manage to build an internally consistent theory of all processes in the micro- and macro-worlds, that still does not mean that their theory is correct.

In the face of the now familiar (and inevitable) dearth of experimental data on particle interactions at energies approaching 10^{19} GeV, and on the structure of the universe on scales $l \gg 10^{28}$ cm, it becomes especially important to guess, if only in broad outline, the true direction that this science will take, a direction that should remain valid even if many specific details of the theory under construction should change. This explains the recent emergence of such unusual terms as *scenario* and *paradigm* in the physicists' lexicon.

The major developments in elementary particle physics over the past two decades can be characterized by a few key words, such as *gauge invariance, unified theories with spontaneous symmetry breaking, supersymmetry,* and *strings*. The term *inflation* has become such a word in the cosmology of the 1980's.

The inflationary universe scenario could only have been created through the joint efforts of cosmologists and elementary particle theorists. The need for and fruitfulness of such a collaboration is now obvious. It should be pointed out that *inflation* is certainly not a magic word that will automatically solve all our problems and open all doors. In some theories of elementary particles, it is difficult to implement the inflationary universe scenario, whereas many other theories fail to lead to a good cosmology, even with the help of inflation. The road to a consistent cosmological theory may yet prove to be a very long one, and we may still find that many details of our present scenarios will be cast off as inessential excess baggage. At the moment, however, it does seem necessary to have something like inflation to obtain a consistent cosmology at peace with particle physics.

Inflationary cosmology continues to develop rapidly. We are witnessing a gradual change in our most general concepts about the evolution of the universe. Just a few years ago, most authorities had virtually no doubt that the universe was born in a unique Big Bang approximately 10–15 billion years in the past. It seemed obvious that space-time was four-dimensional from the very outset, and that it remains four-dimensional all over the universe. It was believed that if the universe were closed, its total size could hardly exceed that of its observable part, $l \sim 10^{28}$ cm, and that in no more than 10^{11} years such a universe would collapse and disappear. If, on the other hand, the universe were open or flat, then it would be infinite, and the general conviction was that it would then exhibit properties everywhere that were almost the same as those in its observable part. Such a universe would exist forever, but after its protons had decayed, as predicted by unified theories of the weak, strong, and electromagnetic interactions, no baryon matter would remain to support life. The only alternatives, then, were a "hot end" with the expected collapse of the universe, and a "cold end" in the infinite void of space.

It now seems more likely that the universe as a whole exists eternally, endlessly spawning ever newer exponentially large regions in which the low-energy laws governing elementary particle interactions, and even the effective dimensionality of space-time itself, may be different. We do not know whether life can evolve forever in each such region , but we do know for certain that life will appear again and again in different regions of the universe, taking on all possible forms. This change in our notions of the global structure of the universe and our place within it is one of the most important consequences to come out of inflationary cosmology.

We have finally come to a newfound appreciation of why it was necessary to write the scenario even though the performance was already over — the show is still going on, and most likely will continue forever. In different parts of the universe, different audiences will observe it in its infinite variations. We cannot witness the whole play in all its grandeur, but we can try to imagine its most essential features — and ultimately perhaps even grasp its meaning.

References

1. S. L. Glashow, Nucl. Phys. **22**, 579 (1961);
   S. Weinberg, Phys. Rev. Lett. **19**, 1264 (1967);
   A. Salam, in: Elementary Particle Theory, edited by N. Svartholm, Almquist and Wiksell, Stockholm (1968), p. 367.
2. G. 't Hooft, Nucl. Phys. **B35**, No. 1, 167–188 (1971);
   B. W. Lee, Phys. Rev. **D5**, No. 4, 823–835 (1972);
   B. W. Lee and J. Zinn–Justin, Phys. Rev. **D5**, No. 12, 3121–3160 (1972);
   G. 't Hooft and M. Veltman, Nucl. Phys. **B50**, No. 2, 318–335 (1972);
   I. V. Tyutin and E. S. Fradkin, Sov. J. Nucl. Phys. **16**, 464 (1972);
   R. E. Kallosh and I. V. Tyutin, Sov. J. Nucl. Phys. **17**, 98 (1973).
3. D. J. Gross and F. Wilczek, Phys. Rev. Lett. **30**, No. 26, 1343–1346 (1973);
   H. D. Politzer, Phys. Rev. Lett. **30**, No. 26, 1346–1349 (1973).
4. H. Georgi and S. L. Glashow, Phys. Rev. Lett. **32**, 438 (1974).
5. D. Z. Friedman, P. van Nieuwenhuizen, and S. Ferrara, Phys. Rev. **D13**, 3214 (1976).
6. Th. Kaluza, Stizungsber. Preuss. Akad. Wiss. Phys., Math., **K1**, 966 (1921);
   O. Klein, Z. Phys. **37**, 895 (1926);
   E. Cremmer and J. Scherk, Nucl. Phys. B **108**, 409;
   E. Witten, Nucl. Phys. B **186**, 412 (1981).
7. M. B. Green and J. H. Schwarz, Phys. Lett. **149B**, 117 (1984);
   Phys. Lett. **151B**, 21 (1984);
   D. J. Gross, J. A. Harvey, E. Martinec, and R. Rohm, Phys. Rev. Lett. **54**, 502 (1985);
   E. Witten, Phys. Lett. **149B**, 351 (1984).

8. A. A. Slavnov and L. D. Faddeev, Gauge Fields: Introduction to Quantum Theory, Benjamin–Cummings, London (1980).
9. J. C. Taylor, Gauge Theories of Weak Interactions, Cambridge University Press, Cambridge (1976).
10. L. B. Okun, Leptons and Quarks, North–Holland, Amsterdam (1982).
11. A. M. Polyakov, Gauge Fields and Strings, Harwood, London (1987).
12. P. Langacker, Phys. Rep. C **72**, 185 (1981).
13. V. I. Ogievetsky and L. Mezincescu, Sov. Phys. Usp. **18**, 960 (1975);
J. Bagger and J. Wess, Supersymmetry and Supergravity, Princeton Univ. Press, Princeton (1983);
P. West, Introduction to Supersymmetry and Supergravity, World Scientific, Singapore (1986).
14. P. van Nieuwenhuizen, Phys. Rep. C **68**, 192 (1981).
15. H. P. Nilles, Phys. Rep. C **110**, 3 (1984).
16. M. J. Duff, B. E. W. Nilsson, and C. N. Pope, Phys. Rep. C **130**, 1 (1986).
17. M. B. Green, J. H. Schwarz, and E. Witten, Superstring Theory, Cambridge University Press, Cambridge (1987).
18. D. A. Kirzhnits, JETP Lett. **15**, 529 (1972).
19. D. A. Kirzhnits and A. D. Linde, Phys. Lett. **42B**, 471 (1972).
20. S. Weinberg, Phys. Rev. **D9**, 3320 (1974);
L. Dolan and R. Jackiw, Phys. Rev. **D9**, 3357 (1974).
21. D. A. Kirzhnits and A. D. Linde, Sov. Phys. JETP **40**, 628 (1974).
22. D. A. Kirzhnits and A. D. Linde, Lebedev Phys. Inst. Preprint No. 101 (1974).
23. D. A. Kirzhnits and A. D. Linde, Ann. Phys. **101**, 195 (1976).
24. A. D. Linde, Rep. Prog. Phys. **42**, 389 (1979).
25. T. D. Lee and G. C. Wick, Phys. Rev. **D9**, 2291 (1974).
26. B. J. Harrington and A. Yildis, Phys. Rev. Lett. **33**, 324 (1974).
27. A. D. Linde, Phys. Rev. **D14**, 3345 (1976);
I. V. Krive, A. D. Linde, and E. M. Chudnovsky, Sov. Phys. JETP **44**, 435 (1976).
28. A. D. Linde, Phys. Lett. **86B**, 39 (1979).
29. I. V. Krive, Sov. Phys. JETP **56**, 477 (1982).
30. A. Salam and J. Strathdee, Nature **252**, 569 (1974);
A. Salam and J. Strathdee, Nucl. Phys. **B90**, 203 (1975).
31. A. D. Linde, Phys. Lett. **62B**, 435 (1976).
32. I. V. Krive, V. M. Pyzh, and E. M. Chudnovsky, Sov. J. Nucl. Phys. **23**, 358 (1976).

33. V. V. Skalozub, Sov. J. Nucl. Phys. **45**, 1058 (1987).
34. Ya. B. Zeldovich and I. D. Novikov, Structure and Evolution of the Universe [in Russian], Nauka, Moscow (1975) [transl. as Relativistic Astrophysics, Vol. II, Univ. of Chicago Press, Chicago (1983)].
35. S. Weinberg, Gravitation and Cosmology: Principles and Applications of the General Theory of Relativity, J. Wiley and Sons, New York (1972).
36. A. D. Sakharov, JETP Lett. **5**, 24 (1967).
37. V. A. Kuzmin, JETP Lett. **12**, 335 (1970).
38. A. Yu. Ignatiev, N. V. Krasnikov, V. A. Kuzmin, and A. N. Tavkhelidze, Phys. Lett. **76B**, 436 (1978);
M. Yoshimura, Phys. Rev. Lett. **41**, 281 (1978);
S. Weinberg,Phys. Rev. Lett. **42**, 850 (1979);
A. D. Dolgov, JETP Lett. **29**, 228 (1979);
W. Kolb and S. Wolfram, Nucl. Phys. B **172**, 224 (1980).
39. Ya. B. Zeldovich, in: Magic without Magic: John Archibald Wheeler, a Collection of Essays in Honor of His 60th Birthday, edited by J. R. Klauder, W. H. Freeman, San Francisco (1972).
40. Ya. B. Zeldovich and M. Yu. Khlopov, Phys. Lett. **79B**, 239 (1978);
J. P. Preskill, Phys. Rev. Lett. **43**, 1365 (1979).
41. Ya. B. Zeldovich, I. Yu. Kobzarev, and L. B. Okun, Phys. Lett. **50B**, 340 (1974).
42. S. Parke and S. Y. Pi, Phys. Lett. **107B**, 54 (1981);
G. Lazarides, Q. Shafi, and T. F. Walsh, Nucl. Phys. B **195**, 157 (1982).
43. P. Sikivie, Phys. Rev. Lett. **48**, 1156 (1982).
44. J. Ellis, A. D. Linde, and D. V. Nanopoulos, Phys. Lett. **128B**, 295 (1983).
45. M. Yu. Khlopov and A. D. Linde, Phys. Lett. **138B**, 265 (1982).
46. J. Polonyi, Budapest Preprint KFKI-93 (1977).
47. G. D. Coughlan, W. Fischler, E. W. Kolb, S. Raby, and G. G. Ross, Phys. Lett. **131B**, 59 (1983).
48. A. S. Goncharov, A. D. Linde, and M. I. Vysotsky, Phys. Lett. **147B**, 279 (1984).
49. J. P. Preskill, M. B. Wise, and F. Wilczek, Phys. Lett. **120B**, 127 (1983);
L. F. Abbott and P. Sikivie, Phys. Lett. **120B**, 133 (1983);
M. Dine and W. Fischler, Phys. Lett. **120B**, 133 (1983).
50. K. Choi and J. E. Kim, Phys. Lett. **154B**, 393 (1985).
51. E. B. Gliner, Sov. Phys. JETP **22**, 378 (1965);

E. B. Gliner, Dokl. Akad. Nauk SSSR **192**, 771 (1970);
E. B. Gliner and I. G. Dymnikova, Sov. Astron. Lett. **1**, 93 (1975).
52. A. A. Starobinsky, JETP Lett. **30**, 682 (1979);
A. A. Starobinsky, Phys. Lett. B **91**, 99 (1980).
53. A. H. Guth, Phys. Rev. **D23**, 347 (1981).
54. A. D. Linde, Phys. Lett. **108B**, 389 (1982).
55. A. Albrecht and P. J. Steinhardt, Phys. Rev. Lett. **48**, 1220 (1982).
56. A. D. Linde, JETP Lett. **38**, 149 (1983);
A. D. Linde, Phys. Lett. **129B**, 177 (1983).
57. A. D. Linde, Mod. Phys. Lett. **1A**, 81 (1986);
A. D. Linde, Phys. Lett. **175B**, 395 (1986);
A. D. Linde, Physica Scripta **T15**, 169 (1987).
58. N. N. Bogolyubov and D. V. Shirkov, Introduction to the Theory of Quantized Fields, Wiley, New York (1980).
59. P. W. Higgs, Phys. Rev. Lett. **13**, 508 (1964);
T. W. B. Kibble, Phys. Rev. **155**, 1554 (1967);
G. S. Guralnik, C. R. Hagen, and T. W. B. Kibble, Phys. Rev. Lett. **13**, 585 (1964);
F. Englert and R. Brout, Phys. Rev. Lett. **13**, 321 (1964).
60. V. L. Ginzburg and L. D. Landau, Sov. Phys. JETP **20**, 1064 (1950).
61. L. D. Landau and E. M. Lifshitz, Statistical Physics, Pergamon, London (1968).
62. A. D. Linde, Phys. Lett. **100B**, 37 (1981);
A. D. Linde, Nucl. Phys. B **216**, 421 (1983).
63. A. Friedmann, Z. Phys. **10**, 377 (1922).
64. H. P. Robertson, Rev. Mod. Phys. **5**, 62 (1933);
A. G. Walker, J. London Math. Soc. **19**, 219 (1944).
65. L. D. Landau and E. M. Lifshitz, The Classical Theory of Fields, Pergamon, Oxford (1975).
66. G. Gamow, Phys. Rev. **74**, 505 (1948).
67. A. G. Doroshkevich and I. D. Novikov, Dokl. Akad. Nauk SSSR **154**, 809 (1964).
68. V. A. Belinsky, E. M. Lifshitz, and I. M. Khalatnikov, Sov. Phys. Usp. **13**, 745 (1971).
69. R. Penrose, "Structure of space-time," in: Battelle Rencontres, 1967 Lectures in Mathematics and Physics, edited by C. M. DeWitt and J. A. Wheeler, Benjamin, New York (1968).
70. S. W. Hawking and G. F. Ellis, The Large Scale Structure of Space-Time, Cambridge University Press, Cambridge (1973).

71. J. A. Wheeler, in: Relativity, Groups, and Topology, edited by B. S. DeWitt and C. M. DeWitt, Gordon and Breach, New York (1964);
S. W. Hawking, Nucl. Phys. B **144**, 349 (1978).
72. C. W. Misner, Phys. Rev. Lett. **28**, 1669 (1972).
73. C. B. Collins and S. W. Hawking, Astrophys. J. **180**, 317 (1973).
74. A. A. Grib, S. G. Mamaev, and V. M. Mostepanenko, Quantum Effects in Strong External Fields [in Russian], Atomizdat, Moscow (1980);
N. Birrell and P. C. Davies, Quantum Fields in Curved Space, Cambridge University Press, Cambridge (1984).
75. E. M. Lifshitz, Sov. Phys. JETP **16**, 587 (1946).
76. Ya. B. Zeldovich, Mon. Not. R. Astron. Soc. **160**, 1 (1970).
77. R. H. Dicke, Nature **192**, 440 (1961);
A. A. Zelmanov, in: Infinity and the Universe [in Russian], Mysl', Moscow (1969), p. 274;
B. Carter, in: Confrontation of Cosmological Theories with Observational Data, edited by M. S. Longair, D. Reidel, Dordrecht (1974);
B. J. Carr and M. J. Rees, Nature **278**, 605 (1979);
I. L. Rozental, Big Bang Big Bounce: How Particles and Fields Drive Cosmic Evolution, Springer–Verlag, Berlin (1988).
78. A. D. Linde, in: 300 Years of Gravitation, edited by S. W. Hawking and W. Israel, Cambridge University Press, Cambridge (1987), p. 604.
79. A. D. Linde, Phys. Today **40**, 61 (1987).
80. S. Weinberg, Phys. Rev. Lett. **36**, 294 (1976).
81. A. Vilenkin, Phys. Rep. **121**, 263 (1985).
82. G. 't Hooft, Nucl. Phys. **B79**, 279 (1974).
83. A. M. Polyakov, JETP Lett. **20**, 430 (1974).
84. T. W. B. Kibble, J. Phys. **9A**, 1387 (1976).
85. Yu. A. Golfand and E. P. Likhtman, JETP Lett. **13**, 323 (1971);
J. Gervais and B. Sakita, Nucl. Phys. **B34**, 632 (1971);
D. V. Volkov and V. P. Akulov, JETP Lett. **16**, 438 (1972);
J. Wess and B. Zumino, Nucl. Phys. **B70**, 39 (1974).
86. J. Ellis and D. V. Nanopoulos, Phys. Lett. **116B**, 133 (1982).
87. A. B. Lahanas and D. V. Nanopoulos, Phys. Rep. **145**, 3 (1987).
88. A. D. Linde, JETP Lett. **19**, 183 (1974);
M. Veltman, Rockefeller University Preprint (1974);
M. Veltman, Phys. Rev. Lett. **34**, 77 (1975);
J. Dreitlein, Phys. Rev. Lett. **33**, 1243 (1975).

89. Ya. B. Zeldovich, Sov. Phys. Usp. **11**, 381 (1968).
90. A. Einstein, Über den gegenwärtigen Stand der Feld-Theorie. Festschrift Prof. Dr. A. Stodola zum 70. (Geburstag, Füssle Verlag, Zurich u. Leipzig, 1929) p. 126.
91. E. S. Fradkin, in: Proceedings of the Quark-80 Seminar, Moscow (1980);
S. Dimopoulos and H. Georgi, Nucl. Phys. B **193**, 150 (1981);
N. Sakai, Z. Phys. C **11**, 153 (1981).
92. N. V. Dragon, Phys. Lett. **113B**, 288 (1982);
P. H. Frampton and T. W. Kephart, Phys. Rev. Lett. **48**, 1237 (1982);
F. Buccella, J. P. Deredinger, S. Ferrara, and C. A. Savoy, Phys. Lett. **115B**, 375 (1982).
93. D. V. Nanopoulos and K. Tamvakis, Phys. Lett. B **110**, 449 (1982);
M. Srednicki, Nucl. Phys. B **202**, 327 (1982).
94. P. Freund, Phys. Lett. **151B**, 387 (1985);
A. Casher, F. Englert, H. Nicolai, and A. Taormina, Phys. Lett. **162B**, 121 (1985).
95. M. J. Duff, B. E. W. Nilsson, and C. N. Pope, Phys. Lett. **163B**, 343 (1985).
96. R. E. Kallosh, Phys. Lett. **176B**, 50 (1986);
R. E. Kallosh, Physica Scripta **T15**, 118 (1987).
97. J. Affleck and M. Dine, Nucl. Phys. **B249**, 361 (1985).
98. A. D. Linde, Phys. Lett. **160B**, 243 (1985).
99. S. Dimopoulos and L. J. Hall, Phys. Lett. **196B**, 135 (1987).
100. W. de Sitter, Proc. Kon. Ned. Akad. Wet. **19**, 1217 (1917);
W. de Sitter, Proc. Kon. Ned. Akad. Wet. **20**, 229 (1917).
101. A. D. Sakharov, Sov. Phys. JETP **22**, 241 (1965).
102. B. L. Altshuler, in: Abstracts from the Third Soviet Conference on Gravitation [in Russian], Erevan (1972), p. 6.
103. L. E. Gurevich, Astrophys. Space Sci. **38**, 67 (1975).
104. A. D. Linde, Phys. Lett. **99B**, 391 (1981).
105. A. D. Dolgov and Ya. B. Zeldovich, Rev. Mod. Phys. **53**, 1 (1981).
106. J. S. Dowker and R. Critchley, Phys. Rev. **D13**, 3224 (1976).
107. V. F. Mukhanov and G. V. Chibisov, JETP Lett. **33**, 523 (1981);
V. F. Mukhanov and G. V. Chibisov, Sov. Phys. JETP **56**, 258 (1982).
108. J. D. Barrow and A. Ottewill, J. Phys. **A16**, 2757 (1983).
109. A. A. Starobinsky, Sov. Astron. Lett. **9**, 302 (1983).
110. L. A Kofman, A. D. Linde, and A. A. Starobinsky, Phys. Lett. **157B**,

36 (1985).
111. V. G. Lapchinsky, V. A. Rubakov, and A. V. Veryaskin, Inst. Nucl. Res. Preprint No. P-0195 (1982).
112. S. W. Hawking, I. G. Moss, and J. M. Stewart, Phys. Rev. D**26**, 2681 (1982).
113. A. H. Guth and E. J. Weinberg, Nucl. Phys. B**212**, 321 (1983).
114. S. W. Hawking, Phys. Lett. **115B**, 295 (1982);
A. A. Starobinsky, Phys. Lett. **117B**, 175 (1982);
A. H. Guth and S. Y. Pi, Phys. Rev. Lett. **49**, 1110 (1982);
J. M. Bardeen, P. J. Steinhardt, and M. S. Turner Phys. Rev. **D28**, 679 (1983).
115. A. D. Linde, Phys. Lett. **132B**, 317 (1983).
116. A. D. Linde, Rep. Prog. Phys. **47**, 925 (1984).
117. V. A. Rubakov, M. V. Sazhin, and A. V. Veryaskin, Phys. Lett. **115B**, 189 (1982).
118. A. D. Linde, Phys. Lett. **162B**, 281 (1985);
A. D. Linde, Suppl. Prog. Theor. Phys. **85**, 279 (1985).
119. I. D. Novikov and V. P. Frolov, The Physics of Black Holes [in Russian], Nauka, Moscow (1986).
120. G. W. Gibbons and S. W. Hawking, Phys. Rev. **D15**, 2738 (1977).
121. S. W. Hawking and I. G. Moss, Phys. Lett. **110B**, 35 (1982).
122. W. Boucher and G. W. Gibbons, in: The Very Early Universe, edited by G. W. Gibbons, S. W. Hawking, and S. Siklos, Cambridge University Press, Cambridge (1983);
A. A. Starobinsky, JETP Lett. **37**, 66 (1983);
R. Wald, Phys. Rev. **D28**, 2118 (1983);
E. Martinez-Gonzalez and B. J. T. Jones, Phys. Lett. **167B**, 37 (1986);
I. G. Moss and V. Sahni, Phys. Lett. **B178**, 159 (1986);
M. S. Turner and L. Widrow, Phys. Rev. Lett. **57**, 2237 (1986);
L. Jensen and J. Stein-Schabes, Phys. Rev. **D34**, 931 (1986).
123. A. D. Dolgov and A. D. Linde, Phys. Lett. **116B**, 329 (1982).
124. L. F. Abbott, E. Farhi, and M. B. Wise, Phys. Lett. **117B**, 29 (1982).
125. L. A. Kofman and A. D. Linde, Nucl. Phys. **B282**, 555 (1987).
126. A. Vilenkin and L. H. Ford, Phys. Rev. **D26**, 1231 (1982).
127. A. D. Linde, Phys. Lett. **116B**, 335 (1982).
128. A. A. Starobinsky, Phys. Lett. **117B**, 175 (1982).
129. V. A. Kuzmin, V. A. Rubakov, and M. E. Shaposhnikov, Phys. Lett. **115B**, 36 (1985).

130. M. E. Shaposhnikov, JETP Lett. **44**, 465 (1986); Nucl. Phys. **B299**, 797 (1988);
L. McLerran, Phys. Rev. Lett. **62**, 1075 (1989).
131. A. Dannenberg and L. J. Hall, Phys. Lett. **198B**, 411 (1987).
132. A. S. Goncharov and A. D. Linde, Sov. Phys. JETP **65**, 635 (1987).
133. A. S. Goncharov, A. D. Linde, and V. F. Mukhanov, Int. J. Mod. Phys. **2A**, 561 (1987).
134. A. A. Starobinsky, in: Fundamental Interactions [in Russian], MGPI, Moscow (1984), p. 55.
135. A. A. Starobinsky, in: Current Trends in Field Theory, Quantum Gravity, and Strings, Lecture Notes in Physics, edited by H. J. de Vega and N. Sanches, Springer-Verlag, Heidelberg (1986).
136. L. P. Grishchuk and Ya. B. Zeldovich, Sov. Astron. **22**, 12 (1978).
137. S. Coleman and E. J. Weinberg, Phys. Rev. **D6**, 1888 (1973).
138. R. Jackiw, Phys. Rev. **D9**, 1686 (1973).
139. A. D. Linde, JETP Lett. **23**, 73 (1976).
140. S. Weinberg, Phys. Rev. Lett. **36**, 294 (1976).
141. A. D. Linde, Phys. Lett. **70B**, 306 (1977).
142. A. D. Linde, Phys. Lett. **92B**, 119 (1980).
143. A. H. Guth and E. J. Weinberg, Phys. Rev. Lett. **45**, 1131 (1980).
144. E. Witten, Nucl. Phys. **B177**, 477 (1981).
145. I. V. Krive and A. D. Linde, Nucl. Phys. **B117**, 265 (1976).
146. A. D. Linde, Trieste Preprint No. IC/76/26 (1976).
147. N. V. Krasnikov, Sov. J. Nucl. Phys. **28**, 279 (1978).
148. P. Q. Hung, Phys. Rev. Lett. **42**, 873 (1979).
149. H. D. Politzer and S. Wolfram, Phys. Lett. **82B**, 242 (1979).
150. A. A. Anselm, JETP Lett. **29**, 645 (1979).
151. N. Cabibbo, L. Maiani, A. Parisi, and R. Petronzio, Nucl. Phys. **B158**, 295 (1979).
152. B. L. Voronov and I. V. Tyutin, Sov. J. Nucl. Phys. **23**, 699 (1976).
153. S. Coleman, R. Jackiw, and H. D. Politzer, Phys. Rev. **D10**, 2491 (1974).
154. L. F. Abbott, J. S. Kang, and H. J. Schnitzer, Phys. Rev. **D13**, 2212 (1976).
155. A. D. Linde, Nucl. Phys. **B125**, 369 (1977).
156. L. D. Landau and I. Ya. Pomeranchuk, Dokl. Akad. Nauk SSSR **102**, 489 (1955).
157. E. S. Fradkin, Sov. Phys. JETP **28**, 750 (1955).
158. A. B. Migdal, Fermions and Bosons in Strong Fields [in Russian], Nauka, Moscow (1978).

159. D. A. Kirzhnits and A. D. Linde, Phys. Lett. **73B**, 323 (1978).
160. J. Fröhlich, Nucl. Phys. **B200**, 281 (1982).
161. C. B. Lang, Nucl. Phys. **B265**, 630 (1986).
162. D. A. Kirzhnits, in: Quantum Field Theory and Quantum Statistics, edited by I. A. Batalin, C. J. Isham, and G. A. Vilkovisky, Adam Hilger Press, Bristol (1987), Vol. 1, p. 349.
163. W. A. Bardeen and M. Moshe, Phys. Rev. **D28**, 1372 (1982).
164. K. Enquist and J. Maalampi, Phys. Lett. **180B**, 14 (1986).
165. L. Smolin, Phys. Lett. **93B**, 95 (1980).
166. E. S. Fradkin, Proc. Lebedev Phys. Inst. **29**, 7 (1965) [English transl.: Consultants Bureau, New York (1967)].
167. V. A. Kuzmin, M. E. Shaposhnikov, and I. I. Tkachev, Z. Phys. Ser. C **12**, 83 (1982).
168. M. B. Kislinger and P. D. Morley, Phys. Rev. **D13**, 2765 (1976).
169. E. V. Shuryak, Sov. Phys. JETP **47**, 212 (1978).
170. A. M. Polyakov, Phys. Lett. **72B**, 477 (1978).
171. A. D. Linde, Phys. Lett. **96B**, 289 (1980).
172. D. J. Gross, R. Pisarski, and L. Yaffe, Rev. Mod. Phys. **53**, 43 (1981).
173. A. D. Linde, Phys. Lett. **96B**, 293 (1980).
174. M. A. Matveev, V. A. Rubakov, A. N. Tavkhelidze, and V. F. Tokarev, Nucl. Phys. B**282**, 700 (1987).
175. D. I. Deryagin, D. Ya. Grigoriev, and V. A. Rubakov, Phys. Lett. **178B**, 385 (1986).
176. O. K. Kalashnikov and H. Perez–Rojas, Nucl. Phys. **B293**, 241 (1987).
177. E. J. Ferrer, V. de la Incera, and A. E. Shabad, Phys. Lett. **185B**, 407 (1987);
E. J. Ferrer and V. de la Incera, Phys. Lett. **205B**, 381 (1988).
178. M. E. Shaposhnikov, Nucl. Phys. **B287**, 767 (1987).
179. M. B. Voloshin, I. B. Kobzarev, and L. B. Okun, Sov. J. Nucl. Phys. **20**, 644 (1974).
180. S. Coleman, Phys. Rev. **D15**, 2929 (1977).
181. C. Callan and S. Coleman, Phys. Rev. **D16**, 1762 (1977).
182. S. Fubini, Nuovo Cimento **34A**, 521 (1976).
183. G. 't Hooft, Phys. Rev. **D14**, 3432 (1976);
R. Rajaraman, Solitons and Instantons in Quantum Field Theory, North-Holland, Amsterdam (1982).
184. N. V. Krasnikov, Phys. Lett. **72B**, 455 (1978).

185. J. Affleck, Nucl. Phys. **B191**, 429 (1981).
186. A. S. Goncharov and A. D. Linde, Sov. J. Part. Nucl. **17**, 369 (1986).
187. A. D. Linde, "The kinetics of phase transitions in grand unified theories" [in Russian], FIAN Preprint No. 266 (1981).
188. R. Flores and M. Sher, Phys. Rev. **D27**, 1679 (1983).
189. M. Sher and H. W. Zaglauer, Phys. Lett. **206B**, 527 (1988).
190. A. A. Abrikosov, Sov. Phys. JETP **32**, 1442 (1957);
H. B. Nielsen and P. Olesen, Nucl. Phys. **B61**, 45 (1973).
191. Ya. B. Zeldovich, Mon. Not. R. Astron. Soc. **192**, 663 (1980).
192. A. Vilenkin, Phys. Rev. Lett. **46**, 1169 (1981).
193. N. Turok and R. Brandenberger, Phys. Rev. **D33**, 2175 (1986).
194. E. N. Parker, Astrophys. J. **160**, 383 (1970).
195. E. W. Kolb, S. A. Colgate, and J. A. Harvey, Phys. Rev. Lett. **49**, 1373 (1982);
S. Dimopoulos, J. Preskill, and F. Wilczek, Phys. Lett. **119B**, 320 (1982);
K. Freese, M. S. Turner, and D. N. Schramm, Phys. Rev. Lett. **51**, 1625 (1983).
196. V. A. Rubakov, Nucl. Phys. **B203**, 311 (1982);
C. G. Callan, Phys. Rev. **D26**, 2058 (1982).
197. Y. Nambu, Phys. Rev. **D10**, 4262 (1974).
198. A. Billoire, G. Lazarides, and Q. Shafi, Phys. Lett. **103B**, 450 (1981).
199. T. A. DeGrand and D. Toussaint, Phys. Rev. **D25**, 526 (1982).
200. V. N. Namiot and A. D. Linde, unpublished.
201. G. W. Gibbons and S. W. Hawking, Phys. Rev. **D15**, 2752 (1977).
202. T. S. Bunch and P. C. W. Davies, Proc. R. Soc. London, Ser. A **A360**, 117 (1978).
203. A. Vilenkin, Nucl. Phys. **B226**, 527 (1983).
204. A. Vilenkin, Phys. Rev. **D27**, 2848 (1983).
205. Yu. L. Klimontovich, Statistical Physics [in Russian], Nauka, Moscow (1982).
206. S.-J. Rey, Nucl. Phys. **B284**, 706 (1987).
207. S. Coleman and F. De Luccia, Phys. Rev. **D21**, 3305 (1980).
208. N. Deruelle, Mod. Phys. Lett. **4A**, 1297 (1989).
209. S. W. Hawking and I. G. Moss, Nucl. Phys. **B224**, 180 (1983).
210. H. A. Kramers, Physica **7**, 240 (1940).
211. A. D. Linde, Phys. Lett. **131B**, 330 (1983).
212. W. Israel, Nuovo Cimento **44B**, 1 (1966).
213. V. A. Berezin, V. A. Kuzmin, and I. I. Tkachev, Phys. Lett. **120B**, 91

(1983);
V. A. Berezin, V. A. Kuzmin, and I. I. Tkachev, Phys. Rev. **D36**, 2919 (1987);
A. Aurilia, G. Denardo, F. Legovini, and E. Spalucci, Nucl. Phys. **B252**, 523 (1985);
P. Laguna-Gastillo, and R. A. Matzner, Phys. Rev. **D34**, 2913 (1986);
S. K. Blau, E. I. Guendelman, and A. H. Guth, Phys. Rev. **D35**, 1747 (1987);
A. Aurilia, R. S. Kissack, R. Mann, and E. Spalucci, Phys. Rev. **D35**, 2961 (1987).

214. E. R. Harrison, Phys. Rev. **D1**, 2726 (1973).
215. L. P. Grishchuk, Sov. Phys. JETP **40**, 409 (1974).
216. V. N. Lukash, Sov. Phys. JETP **52**, 807 (1980);
D. A. Kompaneets, V. N. Lukash, and I. D. Novikov, Sov. Astron. **26**, 259 (1982);
V. N. Lukash and I. D. Novikov, in: The Very Early Universe, edited by G. W. Gibbons, S. W. Hawking, and S. Siklos, Cambridge University Press, Cambridge (1983), p. 311.
217. V. F. Mukhanov and G. V. Chibisov, Mon. Not. R. Astron. Soc. **200**, 535 (1982).
218. V. F. Mukhanov, JETP Lett. **41**, 493 (1985).
219. R. H. Brandenberger, Rev. Mod. Phys. **57**, 1 (1985);
R. H. Brandenberger, Int. J. Mod. Phys. **2A**, 77 (1987).
220. J. M. Bardeen, Phys. Rev. **D22**, 1882 (1980);
G. V. Chibisov and V. F. Mukhanov, Lebedev Phys. Inst. Preprint No. 154 (1983).
221. L. A. Kofman, V. F. Mukhanov, and D. Yu. Pogosyan, Sov. Phys. JETP **66**, 441 (1987).
222. V. F. Mukhanov, L. A. Kofman, and D. Yu. Pogosyan, Phys. Lett. **157B**, 427 (1987).
223. P. J. E. Peebles, Astrophys. J. Lett. **263**, L1 (1982).
224. S. F. Shandarin, A. G. Doroshkevich, and Ya. B. Zeldovich, Sov. Phys. Usp. **26**, 46 (1983).
225. A. A. Starobinsky, Sov. Astron. Lett. **9**, 302 (1983).
226. V. N. Lukash, P. D. Naselskij, and I. D. Novikov, in: Quantum Gravity, edited by M. A. Markov, V. A. Berezin, and V. P. Frolov, World Scientific, Singapore (1984), p. 675;
V. N. Lukash and I. D. Novikov, Nature **316**, 46 (1985);
227. L. A. Kofman and A. A. Starobinsky, Sov. Astron. Lett. **11**, 643

(1985);
L. A. Kofman, D. Yu. Pogosyan, and A. A. Starobinsky, Sov. Astron. Lett. **12**, 419 (1985).

228. A. B. Berlin, E. V. Bulaenko, V. V. Vitkovsky, V. K. Kononov, Yu. N. Parijskij, and Z. E. Petrov, Proc. IAU Symposium 104, edited by G. O. Abell and G. Chincarini, D. Reidel, Dordrecht (1983);
F. Melchiorri, B. Melchiorri, C. Ceccarelli, and L. Pietranera, Astrophys. J. Lett. **250**, L1 (1981);
I. A. Strukov and D. P. Skulachev, Sov. Astron. Lett. **10**, 1 (1984);
J. M. Uson and D. T. Wilkinson, Nature **312**, 427 (1984);
R. D. Davies et al., Nature **326**, 462 (1987);
D. J. Fixen, E. S. Cheng, and D. T. Wilkinson, Phys. Rev. Lett. **50**, 620 (1983).
229. D.H. Lyth, Phys. Lett. **147B**, 403 (1984);
S. W. Hawking, Phys. Lett. **150B**, 339 (1985).
230. C. W. Kim, and P. Murphy, Phys. Lett. **167B**, 43 (1986).
231. S. W. Hawking, in: 300 Years of Gravitation, edited by S. W. Hawking and W. Israel, Cambridge University Press, Cambridge (1987), p. 631.
232. H. Georgi and S. L. Glashow, Phys. Rev. Lett. **28**, 1494 (1982).
233. R. D. Peccei and H. Quinn, Phys. Rev. Lett. **38**, 1440 (1977);
R. D. Peccei and H. Quinn, Phys. Rev. **D16**, 1791 (1977).
234. S. Weinberg, Phys. Rev. Lett. **40**, 223 (1978);
F. Wilczek, Phys. Rev. Lett. **40**, 279 (1978).
235. J. R. Primack, in: Proc. of the International School of Physics "Enrico Fermi" (1984);
M. J. Rees, in: 300 Years of Gravitation, edited by S. W. Hawking and W. Israel, Cambridge University Press, Cambridge (1987).
236. A. G. Doroshkevich, Sov. Astron. Lett. **14**, 125 (1988).
237. Q. Shafi and C. Wetterich, Phys. Lett. **152B**, 51 (1985);
Q. Shafi and C. Wetterich, Nucl. Phys. **B289**, 787 (1987).
238. J. Silk and M. S. Turner, Phys. Rev. **D35**, 419 (1986).
239. A. D. Linde, JETP Lett. **40**, 1333 (1984);
A. D. Linde, Phys. Lett. **158B**, 375 (1985).
240. D. Seckel and M. S. Turner, Phys. Rev. **D32**, 3178 (1985).
241. L. A. Kofman, Phys. Lett. **174B**, 400 (1986).
242. L. A. Kofman and D. Yu. Pogosyan, Phys. Lett. **214B**, 508 (1988).
243. L. A. Kofman, A. D. Linde, and J. Einasto, Nature **326**, 48 (1987).
244. J. Goldstone, Nuovo Cimento **19**, 154 (1961);

J. Goldstone, A. Salam, and S. Weinberg, Phys. Rev. **127**, 965 (1962).
245. T. J. Allen, B. Grinstein, and M. B. Wise, Phys. Lett. **197B**, 66 (1987).
246. Q. Shafi and A. Vilenkin, Phys. Rev. **D29**, 1870 (1984).
247. E. T. Vishniac, K. A. Olive, and D. Seckel, Nucl. Phys. **B289**, 717 (1987).
248. A. D. Dolgov and N. S. Kardashov, Inst. of Space Research Preprint (1987);
A. D. Dolgov, A. F. Illarionov, N. S. Kardashov, and I. D. Novikov, Sov. Phys. JETP **67**, 13 (1988).
249. V. de Lapparent, M. Geller, and J. Huchra, Astrophys. J. **302**, L1 (1987).
250. J. P. Ostriker and L. Cowie, Astrophys. J. Lett. **243**, L127 (1981).
251. I. I. Tkachev, Phys. Lett. **191B**, 41 (1987).
252. M. S. Turner, Phys. Rev. **D28**, 1243 (1983).
253. R. J. Scherrer and M. S. Turner, Phys. Rev. **D31**, 681 (1985).
254. L. A. Kofman and A. D. Linde, to be published.
255. Q. Shafi and C. Wetterich, Nucl. Phys. **B297**, 697 (1988).
256. A. Ringwald, Z. Phys. Ser. C **34**, 481 (1987);
A. Ringwald, Heidelberg Preprint HD-THEP-85-18 (1985).
257. E. W. Kolb and M. S. Turner, Ann. Rev. Nucl. Part. Sci. **33**, 645 (1983).
258. M. S. Turner, in: Architecture of Fundamental Interactions at Short Distances, edited by P. Ramond and R. Stora, Elsevier Science Publishers, Copenhagen (1987).
259. V. A. Kuzmin, M. E. Shaposhnikov, and I. I. Tkachev, Nucl. Phys. **B196**, 29 (1982).
260. B. A. Campbell, J. Ellis, D. V. Nanopoulos, and K. A. Olive, Mod. Phys. Lett. **A1**, 389 (1986);
J. Ellis, D. V. Nanopoulos, and K. A. Olive, Phys. Lett. **B184**, 37 (1987);
J. Ellis, K. Enquist, D. V. Nanopoulos, and K. A. Olive, Phys. Lett. **B191**, 343 (1987).
261. M. Fukugita and T. Yanagida, Phys. Lett. **174B**, 45 (1986).
262. K. Yamamoto, Phys. Lett. **168B**, 341 (1986).
263. R. N. Mohapatra and J. W. F. Valle, Phys. Lett. **186B**, 303 (1987).
264. G. M. Shore, Ann. Phys. **128**, 376 (1980).
265. A. D. Linde, Phys. Lett. **114B**, 431 (1982).

266. P. J. Steinhardt, in: The Very Early Universe, edited by G. W. Gibbons, S. W. Hawking, and S. Siklos, Cambridge University Press, Cambridge (1983).
267. A. D. Linde, "Nonsingular regenerating inflationary universe," Cambridge University Preprint (1982).
268. P. J. Steinhardt and M. S. Turner, Phys. Rev. **D29**, 2162 (1984).
269. A. Albrecht, S. Dimopoulos, W. Fischer, E. W. Kolb, S. Raby, and P. J. Steinhardt, Nucl. Phys. **B229**, 528 (1983).
270. J. Ellis, D. V. Nanopoulos, K. A. Olive, and K. Tamvakis, Nucl. Phys. **B221**, 421 (1983);
D. V. Nanopoulos, K. A. Olive, M. Srednicki, and K. Tamvakis, Phys. Lett. **123B**, 41 (1983).
271. B. Ovrut and P. J. Steinhardt, Phys. Rev. Lett. **53**, 732 (1984);
B. Ovrut and P. J. Steinhardt, Phys. Lett. **B147**, 263 (1984).
272. E. Cremmer, S. Ferrara, L. Girardello, and A. Van Proeyen, Nucl. Phys. **B212**, 413 (1983).
273. A. S. Goncharov and A. D. Linde, Sov. Phys. JETP **59**, 930 (1984);
A. S. Goncharov and A. D. Linde, Phys. Lett. **139B**, 27 (1984).
274. A. S. Goncharov and A. D. Linde, Class. Quant. Grav. **1**, L75 (1984).
275. Q. Shafi and A. Vilenkin, Phys. Rev. Lett. **52**, 691 (1984).
276. S. Y. Pi, Phys. Rev. Lett. **52**, 1725 (1984).
277. A. S. Goncharov, "Phase transitions in gauge theories and cosmology," Candidate's Dissertation, Moscow (1984).
278. L. A. Khalfin, Sov. Phys. JETP **64**, 673 (1986).
279. V. A. Belinsky, L. P. Grishchuk, Ya. B. Zeldovich, and I. M. Khalatnikov, Phys. Lett. **155B**, 232 (1985).
280. V. A. Belinsky, and I. M. Khalatnikov, Sov. Phys. JETP **93**, 784 (1987);
V. A. Belinsky, H. Ishihara, I. M. Khalatnikov, and H. Sato, Progr. Theor. Phys. **79**, 676 (1988).
281. A. D. Linde, Phys. Lett. **202B**, 194 (1988).
282. R. Holman, P. Ramond, and G. G. Ross, Phys. Lett. **137B**, 343 (1984).
283. J. Ellis, A. B. Lahanas, D. V. Nanopoulos, and K. Tamvakis, Phys. Lett. **134B**, 429 (1984);
J. Ellis, C. Kounnas, and D. V. Nanopoulos, Nucl. Phys. **B241**, 406 (1984);
J. Ellis, C. Kounnas, and D. V. Nanopoulos, Nucl. Phys. **B247**, 373 (1984).

284. E. Cremmer, S. Ferrara, C. Kounnas, and D. V. Nanopoulos, Phys. Lett. **133B**, 61 (1983).
285. G. B. Gelmini, C. Kounnas, and D. V. Nanopoulos, Nucl. Phys. **B250**, 177 (1985).
286. D. V. Nanopoulos, K. A. Olive, and M. Srednicki, Phys. Lett. **127B**, 30 (1983).
287. Ya. B. Zeldovich, Sov. Phys. Usp. **133**, 479 (1981).
288. V. Ts. Gurovich and A. A. Starobinsky, Sov. Phys. JETP **50**, 844 (1979).
289. Ya. B. Zeldovich, Sov. Astron. Lett. **95**, 209 (1981).
290. L. P. Grishchuk and Ya. B. Zeldovich, in: Quantum Structure of Space-Time, edited by M. Duff and C. J. Isham, Cambridge University Press, Cambridge (1983), p. 353.
291. K. Stelle, Phys. Rev. **D16**, 953 (1977).
292. A. D. Sakharov, Sov. Phys. JETP **60**, 214 (1984).
293. I. Ya. Aref'eva and I. V. Volovich, Phys. Lett. **164B**, 287 (1985);
I. Ya. Aref'eva and I. V. Volovich, Theor. Mat. Phys. **64**, 866 (1986).
294. M. Reuter and C. Wetterich, Nucl. Phys. **B289**, 757 (1987).
295. M. D. Pollock, Nucl. Phys. **B309**, 513 (1988).
296. M. D. Pollock, Phys. Lett. **215B**, 635 (1988).
297. J. Ellis, K. Enquist, D. V. Nanopoulos, and M. Quiros, Nucl. Phys. **B277**, 233 (1986);
P.Binetruy and M.K. Gaillard, Phys. Rev. **D34**, 3069 (1986);
P. Oh, Phys. Lett. **166B**, 292 (1986);
K. Maeda, M. D. Pollock, and C. E. Vayonakis, Class. Quant. Grav. **3**, L89 (1986);
S. R. Lonsdale and I. G. Moss, Phys. Lett. **189B**, 12 (1987);
M. D. Pollock, Phys. Lett. **199B**, 509 (1987);
J. A. Casas and C. Muñoz, Phys. Lett. **216B**, 37 (1989);
S. Kalara and K. Olive, Phys. Lett. **218B**, 148 (1989).
298. J. A. Wheeler, in: Relativity, Groups, and Topology, edited by C. M. DeWitt and J. A. Wheeler, Benjamin, New York (1968).
299. B. S. DeWitt, Phys. Rev. **160**, 1113 (1967).
300. Quantum Cosmology, edited by L. Z. Fang and R. Ruffini, World Scientific, Singapore (1987).
301. V. N. Ponomarev, A. O. Barvinsky, and Yu. N. Obukhov, Geometrodynamical Methods and the Gauge Approach to the Theory of Gravitational Interactions [in Russian], Energoatomizdat, Moscow

(1985).
302. J. A. Wheeler, in: Foundational Problems in the Special Sciences, edited by R. E. Butts and J. Hintikka, D. Reidel, Dordrecht (1977); also in: Quantum Mechanics, a Half Century Later, edited by J. L. Lopes and M. Paty, D. Reidel, Dordrecht (1977).
303. H. Everett, Rev. Mod. Phys. **29**, 454 (1957).
304. B. S. DeWitt and N. Graham, The Many-Worlds Interpretation of Quantum Mechanics, Princeton University Press, Princeton (1973).
305. B. S. DeWitt, Phys. Today **23**, 30 (1970);
B. S. DeWitt, Phys. Today **24**, 36 (1971).
306. L. Smolin, in: Quantum Theory of Gravity, edited by S. M. Christensen, Adam Hilger, Bristol (1984).
307. D. Deutsch, Int. J. Theor. Phys. **24**, 1 (1985).
308. V. F. Mukhanov, in: Proc. Third Seminar on Quantum Gravity, edited by M. A. Markov, V. A. Berezin, and V. P. Frolov, World Scientific, Singapore (1984).
309. M. A. Markov and V. F. Mukhanov, Phys. Lett. **127A**, 251 (1988).
310. S. W. Hawking, Phys. Rev. **D32**, 2489 (1985).
311. D. N. Page, Phys. Rev. **D32**, 2496 (1985).
312. A. D. Sakharov, Sov. Phys. JETP **49**, 594 (1979);
A. D. Sakharov, Sov. Phys. JETP **52**, 349 (1980).
313. M. A. Markov, Ann. Phys. **155**, 333 (1984).
314. J. B. Hartle and S. W. Hawking, Phys. Rev. **D28**, 2960 (1983).
315. E. P. Tryon, Nature **246**, 396 (1973);
P. I. Fomin, Inst. Teor. Fiz. Preprint No. ITF-73-1379;
P. I. Fomin, Dokl. Akad. Nauk SSSR **9**, 831 (1975).
316. R. Brout, F. Englert, and E. Gunzig, Ann. Phys. **115**, 78 (1978).
317. D. Atkatz and H. Pagels, Phys. Rev. **D25**, 2065 (1982).
318. A. Vilenkin, Phys. Lett. **117B**, 25 (1982).
319. A. D. Linde, Sov. Phys. JETP **60**, 211 (1984);
A. D. Linde, Lett. Nuovo Cimento **39**, 401 (1984).
320. Ya. B. Zeldovich and A. A. Starobinsky, Sov. Astron. Lett. **10**, 135 (1984).
321. V. A. Rubakov, JETP Lett. **39**, 107 (1984);
V. A. Rubakov, Phys. Lett. **148B**, 280 (1984).
322. A. Vilenkin, Phys. Rev. **D30**, 509 (1984);
A. Vilenkin, Phys. Rev. **D33**, 3560 (1986).
323. C. W. Misner, K. S. Thorne, and J. A. Wheeler, Gravitation, W. H. Freeman, San Francisco (1973).

324. S. W. Hawking and D. N. Page, Nucl. Phys. **B264**, 185 (1986);
J. J. Halliwell and S. W. Hawking, Phys. Rev. **D31**, 1777 (1985);
J. J. Halliwell, Phys. Rev. **D38**, 2468 (1988);
A. Vilenkin, Phys. Rev. **D37**, 888 (1988);
A. Vilenkin and T. Vachaspati, Phys. Rev. **D37**, 904 (1988);
L. P. Grishchuk and L. Rozhansky, Phys. Lett. **208B**, 369 (1988);
G. W. Gibbons and L. P. Grishchuk, Nucl. Phys. **B313**, 736 (1989).
325. M. Aryal and A. Vilenkin, Phys. Lett. **199B**, 351 (1987).
326. E. Farhi and A. H. Guth, Phys. Lett. **183B**, 149 (1987).
327. K. Maeda, K. Sato, M. Sasaki, and H. Kodama, Phys. Lett. **108B**, 98 (1982).
328. J. R. Gott, Nature **295**, 304 (1982).
329. S. Weinberg, Phys. Rev. Lett. **48**, 1776 (1982).
330. P. Ehrenfest, Proc. Amsterdam Acad. **20**, 200 (1917).
331. J. D. Barrow and F. J. Tipler, The Anthropic Cosmological Principle, Oxford University Press, Oxford (1986).
332. A. Einstein, B. Podolsky, and N. Rosen, Phys. Rev. **47**, 777 (1935).
333. A. D. Linde and M. I. Zelnikov, Phys. Lett. **215B**, 59 (1988).
334. A. D. Linde, Phys. Lett. **201B**, 437 (1988).
335. R. D. Peccei, J. Sola, and C. Wetterich, Phys. Lett. **195B**, 183 (1987).
336. A. D. Linde, Phys. Lett. **211B**, 29 (1988).
337. S. Coleman and J. Mandula, Phys. Rev. **159**, 1251 (1967).
338. J. Scherk, in: Recent Developments in Gravitation, edited by M. Levy and S. Deser, Plenum, New York (1979), p. 479;
J. Scherk, in: Geometrical Ideas in Physics [Russian translation], Mir, Moscow (1983), p. 201;
M. Gell-Mann, Physica Scripta **T15**, 202 (1987).
339. A. D. Dolgov, in: The Very Early Universe, edited by G. W. Gibbons, S. W. Hawking, and S. Siklos, Cambridge University Press, Cambridge (1983), p. 449.
340. S. W. Hawking, Phys. Lett. **134B**, 403 (1984).
341. T. Banks, Nucl. Phys. **B249**, 332 (1985).
342. L. Abbott, Phys. Lett. **150B**, 427 (1985).
343. S. Barr, Phys. Rev. **D36**, 1691 (1987).
344. A. D. Linde, Phys. Lett. **200B**, 272 (1988).
345. S. Coleman, Nucl. Phys. **B307**, 867 (1988).
346. S. Coleman, Nucl. Phys. **B310**, 643 (1988).
347. T. Banks, Nucl. Phys. **B309**, 493 (1988).
348. S. Weinberg, Phys. Rev. Lett. **59**, 2607 (1987); Rev. Mod. Phys. **61**,

1 (1989).
349. S. Giddings and A. Strominger, Nucl. Phys. **B307**, 854 (1988).
350. S. W. Hawking, Phys. Lett. **195B**, 337 (1987);
S. W. Hawking, Phys. Rev. **D37**, 904 (1988).
351. G. V. Lavrelashvili, V. A. Rubakov, and P. G. Tinyakov, JETP Lett. **46**, 167 (1987);
G. V. Lavrelashvili, V. A. Rubakov, and P. G. Tinyakov, Nucl. Phys. **B299**, 757 (1988).
352. S. Giddings and A. Strominger, Nucl. Phys. **B306**, 890 (1988).
353. I. Antoniadis, C. Bachas, J. Ellis, and D. V. Nanopoulos, Phys. Lett. **211B**, 393 (1988).
354. E. Tomboulis, preprint UCLA/89/TEP/8 (1989).
355. A. D. Linde, in: Proceedings of XXIV International Conference on High Energy Physics, Munich 1988 (Springer-Verlag, Berlin 1989), p. 357
356. J. Polchinski, Phys. Lett. **219B**, 251 (1989).
357. W. Fischler, I. Klebanov, J. Polchinski and L. Susskind, Nucl. Phys. **B327**, 157 (1989).
358. V. A. Rubakov and M.E. Shaposhnikov, Mod. Phys. Lett. **4A**, 107 (1989).
359. A.D. Linde, Phys. Lett. **227B**, 352 (1989).
360. P.B. Arnold, Phys. Rev. **D40**, 613 (1989);
M. Sher, Phys. Rep. **C179**, 274 (1989).
361. A.D. Linde, D.H. Lyth, in preparation.

Index